Classics in Mathematics

C. L. Siegel J. K. Moser **Lectures on Celestial Mechanics**

Carl Ludwig Siegel was born on December 31, 1896 in Berlin. He studied mathematics and astronomy in Berlin and Göttingen and held chairs at the Universities of Frankfurt and Göttingen before moving to the Institute for Advanced Study in Princeton in 1940. He returned to Göttingen in 1951 and died there in 1981.

Siegel was one of the leading mathematicians of the twentieth century, whose work, noted for its depth as well as breadth, ranged over many different fields such as number theory from the analytic, algebraic and geometrical points of view, automorphic functions of several complex variables, symplectic geometry, celestial mechanics.

Jürgen Moser was born on July 4, 1928 in Königsberg, then Germany. After the war he studied in Göttingen, where he received his doctoral degree in 1952 and subsequently was assistant to C. L. Siegel. In 1955 he emigrated to the USA. He held positions at M.I.T., Cambridge and primarily at the Courant Institute of Mathematical Sciences in New York; from 1967 to 1970 he was Director of this institute. In 1980 he moved to the ETH in Zürich where he now is Director of the Mathematical Research Institute.

Moser has worked in various areas of analysis. Besides celestial mechanics and KAM theory (presented in Chapter 3 of this book) he contributed to spectral theory, partial differential equations and complex analysis.

C. L. Siegel J. K. Moser

Lectures on Celestial Mechanics

Reprint of the 1971 Edition

Springer

Carl Ludwig Siegel

Jürgen K. Moser
ETH Zentrum, Mathematik
CH-8092 Zürich
Schweiz

Originally published as Vol. 187 of the
Grundlehren der mathematischen Wissenschaften

Mathematics Subject Classification (1991):
Primary 34-02, 70-02, 70F15
Secondary 34C25, 34C35, 58F05, 70F10, 70F35, 70H05, 70H15, 70H20

ISBN 3-540-58656-3 Springer-Verlag Berlin Heidelberg New York

CIP data applied for

SPIN 10485228 41/3140 - 5 4 3 2 1 0 – Printed on acid-free paper

C. L. Siegel · J. K. Moser

Lectures on Celestial Mechanics

Translation by C. I. Kalme

Springer-Verlag Berlin Heidelberg New York 1971

Carl Ludwig Siegel
Mathematisches Institut der Universität Göttingen

Jürgen K. Moser
Courant Institute of Mathematical Sciences, New York

Charles I. Kalme
University of Southern California, Los Angeles

Revised and enlarged translation of „Vorlesungen über Himmelsmechanik" by C. L. Siegel, 1956 (Grundlehren der mathematischen Wissenschaften, Band 85)

AMS Subject Classifications (1970)
Primary 34-02, 70-02, 70 F 15
Secondary 34 C 25, 34 C 35, 58 F 05, 58 F 10, 70 F 35, 70 H 05, 70 H 15, 70 H 20

ISBN 3-540-05419-7 Springer-Verlag Berlin Heidelberg New York
ISBN 0-387-05419-7 Springer-Verlag New York Heidelberg Berlin

Library of Congress Catalog Card Number 71-155595. Printed in Germany. Printing and binding: Brühlsche Universitätsdruckerei, Gießen

In Memory of Franz Rellich

Preface to the First Edition (Translation)

I have lectured on the questions in celestial mechanics treated in this work at Frankfurt on Main and Baltimore as well as again at Göttingen and Princeton, most fully in a lecture series during the winter semester of 1951/52 at Göttingen. At that time Dr. J. Moser, now in New York, prepared a careful set of notes on which this publication is based.

Not being an astronomer by profession, I have made no attempt to present anew the standard methods for the determination of orbits, for which there certainly are good texts available. My aim was rather to develop some of the ideas and results that have evolved over the period of the past 70 years in the study of solutions to differential equations in the large, in which of course applications to Hamiltonian systems and in particular the equations of motion for the three-body problem occupy an important place. Even here I did not strive for completeness, but have made a selection dictated by personal interest and the hope of stimulating the listener within the framework of a lecture.

After preliminary considerations of the transformation theory for differential equations, our aim in the first chapter is to present the important results of K. F. Sundman on the three-body problem. Although Sundman's theorems are almost 50 years old, they have become known only within a small circle and have hardly effected subsequent developments. Next to Poincaré's work in the theory of differential equations, Sundman's work, despite its specialized character, belongs perhaps among the most significant developments in this area.

Even the still older "Méthodes nouvelles de la mécanique céleste" of Poincaré have not at all had the fruitful effect on the mathematics world at large one might have hoped for in view of the richness of the work. Of the next generation it was Birkhoff who penetrated these methods most deeply, and in addition to a simplified presentation and careful proofs he also contributed interesting new theorems. His book "Dynamical systems" has been a stimulus to me and is closely related to part of the problems dealt with in the remaining two chapters of our work. In the second chapter we treat the various methods for finding periodic solutions for systems of differential equations, wherein also the fixed-point method and the related work of Birkhoff is discussed in detail. In this I have for the most part assumed that we are dealing with analytic differential equations, and have derived the results by means of suitable manipulation of power series, whereby the algebraic conclusions are as far as possible kept separate from the analytic ones. A word of justification is needed as to why the investigation has been carried out only for differential equations that do not contain the independent variable t explicitly, whereas also the case of periodic dependence on t is certainly of particular interest. However, the methods in this general case do not in principle differ from those considered here, which already exhibit all the essential difficulties.

The third chapter deals with the problem of stability, and in addition to the classical result of Liapunov contains above all a discussion of the question of convergence in connection with the normal form of analytic differential equations near an equilibrium and the expansion of the general solution in trigonometric series. It was my desire also to give a complete proof at this time of the often cited theorem of Poincaré about the divergence of these series in celestial mechanics, but in this I have not succeeded. The recurrence theorem treated at the end does not quite fit into the framework of this book, but after the disappointments preceding it, it marks a conciliatory ending.

With regard to a more detailed bibliography, one should refer to Wintner's "Analytic foundations of celestial mechanics". In keeping with the character of the present work, also the list of references at the end is certainly incomplete; it serves only to name for the reader a few supplementary works to this text. The formulas here are numbered consecutively only within the individual sections. By $(a; b)$ in the text one should understand formula (b) from § a, while the symbol $[c]$ refers to the relevant place in the bibliography.

Göttingen, October 1955 Carl L. Siegel

Preface to the English Edition

The present book represents to a large extent the translation of the German „Vorlesungen über Himmelsmechanik" by C. L. Siegel. The demand for a new edition and for an English translation gave rise to the present volume which, however, goes beyond a mere translation. To take account of recent work in this field a number of sections have been added, especially in the third chapter which deals with the stability theory. Still, it has not been attempted to give a complete presentation of the subject, and the basic organization of Siegel's original book has not been altered. The emphasis lies in the development of results and analytic methods which are based on the ideas of H. Poincaré, G. D. Birkhoff, A. Liapunov and, as far as Chapter I is concerned, on the work of K. F. Sundman and C. L. Siegel. In recent years the measure-theoretical aspects of mechanics have been revitalized and have led to new results which will not be discussed here. In this connection we refer, in particular, to the interesting book by V. I. Arnold and A. Avez on "Problèmes Ergodiques de la Mécanique Classique", which stresses the interaction of ergodic theory and mechanics.

We list the points in which the present book differs from the German text. In the first chapter two sections on the triple collision in the three-body problem have been added by C. L. Siegel. Chapter II is essentially unchanged except for the inclusion of the convergence proof for the transformation into Birkhoff's normal form of an area-preserving mapping near a hyperbolic fixed-point. The main additions have been made in Chapter III. Section 26 contains a new and simpler proof for Siegel's theorem on conformal mappings near a fixed-point. Sections 32 to 36 contain a derivation of stability theorems for systems of two degrees of freedom as well as the existence theorem for quasi-periodic solutions, which are based on the work of Kolmogorov, Arnold and the undersigned. The responsibility for the accuracy of these additions rests with the undersigned.

The careful preparation of the translation is due to C. Kalme of the University of Southern California who carried out the difficult task of keeping the gist of the original book, maintaining the accuracy and producing a clear English text. In the proofreading we wish to record the help of R. Churchill and M. Braun. H. Rüssmann suggested a simplification in the proof of Section 33.

If it was possible to preserve the spirit of the original book, it is due to Siegel's close cooperation in reading the entire English manuscript and in checking the proofs.

Princeton, April 1971 Jürgen K. Moser

Table of Contents

Chapter One. The Three-Body Problem

Chapter Two. Periodic Solutions

Chapter Three. Stability

Chapter One

The Three-Body Problem

§ 1. Covariance of Lagrangian Derivatives

Ours, according to Leibniz, is the best of all possible worlds, and the laws of nature can therefore be described in terms of extremal principles. Thus, arising from corresponding variational problems, the differential equations of mechanics have invariance properties relative to certain groups of coordinate transformations. Because this is particularly important for celestial mechanics, in the preliminary sections we will develop as much of the transformation theory for the Euler-Lagrange and the Hamiltonian equations as is desirable for our purposes.

Let n be a natural number and $f = f(x, \dot{x}, t)$ a real valued function of the $2n+1$ independent real variables $x_k, \dot{x}_k, t$, where k ranges over the integers $1, \ldots, n$ and $x, \dot{x}$ are vectors with components $x_k, \dot{x}_k$. We restrict t to a closed interval $t_1 \leqq t \leqq t_2$ and the remaining variables $x, \dot{x}$ to an open set G in $2n$-dimensional space. The function f is to be defined for these values of $x, \dot{x}, t$ and have continuous partial derivatives up to second order. We consider the following problem in the calculus of variations: To determine twice continuously differentiable functions $x_k = x_k(t)$ $(k = 1, \ldots, n)$ of the variable t in the interval $t_1 \leqq t \leqq t_2$, with preassigned boundary values $x_k(t_1) = a_k$, $x_k(t_2) = b_k$, so that the integral

$$I = \int_{t_1}^{t_2} f(x, \dot{x}, t)\, dt$$

takes on an extremum; here $\dot{x}_k = \dfrac{dx_k(t)}{dt}$, while $x, \dot{x}$ are to lie in G.

Assuming that a solution to this extremal problem exists, we embed it into a family of admissible comparison functions $x_k = x_k(\alpha; t)$ $(k = 1, \ldots, n)$ so that $x_k(0; t) = x_k(t)$ becomes the given solution. We require that the $x_k(\alpha; t)$, together with their derivatives with respect to t, be continuously differentiable with respect to the parameter α in the interval $-1 < \alpha < 1$. If these comparison functions are inserted into the integral

$$I(\alpha) = \int_{t_1}^{t_2} f(x(\alpha; t), \dot{x}(\alpha; t), t)\, dt \qquad (-1 < \alpha < 1),$$

the function $I(\alpha)$ attains an extremum I at $\alpha=0$, and therefore its derivative $\frac{dI(\alpha)}{d\alpha}$ vanishes there.

To express this derivative more conveniently, we introduce some abbreviations: Considering f as a function of the $2n+1$ independent variables $x_l, \dot{x}_l, t$, we denote the partial derivatives of f with respect to $x_k, \dot{x}_k$ by $f_{x_k}, f_{\dot{x}_k}$, while differentiation with respect to the parameter α will be indicated by a prime. Then we have

$$I'(\alpha)=\int_{t_1}^{t_2}\sum_{k=1}^{n}\left\{f_{x_k}x_k'(\alpha;t)+f_{\dot{x}_k}\dot{x}_k'(\alpha;t)\right\}dt. \tag{1}$$

On the other hand, since $x_k(\alpha;t_1)=a_k, x_k(\alpha;t_2)=b_k$, the expression

$$s=s(\alpha;t)=\sum_{k=1}^{n} f_{\dot{x}_k}x_k'(\alpha;t)$$

vanishes at the boundary points $t=t_1,t_2$, and therefore

$$0=\int_{t_1}^{t_2}\frac{ds}{dt}dt=\int_{t_1}^{t_2}\sum_{k=1}^{n}\left\{\frac{df_{\dot{x}_k}}{dt}x_k'(\alpha;t)+f_{\dot{x}_k}\dot{x}_k'(\alpha;t)\right\}dt. \tag{2}$$

If we introduce the Lagrangian derivatives

$$L_{x_k}f=f_{x_k}-\frac{df_{\dot{x}_k}}{dt},$$

subtraction of (2) from (1) yields the formula

$$I'(\alpha)=\int_{t_1}^{t_2}\sum_{k=1}^{n}x_k'L_{x_k}f\,dt,$$

where one must insert $x_k=x_k(\alpha;t)$, $\dot{x}_k=\frac{dx_k}{dt}$ for the arguments. By the arbitrariness in the choice of $x_k'(0;t)$ and the continuity of $L_{x_k}f$, the condition $I'(0)=0$ gives the differential equations of Euler and Lagrange, namely

$$L_{x_k}f=0 \quad (k=1,\dots,n),$$

for the solution $x_k=x_k(0;t)=x_k(t)$ $(k=1,\dots,n)$.

We now examine the behavior of Lagrangian derivatives under coordinate transformations. Let new coordinates $\xi_1,\dots,\xi_n$ be introduced in place of $x_1,\dots,x_n$ by means of the substitution

$$x_k=x_k(\xi,t) \quad (k=1,\dots,n). \tag{3}$$

Here it is assumed that the $x_k(\xi,t)$ are twice continuously differentiable functions of the $n+1$ independent variables $\xi_1,\dots,\xi_n,t$ and that in the domain considered the Jacobian determinant of n rows satisfies

$$\left|\frac{\partial x_k}{\partial \xi_l}\right| \neq 0,$$

so that the transformation from ξ to x has a unique inverse. The variable t is not transformed, and the functions $x_k(\xi, t)$ have nothing to do with the $x_k = x_k(\alpha; t)$ introduced previously. If in particular one considers $x_k = x_k(\alpha; t)$, the $\xi_k = \xi_k(\alpha; t)$ also become functions of α, t, and from (3) we have

(4) $$\dot{x}_k = x_{kt} + \sum_{l=1}^{n} x_{k\xi_l} \dot{\xi}_l,$$

where x_{kt}, $x_{k\xi_l}$ denote partial derivatives of $x_k(\xi, t)$. By means of the substitutions (3), (4) also $f(x, \dot{x}, t)$ becomes a function of ξ, $\dot{\xi}$, t, with

$$I'(\alpha) = \int_{t_1}^{t_2} \sum_{k=1}^{n} \xi'_k L_{\xi_k} f \, dt,$$

and this, under the assumptions of continuity, implies the invariance of the expression

$$A = \sum_{k=1}^{n} x'_k L_{x_k} f$$

under coordinate transformations.

Algebraically this invariance can be shown as follows. By (3), (4) we have

$$x_{k\dot{\xi}_l} = 0, \quad \dot{x}_{k\dot{\xi}_l} = x_{k\xi_l}, \quad x'_k = \sum_{l=1}^{n} x_{k\xi_l} \xi'_l,$$

so that

$$f_{\dot{\xi}_l} = \sum_{k=1}^{n} (f_{x_k} x_{k\dot{\xi}_l} + f_{\dot{x}_k} \dot{x}_{k\dot{\xi}_l}) = \sum_{k=1}^{n} f_{\dot{x}_k} x_{k\xi_l}$$

$$\sum_{l=1}^{n} f_{\dot{\xi}_l} \xi'_l = \sum_{k,l=1}^{n} f_{\dot{x}_k} x_{k\xi_l} \xi'_l = \sum_{k=1}^{n} f_{\dot{x}_k} x'_k = s.$$

Hence s, as well as

$$\frac{ds}{dt} = \sum_{k=1}^{n} \left(\frac{df_{\dot{x}_k}}{dt} x'_k + f_{\dot{x}_k} \dot{x}'_k \right),$$

is invariant. The same is true of

$$f' = \sum_{k=1}^{n} (f_{x_k} x'_k + f_{\dot{x}_k} \dot{x}'_k),$$

and therefore finally of

$$f' - \frac{ds}{dt} = A .$$

We observe that in neither of the two proofs was it necessary to assume that $x_k = x_k(t)$ was a solution to the extremal problem.

From the arbitrariness of the x_k' it follows that

$$\sum_{k=1}^{n} x_{k\xi_l} L_{x_k} f = L_{\xi_l} f \qquad (l = 1, \ldots, n),$$

which is true in particular for $\alpha = 0$, and hence for $x_k = x_k(t)$. This determines the behavior of Lagrangian derivatives under coordinate transformations, and shows that the Euler-Lagrange equations remain invariant. Observe that if

$$x_\xi = (x_{k\xi_l}) = \mathfrak{M}$$

denotes the Jacobian matrix of the substitution (3), and $L_x = L_x f$ the row with entries $L_{x_k} f$, then

$$L_x \mathfrak{M} = L_\xi .$$

To formulate the invariance property of A without use of the parameter α we introduce the following new symbol: For φ a function of several independent variables, including t, let $\delta\varphi$ denote the differential of φ with t held constant, i.e. with $dt = 0$. Then if x is the column with entries x_k, as a consequence of (3) we have

$$\delta x = \mathfrak{M} \delta \xi .$$

Hence the L_{x_k} transform contragrediently to the δx_k, and the bilinear form $L_x \delta x$ remains invariant.

Finally, we wish to examine the extent to which the function f is determined by its n Lagrangian derivatives $L_{x_k} f$. Written out explicitly, these derivatives are given by

$$(5) \qquad L_{x_k} f = f_{x_k} - \sum_{l=1}^{n} (f_{\dot{x}_k x_l} \dot{x}_l + f_{\dot{x}_k \dot{x}_l} \ddot{x}_l) - f_{\dot{x}_k t} ,$$

where the right side is to be regarded as a function of the $3n+1$ independent variables $x_l, \dot{x}_l, \ddot{x}_l, t$. If for two functions $g(x, \dot{x}, t)$, $h(x, \dot{x}, t)$ the n Lagrangian derivatives $L_{x_k} g$, $L_{x_k} h$ pairwise agree as functions of these $3n+1$ variables, their difference $f = g - h$ satisfies the equations $L_{x_k} f = 0$ identically in $x, \dot{x}, \ddot{x}, t$. In particular, the coefficient of $\ddot{x}_l$ in (5) vanishes, so that $f_{\dot{x}_k \dot{x}_l} = 0$, and f has the form

$$f(x, \dot{x}, t) = f_0(x, t) + \sum_{k=1}^{n} f_k(x, t)\, \dot{x}_k .$$

Introducing this into (5) and comparing coefficients we obtain

$$f_{0x_k} = f_{kt}, \qquad f_{lx_k} = f_{kx_l} \qquad (k, l = 1, \ldots, n),$$

and these are precisely the necessary and sufficient conditions for the existence of a function $v(x, t)$ whose total differential is

$$dv = f_0\, dt + \sum_{k=1}^{n} f_k\, dx_k .$$

It follows that

$$f = \frac{dv}{dt},$$

$$h(x, \dot{x}, t) = g(x, \dot{x}, t) + \frac{dv(x, t)}{dt},$$

where in the total differentiation with respect to t the x_k are to be regarded as functions of t. Thus the Lagrangian derivatives, as functions of $x, \dot{x}, \ddot{x}, t$, determine f up to an additive function which is the total derivative with respect to t of a function $v(x, t)$ that does not depend on $\dot{x}$. On the other hand, addition of such a function changes the integral I only by a value that, in view of the boundary conditions, does not depend on the choice of the $x_k(t)$.

§ 2. Canonical Transformation

According to (1; 5) the Lagrangian derivatives generally involve second derivatives of the functions $x_k(t)$, and thus the corresponding system of n Euler-Lagrange equations is of second order. To rewrite it as a system of $2n$ first order differential equations we introduce the equations

(1) $$y_k = f_{\dot{x}_k}(x, \dot{x}, t) \qquad (k = 1, \ldots, n)$$

and solve for the $\dot{x}_k$ in terms of $x_1, \ldots, x_n, t$ and the new independent variables $y_1, \ldots, y_n$. This is possible if we assume the determinant of n rows

(2) $$|f_{\dot{x}_k \dot{x}_l}| \neq 0 .$$

Then

(3) $$L_{x_k} f = f_{x_k}(x, \dot{x}, t) - \dot{y}_k ,$$

where the dot over y_k denotes total differentiation with respect to t, and the Euler-Lagrange equations read

(4) $$\dot{y}_k = f_{x_k}(x, \dot{x}, t) \quad (k = 1, \ldots, n).$$

In (1), (4) we have a system of $2n$ first order differential equations for the $2n$ unknown functions $x_k(t)$, $y_k(t)$. To remove the asymmetry of this system we introduce the function

(5) $$E = \sum_{k=1}^{n} \dot{x}_k y_k - f(x, \dot{x}, t),$$

treating the $3n+1$ variables $x_k, y_k, \dot{x}_k, t$ at first as being independent. Then

(6) $$dE = \sum_{k=1}^{n} (\dot{x}_k \, dy_k + y_k \, d\dot{x}_k - f_{x_k} dx_k - f_{\dot{x}_k} d\dot{x}_k) - f_t \, dt.$$

If $\dot{x}$ is now determined from (1) as a function of x, y, t, then $E = E(x, y, t)$ also becomes a function of x, y, t only, and the coefficients of $d\dot{x}_k$ in (6) mutually cancel. We thus obtain

$$dE = \sum_{k=1}^{n} (\dot{x}_k \, dy_k - f_{x_k} dx_k) - f_t \, dt$$

for the total differential of $E(x, y, t)$, and this gives

(7) $$E_{x_k} = -f_{x_k}, \quad E_{y_k} = \dot{x}_k \quad (k = 1, \ldots, n)$$

for the values of the partial derivatives of E as a function of x, y, t, while (3) now becomes

(8) $$L_{x_k} f = -E_{x_k} - \dot{y}_k.$$

Thus (1), (5), and the Euler-Lagrange equations lead to

$$\dot{x}_k = E_{y_k}, \quad \dot{y}_k = -E_{x_k} \quad (k = 1, \ldots, n),$$

and these are the Hamiltonian equations.

By (2) the Jacobian determinant $|y_{k\dot{x}_l}|$ for the y_k considered as functions of the $\dot{x}_l$, as well as the corresponding inverse determinant $|\dot{x}_{k y_l}|$, is different from 0, and therefore by (7) also $|E_{y_k y_l}| \neq 0$. Conversely, suppose $E(x, y, t)$ is a given function with the determinant

(9) $$|E_{y_k y_l}| \neq 0.$$

In analogy with (5), we define

(10) $$f = \sum_{k=1}^{n} \dot{x}_k y_k - E(x, y, t)$$

and again treat the $3n+1$ variables $x_k, y_k, \dot{x}_k, t$ as being independent, whereupon

$$df = \sum_{k=1}^{n} (\dot{x}_k\, dy_k + y_k\, d\dot{x}_k - E_{x_k}\, dx_k - E_{y_k}\, dy_k) - E_t\, dt .$$

The equations

$$\dot{x}_k = E_{y_k} \qquad (k = 1, \ldots, n)$$

are then solved for y as a function of $x, \dot{x}, t$, which is possible because of (9), whereby f becomes a function of $x, \dot{x}, t$ only, while

$$df = \sum_{k=1}^{n} (y_k\, d\dot{x}_k - E_{x_k}\, dx_k) - E_t\, dt .$$

Therefore

$$(11) \qquad f_{x_k} = -E_{x_k}, \quad f_{\dot{x}_k} = y_k \quad (k = 1, \ldots, n),$$

and this again leads to (8). Finally, by (9), the Jacobian determinant $|\dot{x}_{k y_l}|$ for the $\dot{x}_k$ as functions of the y_l, as well as the corresponding inverse determinant $|y_{k \dot{x}_l}|$, is different from 0, and therefore by (11) also $|f_{\dot{x}_k \dot{x}_l}| \neq 0$. Thus from (9) and the Hamiltonian equations we have again derived (1), (2), and the Euler-Lagrange equations.

One half of the $2n$ Hamiltonian equations, namely $\dot{y}_k = -E_{x_k}$, come directly from the Euler-Lagrange equations, while the other half are determined by the substitutions (1), (5). It is interesting to observe that collectively the $2n$ Hamiltonian equations can be interpreted as Euler-Lagrange equations. For this we use the variables x, y in place of x in § 1, and in the function f defined by (10) regard the $4n+1$ variables $x_k, y_k, \dot{x}_k, \dot{y}_k, t$ as being independent. Of course, f is actually independent of the $\dot{y}_k$. From the definition of Lagrangian derivatives we then have

$$(12) \quad L_{x_k} f = f_{x_k} - \frac{df_{\dot{x}_k}}{dt} = -E_{x_k} - \dot{y}_k, \quad L_{y_k} f = f_{y_k} - \frac{df_{\dot{y}_k}}{dt} = \dot{x}_k - E_{y_k},$$

and equating the right sides to 0 we obtain the Hamiltonian equations.

This reformulation allows us to apply our earlier results relating to the transformation of Lagrangian derivatives. We will now investigate substitutions of the form

$$(13) \qquad x_k = x_k(\xi, \eta, t), \quad y_k = y_k(\xi, \eta, t) \quad (k = 1, \ldots, n),$$

whereby $\xi_1, \ldots, \xi_n, \eta_1, \ldots, \eta_n$ become new variables while t remains unchanged. We denote the $2n$ quantities $x_1, \ldots, x_n, y_1, \ldots, y_n$ by $z_1, \ldots, z_{2n}$ and interpret $\zeta_1, \ldots, \zeta_{2n}$ accordingly, so that (13) can be expressed more briefly as

$$z_k = z_k(\zeta, t) \quad (k = 1, \ldots, 2n) .$$

The Jacobian matrices

(14) $\mathfrak{A} = x_\xi = (x_{k\xi_l}), \quad \mathfrak{B} = x_\eta = (x_{k\eta_l}), \quad \mathfrak{C} = y_\xi = (y_{k\xi_l}), \quad \mathfrak{D} = y_\eta = (y_{k\eta_l})$

of n rows, and

$$\mathfrak{M} = z_\zeta = (z_{k\zeta_l})$$

are then related by

$$\mathfrak{M} = \begin{pmatrix} \mathfrak{A} & \mathfrak{B} \\ \mathfrak{C} & \mathfrak{D} \end{pmatrix}, \tag{15}$$

and we assume that $|\mathfrak{M}| \neq 0$. We wish to determine under what conditions on the substitution (13) the Lagrangian derivatives in the new variables again take the special form (12), with an unknown function $\mathsf{E}(\xi, \eta, t)$ in place of $E(x, y, t)$. For this we need

$$L_{\xi_k} f = -\mathsf{E}_{\xi_k} - \dot{\eta}_k, \quad L_{\eta_k} f = \dot{\xi}_k - \mathsf{E}_{\eta_k} \quad (k = 1, \ldots, n) \tag{16}$$

to hold identically in $\zeta, \dot{\zeta}, t$, where the function f is determined by (10). Under these conditions the transformation (13) is said to be canonical.

If we set

$$\phi(\zeta, \dot{\zeta}, t) = \sum_{k=1}^{n} \dot{\xi}_k \eta_k - \mathsf{E}(\zeta, t), \tag{17}$$

the Lagrangian derivatives of ϕ are just the right sides of (16), and by the result of § 1 we must therefore have

$$f = \phi + \frac{dv(\zeta, t)}{dt}$$

identically in $\zeta, \dot{\zeta}, t$, when f is expressed in terms of these variables. By (10), (17) this means that

$$\begin{cases} \dfrac{dv(\zeta, t)}{dt} = \mathsf{E} - E + \displaystyle\sum_{r=1}^{n} \dot{x}_r y_r - \sum_{l=1}^{n} \dot{\xi}_l \eta_l \\ = \mathsf{E} - E + \displaystyle\sum_{l=1}^{n} \left(\sum_{r=1}^{n} x_{r\xi_l} y_r - \eta_l \right) \dot{\xi}_l + \sum_{l=1}^{n} \left(\sum_{r=1}^{n} x_{r\eta_l} y_r \right) \dot{\eta}_l + \sum_{k=1}^{n} x_{kt} y_k , \end{cases} \tag{18}$$

and therefore

$$v_{\xi_l} = \sum_{r=1}^{n} x_{r\xi_l} y_r - \eta_l, \quad v_{\eta_l} = \sum_{r=1}^{n} x_{r\eta_l} y_r, \quad v_t = \mathsf{E} - E + \sum_{k=1}^{n} x_{kt} y_k .$$

This prescribes the $2n+1$ first order partial derivatives of $v(\zeta, t)$. The last of these equations can be thought of as defining the unknown function E, while the remaining ones lead, via the relations $v_{\zeta_k \zeta_l} = v_{\zeta_l \zeta_k}$, to the necessary and sufficient conditions for integrability, namely

$$\left.\begin{aligned}\sum_{r=1}^{n}(x_{r\xi_l\xi_k}y_r+x_{r\xi_l}y_{r\xi_k})&=\sum_{r=1}^{n}(x_{r\xi_k\xi_l}y_r+x_{r\xi_k}y_{r\xi_l})\\ \sum_{r=1}^{n}(x_{r\xi_l\eta_k}y_r+x_{r\xi_l}y_{r\eta_k})-e_{kl}&=\sum_{r=1}^{n}(x_{r\eta_k\xi_l}y_r+x_{r\eta_k}y_{r\xi_l})\\ \sum_{r=1}^{n}(x_{r\eta_l\eta_k}y_r+x_{r\eta_l}y_{r\eta_k})&=\sum_{r=1}^{n}(x_{r\eta_k\eta_l}y_r+x_{r\eta_k}y_{r\eta_l})\end{aligned}\right\}(k,l=1,\ldots,n),$$

where $e_{kl}=1$ for $k=l$ and 0 otherwise. Using the matrices defined in (14), we can express these conditions in the form

$$\mathfrak{C}'\mathfrak{A}=\mathfrak{A}'\mathfrak{C},\quad \mathfrak{D}'\mathfrak{A}-\mathfrak{E}=\mathfrak{B}'\mathfrak{C},\quad \mathfrak{D}'\mathfrak{B}=\mathfrak{B}'\mathfrak{D},\tag{19}$$

where $\mathfrak{E}$ is the $n\times n$ identity matrix and the prime denotes transposition. If we introduce the matrix

$$\mathfrak{J}=\begin{pmatrix}0&\mathfrak{E}\\-\mathfrak{E}&0\end{pmatrix}$$

and use (15), the equations (19) combine into

$$\mathfrak{M}'\mathfrak{J}\mathfrak{M}=\mathfrak{J}.\tag{20}$$

This formula gives necessary and sufficient conditions for the transformation (13) to leave the Hamiltonian form (12) of the Lagrangian derivatives unchanged, and it is evident that the conditions do not depend on the function $E(z,t)$. Solving for $v(\zeta,t)$ from the $2n$ partial derivatives v_{ξ_l}, v_{η_l} alone leaves an additive function of t undetermined, and such an arbitrary additive term then appears in v_t as well as in E. However, it again drops out when we form the derivatives $\mathsf{E}_{\xi_k}, \mathsf{E}_{\eta_k}$, so that the right sides of (16) are completely determined.

A matrix $\mathfrak{M}$ that satisfies equation (20) is called symplectic. In forming the determinant we see that $|\mathfrak{M}|^2\,|\mathfrak{J}|=|\mathfrak{J}|=1$, so that $|\mathfrak{M}|^2=1$. It can actually be shown that $|\mathfrak{M}|=1$, but this will not be needed. In any case, for a symplectic matrix $\mathfrak{M}$ the determinant $|\mathfrak{M}|\neq 0$ and therefore $\mathfrak{M}^{-1}$ exists. From (20) it follows that

$$(\mathfrak{M}^{-1})'\mathfrak{J}\mathfrak{M}^{-1}=(\mathfrak{M}^{-1})'(\mathfrak{M}'\mathfrak{J}\mathfrak{M})\,\mathfrak{M}^{-1}=\mathfrak{J},$$

and consequently $\mathfrak{M}^{-1}$ is also symplectic. Similarly, if $\mathfrak{M}_1$ and $\mathfrak{M}_2$ are symplectic so is $\mathfrak{M}_1\mathfrak{M}_2$. Thus under multiplication the symplectic matrices form a group, the symplectic group. We have shown that a transformation $z=z(\zeta,t)$ is canonical if and only if the Jacobian matrix $z_\zeta=\mathfrak{M}$ is symplectic identically in ζ, t. Hence the inverse transformation is also canonical, and more generally, under suitable assumptions about the ranges of the variables, the canonical transformations form a group.

In particular, canonical transformations take each Hamiltonian system of differential equations into another such system. One can pose the more general problem of exhibiting all invertible transformations having this property. For this, instead of the differential equations, we consider the corresponding Hamiltonian expressions $\dot{x}_k - E_{y_k}$, $\dot{y}_k + E_{x_k}$, which can be combined into the column $\dot{z} - \mathfrak{J}E_z$, where E_z denotes the column with entries E_{x_k}, E_{y_k}. If $z = z(\zeta, t)$ is a substitution with Jacobian matrix $z_\zeta = \mathfrak{M}$, then

$$E_{\zeta_k} = \sum_{l=1}^{2n} E_{z_l} z_{l\zeta_k}, \qquad E_\zeta = \mathfrak{M}' E_z,$$

$$\dot{z} = \mathfrak{M}\dot{\zeta} + z_t, \tag{21}$$

so that

$$\mathfrak{M}^{-1}(\dot{z} - \mathfrak{J}E_z) = \dot{\zeta} + \mathfrak{M}^{-1} z_t - \mathfrak{M}^{-1}\mathfrak{J}\mathfrak{M}'^{-1} E_\zeta .$$

If the right side of the last equation is again to have the Hamiltonian form $\dot{\zeta} - \mathfrak{J}\mathsf{E}_\zeta$ for a suitable function $\mathsf{E}(\zeta, t)$, this function must satisfy the equation

$$\mathsf{E}_\zeta = \mathfrak{J}^{-1}\mathfrak{M}^{-1}\mathfrak{J}\mathfrak{M}'^{-1} E_\zeta - \mathfrak{J}^{-1}\mathfrak{M}^{-1} z_t .$$

With

$$\mathfrak{J}^{-1}\mathfrak{M}^{-1}\mathfrak{J}\mathfrak{M}'^{-1} = \mathfrak{P} = (p_{kl}), \qquad -\mathfrak{J}^{-1}\mathfrak{M}^{-1} = \mathfrak{Q} = (q_{kl}),$$

the integrability conditions become

$$\sum_{r=1}^{2n} (p_{kr} E_{\zeta_r} + q_{kr} z_{rt})_{\zeta_l} = \sum_{r=1}^{2n} (p_{lr} E_{\zeta_r} + q_{lr} z_{rt})_{\zeta_k} \qquad (k, l = 1, \ldots, 2n).$$

If these are to be fulfilled for every choice of the function $E(z, t)$ we must have

$$\sum_{r=1}^{2n} p_{kr} E_{\zeta_r \zeta_l} = \sum_{r=1}^{2n} p_{lr} E_{\zeta_r \zeta_k}, \qquad \sum_{r=1}^{2n} p_{kr\zeta_l} E_{\zeta_r} = \sum_{r=1}^{2n} p_{lr\zeta_k} E_{\zeta_r},$$

and therefore

$$p_{kl} = 0, \qquad p_{kk} = p_{ll} \ (k \neq l), \qquad p_{kr\zeta_l} = p_{lr\zeta_k} \ (k, l, r = 1, \ldots, 2n).$$

Consequently $\mathfrak{P}$ can differ from the identity matrix only by a scalar factor that does not depend on ζ.

We still have to satisfy the remaining conditions

$$\sum_{r=1}^{2n} (q_{kr} z_{rt})_{\zeta_l} = \sum_{r=1}^{2n} (q_{lr} z_{rt})_{\zeta_k} . \tag{22}$$

Abbreviating $\mathfrak{J}z_t = u$ and using the relation

$$\mathfrak{P}^{-1}\mathfrak{Q} = -\mathfrak{M}'\mathfrak{J}^{-1} = (z_{l\zeta_k})\,\mathfrak{J},$$

we can express equations (22) as

$$\sum_{r=1}^{2n} (z_{r\zeta_k\zeta_l} u_r + z_{r\zeta_k} u_{r\zeta_l}) = \sum_{r=1}^{2n} (z_{r\zeta_l\zeta_k} u_r + z_{r\zeta_l} u_{r\zeta_k}),$$

or in matrix form,

$$\mathfrak{M}'\mathfrak{J}\mathfrak{M}_t = (\mathfrak{M}'\mathfrak{J}\mathfrak{M}_t)'.$$

Since $\mathfrak{J}' = -\mathfrak{J}$, this says that

$$\mathfrak{M}'\mathfrak{J}\mathfrak{M}_t + \mathfrak{M}_t'\mathfrak{J}\mathfrak{M} = 0,$$

so that the matrix $\mathfrak{M}'\mathfrak{J}\mathfrak{M}$, and therefore also $\mathfrak{P}$, is independent of t. Thus, as a necessary and sufficient condition for the transformation to have the desired property we have the relation

$$\mathfrak{M}'\mathfrak{J}\mathfrak{M} = \lambda\mathfrak{J}, \tag{23}$$

where $\lambda \neq 0$ is a constant scalar. The appearance of the arbitrary factor λ leads to a generalization of the canonical substitutions and the symplectic group. However, the particular substitution $x = \xi$, $y = \lambda\eta$ satisfies (23), and all such substitutions can be obtained by combining the canonical ones with this trivial one. For this reason we will henceforth restrict ourselves to canonical substitutions.

§ 3. The Hamilton-Jacobi Equation

We now treat the problem of expressing canonical transformations collectively in parametric form. First we consider the case

$$|\mathfrak{B}| = |x_{k\eta_l}| \neq 0. \tag{1}$$

The n equations $x_k = x_k(\xi, \eta, t)$ then determine the η_l as functions of x, ξ, t, and the corresponding Jacobian determinant $|\eta_{kx_l}|$ is also different from 0. We use equation (2; 18), which gives the necessary and sufficient condition for a substitution to be canonical, and express the function $v(\zeta, t)$ in terms of x, ξ, t. Thus setting

$$v = v(\zeta, t) = w(x, \xi, t),$$

we have

$$\frac{dv}{dt} = w_t + \sum_{k=1}^{n} (w_{x_k}\dot{x}_k + w_{\xi_k}\dot{\xi}_k),$$

and comparison of coefficients in (2;18) leads to the relations

$$(2) \qquad y_k = w_{x_k}, \quad \eta_k = -w_{\xi_k} \ (k=1,\dots,n), \quad \mathsf{E} = E + w_t .$$

From $|\eta_{k x_l}| \neq 0$ it also follows that

$$(3) \qquad |w_{\xi_k x_l}| \neq 0 .$$

On the other hand, if w is a function of x, ξ, t satisfying (3), the second equation in (2) can be solved for x as a function of ξ, η, t, and substitution into the first equation of (2) then determines also y as a function of ξ, η, t. Moreover, (1) is also satisfied. Conversely, by (3) the first equation in (2) can be solved for ξ as a function of x, y, t, and substitution into the second equation of (2) then determines also η as a function of x, y, t. If in addition the third equation in (2) is used to define E, then (2;18) is also satisfied, and the resulting transformation is therefore canonical, with $|\mathfrak{B}| \neq 0$. Consequently (2) generates all canonical transformations for which $|\mathfrak{B}| \neq 0$, with the third equation determining E.

As an example, let

$$w = \sum_{k=1}^{n} x_k \xi_k .$$

Then (3) is satisfied, and the resulting canonical transformation is $x_k = -\eta_k$, $y_k = \xi_k$ $(k=1,\dots,n)$ whose Jacobian matrix is $\mathfrak{M} = -\mathfrak{J} = \mathfrak{J}^{-1}$. There are, of course, canonical transformations for which $|\mathfrak{B}| = 0$, as for example the identity transformation $z = \zeta$, for which $\mathfrak{M}$ is the identity matrix. The case

$$|\mathfrak{A}| = |x_{k\xi_l}| \neq 0$$

can be handled in a manner analogous to the above treatment for $|\mathfrak{B}| \neq 0$. However, it is shorter to reduce it to the previous case by the additional substitution $\xi = -\eta^*$, $\eta = \xi^*$. The Jacobian matrix of z as a function of ζ^* then becomes

$$\mathfrak{M}\mathfrak{J}^{-1} = \begin{pmatrix} \mathfrak{B} & -\mathfrak{A} \\ \mathfrak{D} & -\mathfrak{C} \end{pmatrix},$$

so that $-\mathfrak{A}$ appears in place of $\mathfrak{B}$, and the formulas in (2) are valid with ξ^*, η^* in place of ξ, η. It follows that

$$(4) \qquad y_k = w_{x_k}, \quad \xi_k = w_{\eta_k} \ (k=1,\dots,n), \quad \mathsf{E} = E + w_t ,$$

where $w = w(x, \eta, t)$, and

$$|w_{\eta_k x_l}| \neq 0 .$$

This generates all canonical transformations for which $|\mathfrak{A}| \neq 0$. In particular,

$$w = \sum_{k=1}^{n} x_k \eta_k$$

leads to the identity transformation $z = \zeta$. We will not discuss the case when both $|\mathfrak{A}|$ and $|\mathfrak{B}|$ are 0. However, we note that every canonical transformation can be expressed as a composition of two for which either $|\mathfrak{A}| \neq 0$ or $|\mathfrak{B}| \neq 0$.

Our subsequent aim is to simplify a given Hamiltonian system

$$(5) \qquad \dot{x}_k = E_{y_k}, \quad \dot{y}_k = -E_{x_k} \quad (k = 1, \dots, n)$$

as far as possible by means of an appropriate canonical transformation. The simplest case for the transformed system

$$\dot{\xi}_k = \mathsf{E}_{\eta_k}, \quad \dot{\eta}_k = -\mathsf{E}_{\xi_k}$$

would be to have $\mathsf{E}(\xi, \eta, t) = 0$. If we assume that this can be achieved by a transformation with $|\mathfrak{A}| \neq 0$, then (4) must hold, and from it we see that the generating function $w = w(x, \eta, t)$ must satisfy the partial differential equation

$$(6) \qquad E(x, w_x, t) + w_t = 0,$$

subject to the condition $|w_{x_k \eta_l}| \neq 0$. This is the Hamilton-Jacobi equation. Conversely, if one can find a solution $w(x, \eta, t)$ of (6) that depends on n parameters $\eta_1, \dots, \eta_n$ and has the determinant $|w_{x_k \eta_l}|$ different from 0, then (4) determines a canonical transformation which takes the Hamiltonian system (5) into

$$(7) \qquad \dot{\xi}_k = 0, \quad \dot{\eta}_k = 0 \quad (k = 1, \dots, n).$$

These equations can be integrated directly, and the ξ_k, η_k enter as constants of integration. Thus, solving the Hamiltonian system of ordinary differential equations reduces to solving the Hamilton-Jacobi partial differential equation. Observe that it is by no means necessary to obtain the general solution of (6), but only a solution that depends on n parameters $\eta_1, \dots, \eta_n$ and satisfies $|w_{x_k \eta_l}| \neq 0$. This is a simplification also in that a complete solution to the Hamiltonian equations (5) requires $2n$ integration constants.

Does there always exist a canonical transformation that takes a given system (5) into the normal form (7)? The answer, as we will now show, is affirmative.

Consider first a general system of first order differential equations

$$\dot{z}_k = g_k(z, t) \quad (k = 1, \dots, m)$$

for m unknown functions $z_k = z_k(t)$. By known existence theorems, for given initial conditions $t=\tau$, $z=\zeta$ there exists a unique solution $z=z(\zeta, t)$, provided, say, a Lipschitz condition is satisfied and the range of the variables is suitably restricted. Keeping τ fixed, consider ζ, t as independent variables. For $z=z(\zeta, t)$ the differential equations imply that

(8) $$z_{kt} = g_k(z, t),$$

and therefore

$$z_{kt\zeta_l} = \sum_{r=1}^{m} g_{kz_r} z_{r\zeta_l} \qquad (k, l = 1, \ldots, m).$$

Thus, setting

$$z_\zeta = (z_{k\zeta_l}) = \mathfrak{M}, \qquad g_z = (g_{kz_l}) = \mathfrak{G},$$

one obtains the homogeneous system of first order linear differential equations

(9) $$\mathfrak{M}_t = \mathfrak{G}\mathfrak{M}$$

for the Jacobian matrix $\mathfrak{M}$ as a function of t, with ζ fixed. Because $z(\zeta, \tau) = \zeta$ identically in ζ, we have $\mathfrak{M} = \mathfrak{E}$ at $t = \tau$.

For a Hamiltonian system $\dot{z} = \mathfrak{J}E_z$ one has $m = 2n$ and

$$\mathfrak{G} = \mathfrak{J}E_{zz} = \mathfrak{J}(E_{z_k z_l}),$$

so that the matrix

$$\mathfrak{J}\mathfrak{G} = -E_{zz}$$

is symmetric. This, in conjunction with (9), implies that the matrix

$$\mathfrak{M}'\mathfrak{J}\mathfrak{G}\mathfrak{M} = \mathfrak{M}'\mathfrak{J}\mathfrak{M}_t$$

is also symmetric, and therefore

$$\mathfrak{M}'\mathfrak{J}\mathfrak{M}_t = (\mathfrak{M}'\mathfrak{J}\mathfrak{M}_t)' = \mathfrak{M}_t'\mathfrak{J}'\mathfrak{M} = -\mathfrak{M}_t'\mathfrak{J}\mathfrak{M}.$$

Consequently $\mathfrak{M}'\mathfrak{J}\mathfrak{M}$ is independent of t, and, since $\mathfrak{M} = \mathfrak{E}$ for $t = \tau$, it follows that

$$\mathfrak{M}'\mathfrak{J}\mathfrak{M} = \mathfrak{J}.$$

Therefore the transformation $z = z(\zeta, t)$ is canonical, and moreover, as a consequence of (2;21) and (8), it takes the given system into $\dot{\zeta} = 0$. Furthermore, the determinant $|\mathfrak{A}| = |x_{k\xi_l}|$, being equal to 1 for $t = \tau$, will not vanish for t sufficiently near τ. Hence the transformation $z = z(\zeta, t)$ can actually be obtained by solving the Hamilton-Jacobi equation (6) and using (4).

Even if it is not possible to find an integral $w(x, \eta, t)$ of the Hamilton-Jacobi equation directly, the problem can sometimes be simplified as follows. Suppose that the Hamiltonian function $E(x, y, t)$ can be decomposed as a sum of two terms

$$E(x, y, t) = F(x, y, t) + G(x, y, t),$$

and that for the first term an integral $w(x, \eta, t)$, with $|w_{x_k \eta_l}| \neq 0$, of the Hamilton-Jacobi equation

$$F(x, w_x, t) + w_t = 0$$

is known. For the corresponding canonical transformation (4) we then have

$$\mathsf{E} = E + w_t = F + G + w_t = G,$$

so that, with G regarded as a function of ξ, η, t, the original system (5) becomes

$$\dot{\xi}_k = G_{\eta_k}, \quad \dot{\eta}_k = -G_{\xi_k} \quad (k = 1, \ldots, n).$$

In certain cases this system can be solved directly, or, if not, the function G can possibly again be split up and the same reduction process carried out. In this way we can perhaps decompose $E(x, y, t)$ into a finite or infinite sum, such that the corresponding Hamilton-Jacobi equation can be solved at each step. Even if the process should have infinitely many steps and fail to converge, by terminating it at a suitable place one may still obtain a useful approximate solution.

In the preceding analysis we have repeatedly used the implicit function theorem, emphasizing only the condition that certain Jacobian determinants do not vanish, and not giving precise ranges for the corresponding variables. Thus the results have only local character, and for behavior in the large each individual case must be treated separately.

§ 4. The Cauchy Existence Theorem

We consider a system of first order differential equations

(1) $$\dot{x}_k = f_k(x) \quad (k = 1, \ldots, m),$$

where the m functions $f_k = f_k(x)$ depend on $x_1, \ldots, x_m$, but not on t itself. It is known that if the f_k satisfy a Lipschitz condition in a real neighborhood of $x = \xi$, then for given initial values $x(\tau) = \xi$ the system (1) has a unique solution. We will even assume that the f_k are regular analytic functions of the variables $x_1, \ldots, x_m$ in a complex neighborhood of $x = \xi$, and our aim is to prove the following theorem:

If the functions $f_1, \ldots, f_m$ are regular and bounded in absolute value by M in the domain $|x_k - \xi_k| < r$ $(k = 1, \ldots, m)$, then in the complex neighborhood

$$|t - \tau| < \frac{r}{(m+1)M}$$

of τ the solution $x_k(t)$ of (1), determined by the initial conditions $x_k(\tau)=\xi_k$ $(k=1,\ldots,m)$, is a regular analytic function of t and satisfies

$$|x_k(t)-\xi_k|<r \quad (k=1,\ldots,m).$$

If the quantities x_k, f_k, t are replaced by ξ_k+rx_k, Mf_k, $\tau+M^{-1}rt$, the system (1) remains unchanged, while the constants ξ_k, r, M, τ are replaced by 0, 1, 1, 0. Therefore it is sufficient to prove the theorem for these particular values. To solve (1) with initial conditions $x_k(0)=0$, we set

$$x_k=x_k(t)=\sum_{n=1}^{\infty}\alpha_{k,n}t^n \quad (k=1,\ldots,m)$$

and determine the coefficients by inserting this expression into the differential equation. For convenience we introduce the following notation. If

$$\phi=\sum_{k=0}^{\infty}c_k t^k$$

is a formal power series, irrespective of convergence, let

$$\phi_n=\sum_{k=0}^{n}c_k t^k, \qquad (\phi)_n=c_n \quad (n=0,1,\ldots).$$

Then $(\phi)_n=(\phi_n)_n$ and, if ψ is another formal power series in t, also $(\phi\psi)_n=(\phi_n\psi_n)_n$, $(\phi\pm\psi)_n=(\phi_n\pm\psi_n)_n$. Suppose now that

$$f_k=\sum_l a_{k,l_1\ldots l_m}x_1^{l_1}\ldots x_m^{l_m} \quad (k=1,\ldots,m),$$

where the sum is taken over all m-tuples l of nonnegative integers $l_1,\ldots,l_m$. Then by comparison of coefficients (1) gives

$$(2) \qquad \begin{cases} (n+1)\,\alpha_{k,n+1}=\sum_l a_{k,l_1\ldots l_m}(x_1^{l_1}\ldots x_m^{l_m})_n \\ \qquad\qquad =\sum_l a_{k,l_1\ldots l_m}(x_{1n}^{l_1}\ldots x_{mn}^{l_m})_n \quad (n=0,1,\ldots), \end{cases}$$

and it follows by induction that the $\alpha_{k,n}$ are polynomials in the $a_{r,l_1\ldots l_m}$ $(r=1,\ldots,m)$ with nonnegative rational coefficients.

To prove convergence, we use the method of majorants. If

$$f=\sum_l a_{l_1\ldots l_m}x_1^{l_1}\ldots x_m^{l_m}, \quad g=\sum_l b_{l_1\ldots l_m}x_1^{l_1}\ldots x_m^{l_m}$$

are two formal power series, which need not converge, then g is said to be a majorant of f, symbolically $f\prec g$, if

$$|a_{l_1\ldots l_m}|\leqq b_{l_1\ldots l_m}$$

for all the coefficients. In particular, the coefficients of g must be real and nonnegative. Suppose now that $f_k\prec g_k$ $(k=1,\ldots,m)$ and, together with (1), consider the majorant system

$$\dot{y}_k = g_k(y) \quad (k=1, \ldots, m) \tag{3}$$

with initial conditions $y_k(0)=0$. This can again be solved formally by a power series

$$y_k = y_k(t) = \sum_{n=1}^{\infty} \beta_{k,n} t^n,$$

and we claim that $y_k(t)$ is also a majorant of $x_k(t)$, or that

$$|\alpha_{k,\nu}| \leqq \beta_{k,\nu} \quad (k=1, \ldots, m;\ \nu = 1, 2, \ldots). \tag{4}$$

Indeed, since the coefficients $b_{k, l_1 \ldots l_m}$ of g_k are nonnegative, the recursion formulas

$$(n+1)\, \beta_{k,n+1} = \sum_l b_{k, l_1 \ldots l_m} (y_{1n}^{l_1} \ldots y_{mn}^{l_m})_n \quad (n=0, 1, \ldots)$$

corresponding to (2) imply that the $\beta_{k,\nu}$ are all nonnegative real numbers, and assertion (4) can now be proved by induction. We assume this assertion to hold for $\nu \leqq n$, an assumption which is vacuous for $n=0$. Then we have $x_{kn} \prec y_{kn}$ and therefore

$$\begin{aligned} (n+1)\, |\alpha_{k,n+1}| &\leqq \sum_l |a_{k, l_1 \ldots l_m}|\, |(x_{1n}^{l_1} \ldots x_{mn}^{l_m})_n| \\ &\leqq \sum_l b_{k, l_1 \ldots l_m} (y_{1n}^{l_1} \ldots y_{mn}^{l_m})_n = (n+1)\, \beta_{k,n+1}\,. \end{aligned}$$

This verifies (4) for $\nu = n+1$, and proves that $x_k \prec y_k$.

Consequently, it is sufficient to find a majorant g_k of f_k for which one can solve the corresponding system (3) explicitly, and prove the estimates claimed in the theorem for its solution. As in the case of one variable, the assumption $|f_k| \leqq 1$ for $|x_1| < 1, \ldots, |x_m| < 1$ implies by Cauchy's formula that

$$|a_{k, l_1 \ldots l_m}| \leqq 1\,,$$

and therefore the particular power series

$$g(x) = g_k(x) = \sum_l x_1^{l_1} \ldots x_m^{l_m} = \prod_{r=1}^{m} (1 - x_r)^{-1}$$

majorizes all the $f_k(x)$, independently of k. Since the right sides of the equations

$$\dot{y}_k = g(y)\,, \quad y_k(0) = 0 \quad (k = 1, \ldots, m)$$

do not depend on k, the solutions $y_k(t)$ are all given by a single power series $y = y(t)$ satisfying

$$\dot{y} = (1-y)^{-m}, \quad y(0) = 0\,,$$

and direct integration gives

$$1-(1-y)^{m+1}=(m+1)\,t$$

$$y=1-\{1-(m+1)\,t\}^{\frac{1}{m+1}}.$$

This function, on the other hand, has a convergent power series for $|t|<(m+1)^{-1}$ and in this circle satisfies

$$|y|\leqq 1-\{1-(m+1)\,|t|\}^{\frac{1}{m+1}}<1,$$

so that certainly also $|x_k(t)|<1$. Thus the power series obtained recursively for the $x_k(t)$, which satisfy the system (1) formally, converge for $|t|<(m+1)^{-1}$, and the functions defined by these series are therefore solutions to the given system of differential equations. This proves the theorem.

From the recursion formulas for the coefficients in the expansion of the solution $x_k(t)$ in powers of $t-\tau$, it follows that if the initial values ξ_k $(k=1,\ldots,m)$ and the coefficients of $f_k(x)$ are real, then so are the coefficients of $x_k(t)$. From now on we make this assumption, and also take τ to be real. Since we are interested in the solution $x_k(t)$ of (1) for real $t\geqq\tau$, let us assume that all the functions $x_k(t)$ $(k=1,\ldots,m)$ are regular in the interval $\tau\leqq t<t_1$ and, moreover, that for $\tau\leqq t<t_1$ the curve $x=x(t)$ lies in a closed bounded point-set P in which the m functions $f_k(x)$ are regular in the complex variables $x_1,\ldots,x_m$. We will show that then, as a consequence of the existence theorem, the $x_k(t)$ are regular also at $t=t_1$.

Since the $f_k(x)$ are regular in the compact set P, there exists a positive number ϱ, independent of ξ, such that for all ξ in P the functions $f_k(x)$ remain regular in the closed complex domain $|x_l-\xi_l|\leqq\varrho$ $(l=1,\ldots,m)$. Furthermore, there exists a positive number M, also independent of ξ, such that $|f_k(x)|\leqq M$ in this domain. Hence, if we choose in the interval $\tau\leqq t<t_1$ a number $t=t_2$ satisfying

$$t_1-t_2<\frac{\varrho}{(m+1)\,M}$$

and apply the existence theorem to (1) with $\xi_k=x_k(t_2)$ $(k=1,\ldots,m)$, $\tau=t_2$, $r=\varrho$, it follows that the $x_k(t)$ are regular in the circle

$$|t-t_2|<\frac{\varrho}{(m+1)\,M},$$

and therefore in particular at $t=t_1$, as claimed.

In the case of a Hamiltonian system

$$\dot{x}_k=E_{y_k}(x,y),\quad \dot{y}_k=-E_{x_k}(x,y)\quad (k=1,\ldots,n) \tag{5}$$

it will be useful for later applications to recast the previous results in a more convenient form by expressing the required estimate on the partial derivatives of E in terms of an estimate on E itself. To do this we use a classical argument in function theory. Namely, if $g(z)$ is a regular function of a complex variable z in the circle $|z-\zeta|<2\varrho$ where it satisfies $|g(z)|\leqq M$, then for $|z-\zeta|<\varrho$ Cauchy's integral formula gives

$$g'(z)=\frac{1}{2\pi i}\int_C \frac{g(u)\,du}{(u-z)^2},$$

where the path of integration C can be taken as the circle $|u-z|=\varrho$. Since the circle lies within $|u-\zeta|<2\varrho$, this leads to the estimate

$$|g'(z)|\leqq M\varrho^{-1} \quad (|z-\zeta|<\varrho).$$

Suppose now that the Hamiltonian function $E(x, y)$ is analytic in each of the $2n$ variables $x_1, \ldots, y_n$ for $|x_k-\xi_k|<2\varrho$, $|y_k-\eta_k|<2\varrho$ and satisfies $|E(x, y)|\leqq M$. The above argument then shows that in the domain $|x_k-\xi_k|<\varrho, |y_k-\eta_k|<\varrho$ $(k=1, \ldots, n)$ we have the desired estimates

$$|E_{x_l}|\leqq M\varrho^{-1}, \quad |E_{y_l}|\leqq M\varrho^{-1} \qquad (l=1, \ldots, n),$$

and the existence theorem now says that the solutions $x_k(t)$, $y_k(t)$ of the Hamiltonian system (5) with initial values $x_k(\tau)=\xi_k$, $y_k(\tau)=\eta_k$ are regular in the disc

$$|t-\tau|<\frac{\varrho^2}{(2n+1)M}$$

and satisfy

$$|x_k(t)-\xi_k|<\varrho, \quad |y_k(t)-\eta_k|<\varrho.$$

Accordingly, for Hamiltonian systems the previous statement on analytic continuation of solutions along the real t-axis takes the following form: If the solutions $x_k(t)$, $y_k(t)$ $(k=1, \ldots, n)$ of (5) are regular for $\tau\leqq t<t_1$, and the corresponding arc of the solution curve in $2n$-dimensional (x, y)-space remains in a closed bounded point-set P in which the Hamiltonian function $E(x, y)$ is regular, then the $x_k(t)$, $y_k(t)$ are regular also at $t=t_1$. This fact will be used in dealing with the three-body problem.

§ 5. The n-Body Problem

We consider the motion of n particles $P_k (k=1, \ldots, n)$ in three-dimensional Euclidean space, where $n>1$. Let x_k, y_k, z_k denote the coordinates of P_k in a fixed Cartesian coordinate system, $m_k>0$ its mass, and r_{kl} the distance between P_k and P_l, so that

$$r_{kl}^2=(x_k-x_l)^2+(y_k-y_l)^2+(z_k-z_l)^2 \quad (k,l=1,\ldots,n). \tag{1}$$

As a notational convenience, we will frequently use q_k to denote x_k, y_k, or z_k when referring to projections on a particular coordinate axis, and use q to denote one of the $3n$ possible coordinates q_k $(k=1,\ldots,n)$. Moreover, m will then always denote the mass m_k of the particle referred to by q. In an appropriate system of units the potential function for Newton's law of attraction is

$$U=\sum_{k<l}\frac{m_k m_l}{r_{kl}}, \tag{2}$$

and the equations of motion for the n-body problem take the abbreviated form

$$m\ddot{q}=U_q, \tag{3}$$

where U_q is the partial derivative of U with respect to q. They can also be written in the form

$$\dot{q}=v, \quad \dot{v}=m^{-1}U_q, \tag{4}$$

which is a system of $6n$ first order differential equations for the $6n$ quantities q, v as functions of t. Denoting the initial values at a real time $t=\tau$ by the subscript τ, which consequently will not refer to partial differentiation, we prescribe the $6n$ real values $q=q_\tau$, $v=v_\tau$ subject only to the restriction

$$r_{kl\tau}=\varrho_{kl}>0 \quad (k\neq l;\, k,l=1,\ldots,n).$$

The n-body problem consists of describing the complete behavior of all solutions to these equations of motion for arbitrary preassigned initial conditions. Despite efforts by outstanding mathematicians for over 200 years, the problem for $n>2$ remains unsolved to this day.

In 1858 Dirichlet [1] told his friend Kronecker of having discovered a general method for treating problems of mechanics, the method consisting not of a direct integration of the differential equations of motion, but rather of a step by step approximation of the solution to the problem. He also said in another conversation that he had succeeded in proving stability of the planetary system. However, Dirichlet died soon after, without leaving behind a written record of these discoveries, and nothing more is known of his method. Weierstrass suspected that it was a question of using power series expansions, and he endeavored to find a corresponding solution to the n-body problem, directing also his students S. Kovalevski and G. Mittag-Leffler to that goal [2]. At the suggestion of Mittag-Leffler, the King of Sweden and Norway established a prize for the solution to the problem of finding a series expansion for the coordinates of the n bodies valid for all time. In 1889 the prize was awarded to Poincaré, although he too had not solved the given problem. Still, his prize essay [3] contained an abundance of original ideas that were of great importance for the further development of mechanics, and served as a stimulus for other branches of mathematics as well. Finally, 20 years later, the proposed problem was solved by Sundman [4] for the case $n=3$. The main difficulty is that one has not succeeded until now in excluding collisions of two bodies by appropriate restrictions on the initial conditions. Sundman circumvented this difficulty by introducing a new variable ω in place of the time t, so that t and the coordinates q remain regular functions of ω when the two bodies collide. In this way he obtained series expansions for t and q in powers of ω that describe the entire motion. This beautiful and important result will be derived completely in the further course of chapter one. Unfortunately no corresponding result is known for $n>3$.

To begin with, let us construct the 10 classical integrals of the n-body problem for arbitrary $n>1$. By (1), (2) we have

$$U_{q_k} = \sum_{l \neq k} \frac{m_k m_l}{r_{kl}^3}(q_l - q_k) \quad (k=1, \dots, n), \tag{5}$$

hence

$$\sum_{k=1}^{n} U_{q_k} = 0, \tag{6}$$

and the equations of motion (4) give

$$\sum_{k=1}^{n} m_k \dot{v}_k = 0, \quad v_k = \dot{q}_k .$$

This leads to the six center of mass integrals

$$\left\{\begin{aligned} &\sum_{k=1}^{n} m_k \dot{x}_k = a, && \sum_{k=1}^{n} m_k \dot{y}_k = b, && \sum_{k=1}^{n} m_k \dot{z}_k = c,\\ &\sum_{k=1}^{n} m_k (x_k - t\dot{x}_k) = a^*, && \sum_{k=1}^{n} m_k (y_k - t\dot{y}_k) = b^*, && \sum_{k=1}^{n} m_k (z_k - t\dot{z}_k) = c^*, \end{aligned}\right. \tag{7}$$

or

$$\sum_{k=1}^{n} m_k x_k = at + a^*, \quad \sum_{k=1}^{n} m_k y_k = bt + b^*, \quad \sum_{k=1}^{n} m_k z_k = ct + c^*,$$

with six constants of integration a, a^*, b, b^*, c, c^*.

Moreover, if p_k, like q_k, also represents one of the variables x_k, y_k, z_k for $k \doteq 1, \dots, n$, then from (5) we have

$$U_{q_k} p_k - U_{p_k} q_k = \sum_{l \neq k} \frac{m_k m_l}{r_{kl}^3} (q_l p_k - p_l q_k) \quad (k = 1, \dots, n),$$

so that

$$\sum_{k=1}^{n} (U_{q_k} p_k - U_{p_k} q_k) = 0,$$

and the equations of motion (4) give

$$\sum_{k=1}^{n} m_k (\dot{v}_k p_k - \dot{u}_k q_k) = 0, \quad v_k = \dot{q}_k, \quad u_k = \dot{p}_k.$$

This leads to the 3 angular momentum integrals

$$(8) \quad \begin{cases} \sum_{k=1}^{n} m_k (y_k \dot{z}_k - z_k \dot{y}_k) = \alpha, \quad \sum_{k=1}^{n} m_k (z_k \dot{x}_k - x_k \dot{z}_k) = \beta, \\ \sum_{k=1}^{n} m_k (x_k \dot{y}_k - y_k \dot{x}_k) = \gamma, \end{cases}$$

with three constants of integration α, β, γ. Finally, upon summing over all the coordinates, from (4) we obtain

$$\sum_q (m v \dot{v} - U_q \dot{q}) = 0,$$

and this leads to the energy integral

$$(9) \qquad T - U = h,$$

with one integration constant h. Here

$$(10) \qquad T = \frac{1}{2} \sum_q m \dot{q}^2 = \frac{1}{2} \sum_v m v^2$$

is the kinetic energy of the system of particles. With the aid of the ten integrals (7), (8), (9) one can eliminate ten coordinates q, v from the equations of motion (4) and thereby reduce the system to one of $6n - 10$ first order differential equations.

It is interesting to note that the left sides of (7), (8), (9) are actually algebraic functions of the $6n + 1$ variables q, v, t, and one may ask if there are any additional integrals of this kind. To make this question clearer, let us define the concept of an integral more precisely. Let

$$(11) \qquad \dot{x}_k = f_k(x, t) \quad (k = 1, \dots, m)$$

again be a system of m first order differential equations, whose right sides, however, in addition to $x_1, \ldots, x_m$, may also depend on t. A continuously differentiable function $g(x, t)$ of the $m+1$ independent variables $x_1, \ldots, x_m, t$ is said to be an integral of the system (11) if it is constant along each solution $x = x(t)$ of (11). This is easily seen to be equivalent to the requirement that it satisfy the homogeneous first order linear partial differential equation

$$g_t + \sum_{k=1}^{m} f_k(x, t)\, g_{x_k} = 0$$

identically in the variables x, t. Given l integrals $g = g_1, \ldots, g_l$ of (11), they are said to be independent if the Jacobian matrix formed with the $m+1$ partial derivatives with respect to x_k, t has rank l. Finally, an integral is said to be algebraic if it is an algebraic function of the x_k, t. Thus, for each $n > 1$, we have constructed ten algebraic integrals of the system (4) of differential equations for the n-body problem, and they are easily seen to be independent. Bruns [5] has proved the interesting theorem that there are no additional algebraic integrals of (4) independent of these ten, and it follows from this that every algebraic integral of (4) is an algebraic function of these ten known integrals. On the other hand, we know from the existence theorems that the system (4) has altogether $6n$ independent integrals, and consequently, since $6n > 10$, they cannot all be algebraic. Unfortunately, the proof of Bruns' theorem, because of its length, cannot be reproduced here.

We now apply to the system (4) the existence theorem of Cauchy derived in the preceding section. Here τ and the initial values $q = q_\tau$, $v = v_\tau$ are real, but in determining the positive constants r, M appearing in the existence theorem we allow the values of the variables to be complex. By assumption, the initial values satisfy $r_{kl\tau} = \varrho_{kl} > 0\ (k \neq l)$. Let A be an upper bound of U for $t = \tau$, so that

$$U_\tau \leqq A,$$

and let

$$\min_{k \neq l} \varrho_{kl} = \varrho, \quad \min_k m_k = \mu. \tag{12}$$

By (2) we have

$$\frac{\mu^2}{\varrho} \leqq U_\tau \leqq A,$$

and hence

$$\varrho \geqq \mu^2 A^{-1}. \tag{13}$$

To obtain an upper bound for the absolute values of the derivatives U_q we first estimate the values $|r_{kl}|$ $(k \neq l)$ from below, for the q complex and near q_τ. Denoting the three expressions $(q_k - q_{k\tau}) - (q_l - q_{l\tau})$ for $q = x, y, z$ by ϕ, ψ, χ, and using the Schwarz inequality, from (1) we obtain the estimate

$$|r_{kl}|^2 \geqq \varrho_{kl}^2 - 2\varrho_{kl}(|\phi|^2 + |\psi|^2 + |\chi|^2)^{\frac{1}{2}} - (|\phi|^2 + |\psi|^2 + |\chi|^2). \tag{14}$$

Hence, if we have

$$|q - q_\tau| < \frac{\varrho}{14} \tag{15}$$

for all q, so that ϕ, ψ, χ are less than $\varrho/7$ in absolute value, and therefore

$$|\phi|^2 + |\psi|^2 + |\chi|^2 < 3\frac{\varrho^2}{49} < \frac{\varrho^2}{16},$$

then (12), (14) combine to give

$$|r_{kl}|^2 \geqq \varrho_{kl}^2 - \frac{1}{2}\varrho_{kl}\varrho - \frac{1}{16}\varrho^2 > \frac{1}{4}\varrho_{kl}^2, \quad |r_{kl}| > \frac{1}{2}\varrho_{kl}. \tag{16}$$

In view of (13), we can certainly satisfy (15) by requiring that

$$|q - q_\tau| < \frac{\mu^2}{14A}, \tag{17}$$

in which case

$$|q_l - q_k| \leqq |q_l - q_{l\tau}| + |q_k - q_{k\tau}| + |q_{l\tau} - q_{k\tau}| < \frac{\varrho}{7} + \varrho_{kl} \leqq \frac{8}{7}\varrho_{kl}.$$

By (13), (16) we then have

$$\left|\frac{q_l - q_k}{r_{kl}^3}\right| \leqq \left(\frac{2}{\varrho_{kl}}\right)^3 \frac{8}{7}\varrho_{kl} = \frac{64}{7}\varrho_{kl}^{-2} \leqq \frac{64}{7}A^2\mu^{-4} \qquad (k \neq l)$$

$$|m_k^{-1}U_{q_k}| = \left|\sum_{l \neq k} \frac{m_l}{r_{kl}^3}(q_l - q_k)\right| < c_1 A^2 \qquad (k = 1, \ldots, n),$$

where the positive constant c_1 depends only on the given masses. Moreover, with $\dot{q} = v$, the energy integral gives

$$\frac{m}{2}v_\tau^2 \leqq T_\tau = U_\tau + h \leqq A + h$$

$$|v_\tau| \leqq c_2\sqrt{A + h},$$

where also c_2 depends only on the masses. Setting

$$(18) \qquad \frac{\mu^2}{14A} = r$$

and, in addition to (17), assuming that

$$|v - v_\tau| < r,$$

we finally have

$$|v| \leqq |v - v_\tau| + |v_\tau| < r + c_2\sqrt{A+h}.$$

Consequently, if the constants r, M in the existence theorem are defined by (18) and

$$M = c_1 A^2 + \frac{\mu^2}{14A} + c_2\sqrt{A+h},$$

it follows that the solutions $q(t)$, $v(t) = \dot{q}(t)$ of (4) are regular in the circle

$$|t - \tau| < \frac{r}{(6n+1)M} = \delta,$$

and in particular in the interval $\tau \leqq t < \tau + \delta$. Moreover, the radius δ depends only on A, h, and the masses.

Let us now look at the solution curve in the coordinate space of the $6n$ variables q, $v = \dot{q}$ for $t \geqq \tau$. As long as $t - \tau < \delta$, the functions $q(t)$ remain regular in t, and in particular, no collision can occur. Indeed, otherwise U would become infinite, and by the energy relation also T would become infinite, as would at least one of the velocity components $\dot{q}(t)$, contrary to the established regularity of the $q(t)$. If the solution curve is continued analytically along the real t-axis for $t > \tau$, then either the $6n$ coordinates remain regular for all finite real time $t \geqq \tau$, or there is a first time $t_1 > \tau$ which is a singularity for at least one $q(t)$, with all the $q(t)$ remaining regular for $\tau \leqq t < t_1$. We contend that as $t \to t_1$ $(t < t_1)$ the positive function U tends to infinity. If this were not so, there would exist a positive constant A and a sequence of times $\tau_1 > \tau$ converging to t_1 from below, such that

$$U_{\tau_1} \leqq A.$$

Hence if we chose $\tau_1 > t_1 - \delta$, with δ the quantity defined in the preceding paragraph, and applied the existence theorem with τ_1 in place of τ, it would follow that all the $q(t)$ remain regular at $t = t_1$, contrary to the definition of t_1. This shows that $U \to \infty$ as $t \to t_1$, and from (2) we see that the minimum of the $\dfrac{n(n-1)}{2}$ distances r_{kl} $(k < l)$ must then tend to 0 as $t \to t_1$.

§ 6. Collision

From the six center of mass integrals (5;7) we see that the center of mass P_0 of the n bodies moves in a straight line at a constant speed. If we introduce, by parallel translation, a moving coordinate system with P_0 as the origin, the equations of motion (5;3) remain invariant. Consequently we will assume in the following that P_0 is at rest at the origin O.

The following important theorem was first proved by Sundman. Evidently it was known already to Weierstrass, who, however, did not give a proof. The theorem states:

If at time $t=t_1$ all n bodies collide at one point, then the three angular momentum constants α, β, γ must all be 0.

Since the center of mass rests at the origin, a collision of all n bodies can occur only at O. Introducing the expression

$$I=\sum_q mq^2=\sum_{k=1}^{n} m_k(x_k^2+y_k^2+z_k^2),$$

for $\tau \leqq t < t_1$ we have

$$\text{(1)} \qquad \tfrac{1}{2}\dot{I}=\sum_q mq\dot{q},$$

$$\tfrac{1}{2}\ddot{I}=\sum_q m(\dot{q}^2+q\ddot{q})=2T+\sum_q qU_q.$$

Since U is a homogeneous function of degree -1 in the coordinates q, it follows from a well known theorem of Euler that

$$\sum_q qU_q=-U.$$

This leads to Lagrange's formula

$$\tfrac{1}{2}\ddot{I}=2T-U,$$

which can be expressed with the aid of the energy integral $T-U=h$ in the form

$$\text{(2)} \qquad \tfrac{1}{2}\ddot{I}=T+h=U+2h.$$

At first we do not assume that all particles collide at time t_1, but, as in the previous section, only that t_1 is a singularity for at least one coordinate $q(t)$. As we have seen, U then tends to ∞ as $t\to t_1$, and therefore, if $t_2<t_1$ is sufficiently near t_1, the right side of (2) will be positive throughout the interval $t_2 \leqq t < t_1$. Hence

$$\ddot{I}>0 \qquad (t_2 \leqq t < t_1),$$

and the function $\dot{I}$ is monotonically increasing. We may also assume that $\dot{I}$ is always positive or always negative in this interval, for should it change sign there, say at t_3, we need only replace t_2 by a number between t_3 and t_1. Thus the positive function I is monotonic in the interval $t_2 \leqq t < t_1$, and therefore has a limit as $t \to t_1$. From the definition of I this limit is 0 if and only if all n bodies collide at the origin at time t_1.

Using the algebraic identity

$$\sum_{k=1}^{g} \xi_k^2 \sum_{k=1}^{g} \eta_k^2 = \left(\sum_{k=1}^{g} \xi_k \eta_k \right)^2 + \sum_{k<l} (\xi_k \eta_l - \xi_l \eta_k)^2$$

with $g = 3n$, $\xi_k = q\sqrt{m}$, $\eta_k = \dot{q}\sqrt{m}$, we obtain from (1) the formula

$$2IT = \tfrac{1}{4}\dot{I}^2 + \sum_{k<l} (\xi_k \eta_l - \xi_l \eta_k)^2, \tag{3}$$

and retaining in the last sum only those terms in which ξ_k and η_l refer to the same particle, we are led to the estimate

$$2IT \geqq \tfrac{1}{4}\dot{I}^2 + \sum_{k=1}^{n} m_k^2 \{(y_k \dot{z}_k - z_k \dot{y}_k)^2 + (z_k \dot{x}_k - x_k \dot{z}_k)^2 + (x_k \dot{y}_k - y_k \dot{x}_k)^2\}.$$

On the other hand, (5;8) in conjunction with the Schwarz inequality implies that

$$\alpha^2 = \left\{ \sum_{k=1}^{n} m_k (y_k \dot{z}_k - z_k \dot{y}_k) \right\}^2 \leqq n \sum_{k=1}^{n} m_k^2 (y_k \dot{z}_k - z_k \dot{y}_k)^2,$$

with analogous inequalities for β, γ, so that the previous estimate gives

$$2IT \geqq \frac{1}{4}\dot{I}^2 + \eta, \quad \eta = \frac{\alpha^2 + \beta^2 + \gamma^2}{n}. \tag{4}$$

For the proof of our theorem it is sufficient to consider, instead of (4), the weaker inequality

$$2IT \geqq \eta,$$

which by (2) is equivalent to

$$\ddot{I} \geqq \eta I^{-1} + 2h. \tag{5}$$

If I is monotonically decreasing in the interval $t_2 \leqq t < t_1$, multiplying (5) by the positive quantity $-2\dot{I}$ and integrating from t_2 to t, we obtain

$$\dot{I}_2^2 - \dot{I}^2 \geqq 2\eta \log \frac{I_2}{I} + 4h(I_2 - I),$$

where I_2, $\dot{I}_2$ denote the values of $I, \dot{I}$ at $t = t_2$. Therefore certainly

$$2\eta \log \frac{I_2}{I} \leqq \dot{I}_2^2 + 4|h|\, I_2 \quad (t_2 \leqq t < t_1), \tag{6}$$

and when $\eta > 0$, i.e. when α, β, γ are not all 0, this results in a positive lower bound for I. If, on the other hand, I is monotonically increasing for $t_2 \leqq t < t_1$, then trivially we have $I \geqq I_2$ in this interval. Thus in both cases, if $\eta > 0$, the function I has a positive lower bound in the interval $t_2 \leqq t < t_1$. Denoting the distance from the center of mass, the origin of our coordinate system, to the point P_k by ϱ_k, we have

$$I = \sum_{k=1}^{n} m_k \varrho_k^2 .$$

Moreover,

$$\sum_{k=1}^{n} m_k q_k = 0 ,$$

and therefore

$$\sum_{k=1}^{n} m_k (q_l - q_k) = M q_l , \quad M = \sum_{k=1}^{n} m_k$$

$$M q_l^2 \leqq \sum_{k=1}^{n} m_k (q_l - q_k)^2 \quad (l = 1, \ldots, n)$$

$$M I \leqq 2 \sum_{k<l} m_k m_l r_{kl}^2 .$$

Consequently, if $\eta > 0$, the maximum of the $\frac{n(n-1)}{2}$ distances $r_{kl}\,(k<l)$ has a positive lower bound for $t_2 \leqq t < t_1$, and therefore also for $\tau \leqq t < t_1$, and it follows that collision of all n bodies at O cannot occur at time t_1. This proves the theorem.

In the remainder of this chapter we will consider only the case $n = 3$. We will show now that the angular momentum constants α, β, γ can all be 0 only if the three bodies move in a fixed plane. Since the expression $\alpha^2 + \beta^2 + \gamma^2$ and the equations of motion are invariant under orthogonal coordinate transformations and since the center of mass P_0 lies at O, we may assume that at $t = \tau$ the three particles lie in the plane $z = 0$. By (5;8), the vanishing of α, β, γ then implies that

$$\sum_{k=1}^{3} m_k y_k \dot{z}_k = 0 , \quad \sum_{k=1}^{3} m_k x_k \dot{z}_k = 0 \quad (t = \tau) ,$$

and because the center of mass is at rest, we also have

$$\sum_{k=1}^{3} m_k \dot{z}_k = 0 .$$

These three homogeneous linear equations for the quantities $m_k \dot{z}_k$ $(k = 1, 2, 3)$ at $t = \tau$ imply that either the $\dot{z}_k$ all vanish at $t = \tau$, or

$$\begin{vmatrix} x_1 & x_2 & x_3 \\ y_1 & y_2 & y_3 \\ 1 & 1 & 1 \end{vmatrix} = 0 \quad (t = \tau) .$$

In the first case, the initial directions of motion for the three bodies lie in the plane $z=0$, and by the uniqueness theorem for differential equations, the equations of motion imply that the bodies remain in this plane permanently. In the second case, the three bodies lie initially on a line, and the coordinate axes can still be rotated so that $\dot{z}_3$ vanishes at $t=\tau$. If we then exclude the case $\dot{z}_k=0$ $(k=1, 2, 3)$ already treated, the above equations give

$$y_1 = y_2, \quad x_1 = x_2 \quad (t=\tau),$$

implying that, contrary to earlier assumptions, the points P_1, P_2 coincide at $t=\tau$. This proves our assertion.

In particular, one now sees that if all three bodies are to collide, the motion must necessarily take place in a fixed plane. The case of triple collision, characterized by the condition

$$\lim_{t \to t_1} I = 0,$$

will be investigated further in § 12 and § 13.

We will assume from now until the end of § 11 that

$$\lim_{t \to t_1} I > 0,$$

from which it follows that I lies above a positive lower bound for $\tau \leqq t < t_1$. By our earlier result, the maximum of the three lengths r_{12}, r_{23}, r_{31} of the sides of the triangle with the particles as vertices must also lie above a positive bound ε for $\tau \leqq t < t_1$. On the other hand, as we saw in the previous section, the minimum length approaches 0 as $t \to t_1$. Denoting this minimum by r, we can choose $t_2 < t_1$ near t_1 so that

$$r < \frac{\varepsilon}{2} \quad (t_2 \leqq t < t_1).$$

During this time interval a fixed side must remain the shortest one, for otherwise continuity would imply that at some t in this interval two of the sides would become shorter than $\frac{\varepsilon}{2}$, making the third side shorter than ε, and this would contradict the choice of ε. Thus the length of a definite side, say r_{13}, remains below $\frac{\varepsilon}{2}$ for $t_2 \leqq t < t_1$, and we have

$$r_{13} < \frac{\varepsilon}{2}, \quad r_{12} > \frac{\varepsilon}{2}, \quad r_{23} > \frac{\varepsilon}{2} \quad (t_2 \leqq t < t_1). \tag{7}$$

Consequently, as $t \to t_1$, the distance $r_{13}=r$ tends to 0 and the points P_1, P_3 collide, while the other distances remain above a positive lower bound.

We will now show that the collision occurs at a definite point in space. The equation of motion

$$\ddot{q}_2 = \frac{m_1}{r_{12}^3}(q_1 - q_2) + \frac{m_3}{r_{23}^3}(q_3 - q_2)$$

leads to the estimate

$$|\ddot{q}_2| \leqq \frac{m_1}{r_{12}^2} + \frac{m_3}{r_{23}^2},$$

so that by (7) the function $\ddot{q}_2$ is bounded in the interval $t_2 \leqq t < t_1$. Integrating twice we obtain definite limits for $\dot{q}_2, q_2$ as $t \to t_1$, and therefore the point P_2 as well as the corresponding velocity components have definite limits as $t \to t_1$. Since $q_1 - q_3$ tends to 0 as $t \to t_1$ and the center of mass of the three bodies rests at the origin, the center of mass integral $m_1 q_1 + m_2 q_2 + m_3 q_3 = 0$ then implies that also q_1 and q_3 have limits as $t \to t_1$. Thus P_1 and P_3 actually collide at a definite point in space. In addition, it follows that the monotone function I is bounded in the interval $t_2 \leqq t < t_1$, and therefore has a finite limit as $t \to t_1$.

On the other hand, one would expect the velocities of P_1 and P_3 to become infinite as $t \to t_1$. To examine this more closely, we denote the velocity of P_k by V_k $(k = 1, 2, 3)$, and from the energy integral obtain the relation

$$\frac{1}{2} \sum_{k=1}^{3} m_k V_k^2 = T = U + h.$$

Since $r = r_{13}$ approaches 0 as $t \to t_1$, while the lengths of the other two sides of the triangle have positive lower bounds, the product rU tends to $m_1 m_3$, and we have

$$r \sum_{k=1}^{3} m_k V_k^2 \to 2 m_1 m_3 \qquad (t \to t_1). \tag{8}$$

In particular, the expressions $r V_k^2, r \dot{q}_k^2, r^{\frac{1}{2}} \dot{q}_k$ $(k = 1, 2, 3)$ remain bounded at collision. Furthermore, the center of mass relation $m_1 \dot{q}_1 + m_2 \dot{q}_2 + m_3 \dot{q}_3 = 0$ gives

$$r(m_1 \dot{q}_1)^2 - r(m_3 \dot{q}_3)^2 = m_2 r^{\frac{1}{2}} \{ m_2 r^{\frac{1}{2}} \dot{q}_2^2 + 2 m_3 \dot{q}_2 (r^{\frac{1}{2}} \dot{q}_3) \},$$

where the individual terms in the curved brackets are bounded, while the factor $r^{\frac{1}{2}}$ in front tends to 0. Consequently

$$r(m_1 \dot{q}_1)^2 - r(m_3 \dot{q}_3)^2 \to 0, \quad r(m_1 V_1)^2 - r(m_3 V_3)^2 \to 0,$$

and since $r m_2 V_2^2$ also goes to 0, it follows from (8) that

$$r V_1^2 \to \frac{2 m_3^2}{m_1 + m_3} \qquad (t \to t_1). \tag{9}$$

This determines the asymptotic behavior of V_1 and V_3 at $t=t_1$.

Although the function r^{-1} becomes infinite as $t \to t_1$, we will show that the integral

$$\int_{\tau}^{t_1} \frac{dt}{r} = \lim_{t \to t_1} \int_{\tau}^{t} \frac{dt}{r} \tag{10}$$

nevertheless converges. For this we use Lagrange's formula (2), namely

$$\frac{1}{2}\ddot{I} = U + 2h.$$

Since by (7) the difference $U - m_1 m_3 r_{13}^{-1}$ is bounded for $\tau \leqq t < t_1$, it is obviously enough to show that $\dot{I}$ has a finite limit as $t \to t_1$. We have already shown in the beginning of this section that $\dot{I}$ is monotonic in the interval $t_2 \leqq t < t_1$, and therefore it remains only to show that $\dot{I}$ is bounded. Using the center of mass integral, one verifies that

$$\begin{aligned} \frac{1}{2}\dot{I} &= \sum_{k=1}^{3} m_k (x_k \dot{x}_k + y_k \dot{y}_k + z_k \dot{z}_k) \\ &= \sum_{k=1}^{2} m_k \{(x_k - x_3)\dot{x}_k + (y_k - y_3)\dot{y}_k + (z_k - z_3)\dot{z}_k\}, \end{aligned}$$

so that by the Schwarz inequality we have

$$\frac{1}{2}|\dot{I}| \leqq \sum_{k=1}^{2} m_k r_{k3} V_k .$$

In view of (9), the expression $r_{13} V_1$ tends to 0, while r_{23} and V_2 remain bounded. This proves the existence of the limit in (10).

In the next two sections we will investigate the nature of the singularity of q and $\dot{q}$ at $t=t_1$ more closely. Before doing this, however, we interject a preliminary heuristic analysis. As we already know, at least one of the nine derivatives $\dot{q}(t)$ is unbounded as $t \to t_1$, while the functions $q(t)$ themselves all have finite limits. From this it already follows that none of the $\dot{q}(t)$ can have a pole at $t=t_1$, for otherwise $q(t)$ would become infinite there. We now make the assumption, as yet unproved, that none of the $q(t)$ has an essential singularity at $t=t_1$, but at most a branch point of finite order. For l a natural number let

$$s = (t_1 - t)^{\frac{1}{l}}$$

be a local uniformizer common to all the $q(t)$, meaning that the $q(t)$ all have regular power series expansions in s which converge for s sufficiently small in absolute value. Moreover, let s^{μ} be the smallest positive power of s actually appearing in a $q(t)$, so that

(11) $$q(t)=q(t_1)+c_1 s^{\mu}+\cdots,$$

with c_1 different from 0 for at least one coordinate q. Then

(12) $$\dot{q}(t)=-\frac{\mu}{l}c_1 s^{\mu-l}+\cdots,$$

and in particular,

$$V_1^2=c_2 s^{2(\mu-l)}+\cdots,\quad c_2>0.$$

Moreover, by (11) we have

$$r=c_3 s^{\mu}+\cdots,\quad c_3\geqq 0,$$

and if in addition $c_3>0$, then (9) implies that $3\mu-2l=0$ and $\frac{\mu}{l}=\frac{2}{3}$.

This suggests that $l=3$, and

$$s=(t_1-t)^{\frac{1}{3}}$$

will serve as a local uniformizer. In § 8 we will show that this is actually so. By (12), the derivative $\dot{q}(t)$ as a function of s must then either remain regular at $t=t_1$, or have a pole of first order there. The integral

$$\lambda=\int_t^{t_1}\frac{dt}{r}\quad(\tau\leqq t<t_1),$$

whose existence we have already established, will then have a series expansion of the form

$$\lambda=\frac{3}{c_3}s+\cdots,$$

and therefore λ can also serve as a local uniformizer. This choice is suggested by the classical theory of the two-body problem, where λ is just the eccentric anomaly.

In the next section we will lay the groundwork for proving these heuristic conclusions. One cannot expect to cover the singular point $t=t_1$ by a direct application of Cauchy's existence theorem to the equations of motion obtained by introducing λ as independent variable in place of t. Indeed, this is already seen from the fact that at least one $\dot{q}$ is necessarily singular at $\lambda=0$. If, however, our heuristic reasoning is correct, the new variables obtained from $\dot{x}_1, \dot{y}_1, \dot{z}_1$ and $\dot{x}_3, \dot{y}_3, \dot{z}_3$ by radial inversion will still be regular at $\lambda=0$. By introducing these new variables into the equations of motion, it is possible to achieve the desired goal. To simplify the calculations we will follow the precedent of Levi-Civita [1], and first bring the equations of motion into Hamiltonian form, after which we will determine a canonical transformation that has the desired effect on the $\dot{q}$.

§ 7. The Regularizing Transformation

We will now express the equations of motion for the three-body problem in Hamiltonian form, without assuming, at first, that the center of mass P_0 rests at the origin. We denote the coordinates of the points P_1, P_2, P_3 by $q_1, \ldots, q_9$, so that x_k, y_k, z_k $(k=1,2,3)$ become $q_{3k-2}, q_{3k-1}, q_{3k}$, and accordingly denote the coordinates $m_k\dot{x}_k, m_k\dot{y}_k, m_k\dot{z}_k$ of the momenta by $p_{3k-2}, p_{3k-1}, p_{3k}$. Hence

$$T=\frac{1}{2}\sum_{k=1}^{3}\left(\frac{p_k^2}{m_1}+\frac{p_{k+3}^2}{m_2}+\frac{p_{k+6}^2}{m_3}\right). \tag{1}$$

The equations of motion (5; 4), with $n=3$, then take the Hamiltonian form

$$\dot{q}_k=E_{p_k}, \quad \dot{p}_k=-E_{q_k} \quad (k=1,\ldots,9), \tag{2}$$

where $E=T-U$ is to be regarded as a function of the 18 independent variables p_k, q_k. We will use the center of mass integrals to eliminate three pairs of variables p_k, q_k from these 18 differential equations, and we will achieve this by taking all the p_k, q_k into new variables x_k, y_k via a suitable canonical transformation. Following the pattern in (3; 4), we set

$$p_k=w_{q_k}, \quad x_k=w_{y_k} \quad (k=1,\ldots,9), \tag{3}$$

where $w(q,y)$ is a generating function whose Jacobian determinant $|w_{y_k q_l}|$ is not 0. We wish to set up the canonical transformation so that $x_1, \ldots, x_6$ become the coordinates of P_1, P_2 relative to P_3, while x_7, x_8, x_9 remain as the coordinates of P_3, i.e. so that

$$x_k=q_k-q_{k+6}, \quad x_{k+3}=q_{k+3}-q_{k+6}, \quad x_{k+6}=q_{k+6} \quad (k=1,2,3). \tag{4}$$

This requirement is evidently consistent with the second equation of (3) if we choose

$$w=\sum_{k=1}^{3}\{(q_k-q_{k+6})y_k+(q_{k+3}-q_{k+6})y_{k+3}+q_{k+6}y_{k+6}\}.$$

Then $|w_{y_k q_l}|=1$, and (3) does indeed give rise to a canonical transformation. The first equation of (3) gives

$$p_k=y_k, \quad p_{k+3}=y_{k+3}, \quad p_{k+6}=y_{k+6}-y_{k+3}-y_k \quad (k=1,2,3),$$

so that

$$(5)\qquad y_k = p_k,\quad y_{k+3} = p_{k+3},\quad y_{k+6} = p_k + p_{k+3} + p_{k+6}\qquad (k=1,2,3),$$

and in (4), (5) we have the desired transformation, which we see is linear. Since, in addition, it does not depend on t, the new equations of motion read

$$(6)\qquad \dot{x}_k = E_{y_k},\quad \dot{y}_k = -E_{x_k}\qquad (k=1,\ldots,9),$$

where $E = T - U$ is now regarded as a function of the x_k, y_k. Then

$$T = \frac{1}{2}\sum_{k=1}^{3}\{m_1^{-1}y_k^2 + m_2^{-1}y_{k+3}^2 + m_3^{-1}(y_{k+6} - y_k - y_{k+3})^2\},$$

$$(7)\qquad \begin{cases} U = m_1 m_3 (x_1^2 + x_2^2 + x_3^2)^{-\frac{1}{2}} + m_2 m_3 (x_4^2 + x_5^2 + x_6^2)^{-\frac{1}{2}} \\ \qquad + m_1 m_2 \{(x_1 - x_4)^2 + (x_2 - x_5)^2 + (x_3 - x_6)^2\}^{-\frac{1}{2}}, \end{cases}$$

so that E does not depend on x_7, x_8, x_9. By (6) this means that y_7, y_8, y_9 remain constant throughout the motion, which leads directly to the center of mass integrals. If one considers the differential equations (6) for $k = 1, \ldots, 6$ only, one obtains a Hamiltonian system for the first six pairs x_k, y_k alone, since x_7, x_8, x_9 do not appear in E, while y_7, y_8, y_9 remain constant. Having solved this system, one can obtain the remaining variables x_7, x_8, x_9 from

$$\dot{x}_k = E_{y_k}\qquad (k=7,8,9)$$

by quadrature. If we assume from now on that the center of mass rests at the origin, then $y_7 = y_8 = y_9 = 0$ and

$$0 = m_1 q_k + m_2 q_{k+3} + m_3 q_{k+6} = m_1(x_k + x_{k+6}) + m_2(x_{k+3} + x_{k+6}) + m_3 x_{k+6},$$

so that

$$x_{k+6} = -\frac{m_1 x_k + m_2 x_{k+3}}{m_1 + m_2 + m_3}\qquad (k=1,2,3).$$

Therefore we only have to consider the system

$$(8)\qquad \dot{x}_k = E_{y_k},\quad \dot{y}_k = -E_{x_k}\qquad (k=1,\ldots,6)$$

with

$$(9)\quad T = \frac{1}{2}(m_1^{-1} + m_3^{-1})\sum_{k=1}^{3} y_k^2 + \frac{1}{2}(m_2^{-1} + m_3^{-1})\sum_{k=1}^{3} y_{k+3}^2 + m_3^{-1}\sum_{k=1}^{3} y_k y_{k+3}.$$

Using the results of the previous section, we will now take a closer look at the behavior of the solutions x_k, y_k when P_1 and P_3 collide at time t_1. If we abbreviate

$$(10)\qquad x_1^2 + x_2^2 + x_3^2 = x^2,\quad y_1^2 + y_2^2 + y_3^2 = y^2,$$

then $x>0$ for $\tau \leqq t < t_1$, $x \to 0$ as $t \to t_1$, and (6; 9) becomes

$$xy^2 \to \frac{2(m_1 m_3)^2}{m_1+m_3} \quad (t \to t_1). \tag{11}$$

Furthermore, we see from earlier results that x_k, y_k $(k=4,5,6)$ have limits as $t \to t_1$, with

$$xU \to m_1 m_3 \quad (t \to t_1).$$

In accordance with the heuristic reasoning of the previous section, we replace t by the new independent variable

$$s = \int_{\tau}^{t} \frac{dt}{x(t)} \quad (\tau \leqq t < t_1), \tag{12}$$

where $x = x(t)$ is the function defined by (6), (10). Then s is a regular function of t in the interval $\tau \leqq t < t_1$, and increases there from 0 to the finite value

$$s_1 = \int_{\tau}^{t_1} \frac{dt}{x(t)}. \tag{13}$$

From $\dot{s} = x^{-1}$ we see that the inverse function t is also regular in s, and increases monotonically from τ to t_1 for $0 \leqq s < s_1$. If differentiation with respect to s is denoted by a prime, the equations of motion in terms of this new independent variable become

$$x_k' = xE_{y_k}, \quad y_k' = -xE_{x_k} \quad (k=1, \ldots, 6), \tag{14}$$

which, as can be seen, are no longer in Hamiltonian form.

To restore (14) to Hamiltonian form we use a device introduced by Poincaré. Being an integral, the function E has a constant value h along each solution of (8), and therefore the function

$$F = x(E-h) = x(T-U-h) \tag{15}$$

of x_k, y_k $(k=1, \ldots, 6)$ satisfies

$$F_{x_k} = xE_{x_k}, \quad F_{y_k} = xE_{y_k}$$

along the corresponding solution, since the other term arising from differentiating $x(E-h)$ contains a factor $E-h$ that vanishes there. Consequently (14) may be expressed as the Hamiltonian system

$$x_k' = F_{y_k}, \quad y_k' = -F_{x_k} \quad (k=1, \ldots, 6), \tag{16}$$

which is indeed satisfied by all solutions of the original equations of motion for which E has the prescribed value h, hence F the value 0.

Because the new Hamiltonian function F does not explicitly contain the independent variable s, the derivative

$$F' = \sum_{k=1}^{6} (F_{x_k} x_k' + F_{y_k} y_k') = 0$$

along each solution of (16), so that F is constant. Conversely, if $F=0$ and $x \neq 0$, then (14) follows from (16), while if $F \neq 0$ the solutions of (16) bear no relation to the three-body problem. An obvious advantage in introducing $F = xT - xU - hx$ is that the terms xT, xU do not become infinite but have finite limits as $t \to t_1$. Of course, because y becomes infinite, the derivatives F_{x_k} do not all remain bounded, so that (16) is still not suitable for a more precise investigation of the singularity $t = t_1$. We will therefore undertake a canonical transformation whereby y_1, y_2, y_3 are transformed by means of radial inversion.

To find an expression for this transformation we begin with a two-body problem. It is natural to suppose that as $t \to t_1$ the particle P_2 will no longer exert any real influence on the behavior of P_1, P_3. Ignoring P_2 completely, we then consider P_1, P_3 as the particles in a two-body problem, with the center of mass resting at O. Then

$$T = \tfrac{1}{2}(m_1^{-1} + m_3^{-1})(y_1^2 + y_2^2 + y_3^2), \; U = m_1 m_3 (x_1^2 + x_2^2 + x_3^2)^{-\frac{1}{2}} = m_1 m_3 x^{-1},$$
$$F = \tfrac{1}{2}(m_1^{-1} + m_3^{-1})\, x y^2 - m_1 m_3 - hx.$$

Restricting ourselves to the particular case $h=0$, $\frac{1}{2}(m_1^{-1} + m_3^{-1}) = 1$ and disregarding the additive constant $-m_1 m_3$, we obtain the simplified Hamiltonian system

$$(17) \qquad x_k' = F_{y_k}, \quad y_k' = -F_{x_k} \quad (k = 1, 2, 3),$$

with

$$F = F(x_k, y_k) = (x_1^2 + x_2^2 + x_3^2)^{\frac{1}{2}} (y_1^2 + y_2^2 + y_3^2).$$

According to the Hamilton-Jacobi theory developed in § 3, the complete solution of (17) can be obtained if one can set up a solution $w(x_k, \xi_k, s)$ of the partial differential equation

$$(18) \qquad F(x_k, w_{x_k}) + w_s = 0$$

that depends on three parameters ξ_1, ξ_2, ξ_3, subject to the condition $|w_{x_k \xi_l}| \neq 0$; in fact, then

$$(19) \qquad y_k = w_{x_k}, \quad \eta_k = -w_{\xi_k} \quad (k = 1, 2, 3),$$

with six constants of integration ξ_k, η_k. Since F is free of the independent variable s, we seek a solution to (18) of the form

$$(20) \qquad w(x_k, \xi_k, s) = v(x_k, \xi_k) - \lambda(\xi_k)\, s.$$

It would, of course, be even simpler to seek w independent of s, but then, by (19), the general solution x_k, y_k of (17) would also have to be independent of s, which is absurd. Substituting (20) into (18), we are led to

$$F(x_k, v_{x_k}) = \lambda(\xi_k), \quad |v_{x_k \xi_l}| \neq 0, \tag{21}$$

with (19) becoming

$$y_k = v_{x_k}, \quad \eta_k = \lambda_{\xi_k} s - v_{\xi_k} \quad (k = 1, 2, 3). \tag{22}$$

Before actually solving (21), we wish to change notation and instead of (22) introduce the canonical transformation

$$y_k = v_{x_k}, \quad \eta_k = -v_{\xi_k} \quad (k = 1, 2, 3), \tag{23}$$

which does not depend on s. By (21), then

$$F(x_k, y_k) = F(x_k, v_{x_k}) = \lambda(\xi_k),$$

and according to our theory the transformation (23) takes the Hamiltonian system (17) into the system

$$\xi_k' = \lambda_{\eta_k} = 0, \quad \eta_k' = -\lambda_{\xi_k}.$$

Consequently, ξ_k and $\eta_k + \lambda_{\xi_k} s = \zeta_k$ $(k = 1, 2, 3)$ appear as constants of integration.

To solve (21) we will first consider the analogous problem in the plane, and then generalize the resulting solution in a natural way to three dimensions. In the reduced differential equation

$$(x_1^2 + x_2^2)^{\frac{1}{2}}(v_{x_1}^2 + v_{x_2}^2) = \lambda(\xi_k), \quad |v_{x_k \xi_l}| \neq 0 \tag{24}$$

we set $x_1 + i x_2 = z$ and seek v as the imaginary part of an analytic function $\phi(z) = u + iv$. By the Cauchy-Riemann equations we have

$$u_{x_1} = v_{x_2}, \quad v_{x_1}^2 + v_{x_2}^2 = u_{x_1}^2 + v_{x_1}^2 = |\phi_z|^2,$$

and therefore

$$|z\, \phi_z^2| = \lambda(\xi_k)$$

must be constant in z. Since the function $z\,\phi_z^2$ is analytic, it must itself be constant. Setting

$$z\phi_z^2 = \bar{\zeta} = \xi_1 - i\xi_2, \quad \phi_z = \left(\frac{\bar{\zeta}}{z}\right)^{\frac{1}{2}},$$

with $\zeta = \xi_1 + i\xi_2$ a complex constant, we find upon integrating that

$$\phi(z) = 2\sqrt{\bar{\zeta} z}$$

is a solution to the given problem. Thus

$$iv = \sqrt{\bar{\zeta} z} - \sqrt{\zeta \bar{z}}$$

$$v^2 = 2|\zeta z| - \bar{\zeta} z - \zeta \bar{z} = 2\{(\xi_1^2 + \xi_2^2)^{\frac{1}{2}} (x_1^2 + x_2^2)^{\frac{1}{2}} - (\xi_1 x_1 + \xi_2 x_2)\},$$

and an easy computation shows that if $\zeta z \neq 0$ then

$$|v_{x_k \xi_l}| = \frac{1}{4|\zeta z|} \neq 0,$$

so that also the second requirement in (24) is fulfilled.

It is now natural to generalize this solution to three dimensions by setting

$$(25) \quad v^2 = 2\left(\xi x - \sum_{k=1}^{3} \xi_k x_k\right), \quad \xi = (\xi_1^2 + \xi_2^2 + \xi_3^2)^{\frac{1}{2}}, \quad x = (x_1^2 + x_2^2 + x_3^2)^{\frac{1}{2}}.$$

We thus have to show that if

$$F(x_k, y_k) = xy^2 = x(y_1^2 + y_2^2 + y_3^2),$$

then indeed (25) results in (21) for a suitable $\lambda(\xi_k)$. First of all, (25) gives

$$(26) \quad v v_{x_k} = \frac{x_k}{x} \xi - \xi_k \quad (x \neq 0), \qquad v v_{\xi_k} = \frac{\xi_k}{\xi} x - x_k \quad (\xi \neq 0).$$

Multiplying the first equation in (26) by x, squaring, and summing over $k = 1, 2, 3$, we find that

$$x^2 v^2 \sum_{k=1}^{3} v_{x_k}^2 = 2\xi^2 x^2 - 2\xi x \sum_{k=1}^{3} \xi_k x_k = \xi x v^2$$

$$(27) \qquad x \sum_{k=1}^{3} v_{x_k}^2 = \xi \quad (x v^2 \neq 0),$$

while the Jacobian determinant becomes

$$|v_{x_k \xi_l}| = \frac{-1}{4\xi x v} \quad (\xi x v \neq 0).$$

This verifies (21) for

$$\lambda(\xi_k) = \xi.$$

To satisfy the requirement $\xi x v \neq 0$, one still has to assume that the two real vectors $(\xi_1, \xi_2, \xi_3) = (\xi_k)$ and $(x_1, x_2, x_3) = (x_k)$ are linearly independent. The initial formulation (20) is thereby justified, and it only remains to express the canonical transformation (23) generated by $v(x_k, \xi_k)$.

Multiplying the first equation in (26) by x and the second by ξ gives

$$x v v_{x_k} = \xi x_k - \xi_k x = -\xi v v_{\xi_k},$$

which by (23) leads to

$$x y_k = \xi \eta_k \quad (k=1,2,3). \tag{28}$$

By (23), (27) we have

$$x y^2 = \xi,$$

while by (26), in analogy to (27), we have

$$\xi \sum_{k=1}^{3} v_{\xi_k}^2 = x,$$

which by (23), with the abbreviation $\eta_1^2+\eta_2^2+\eta_3^2=\eta^2$, gives

$$\xi \eta^2 = x. \tag{29}$$

Multiplying (28) by y^2, one obtains

$$\eta_k = \frac{y_k}{y^2} \quad (k=1,2,3), \tag{30}$$

while analogously

$$y_k = \frac{\eta_k}{\eta^2} \quad (k=1,2,3). \tag{31}$$

From $x\xi \neq 0$, we observe that also $y \neq 0$, $\eta \neq 0$. Thus, in view of (30), (31), the triples y_1, y_2, y_3 and η_1, η_2, η_3 transform into one another by means of radial inversion.

Finally, we derive explicit expressions for the transformation of the x_k. By (23), (26) one has

$$v y_k = \frac{x_k}{x}\xi - \xi_k,$$

and multiplying this by x_k, summing over k, and using the abbreviations

$$g = \sum_{k=1}^{3} x_k y_k, \quad \gamma = \sum_{k=1}^{3} \xi_k \eta_k,$$

we obtain

$$v g = x\xi - \sum_{k=1}^{3} x_k \xi_k = \frac{v^2}{2},$$

or

$$g = \frac{v}{2}.$$

Analogously,

$$v\eta_k = x_k - \frac{\xi_k}{\xi} x$$

leads to

$$y = -\frac{v}{2}.$$

Furthermore, the next to last equation gives

$$x_k = \frac{\xi_k}{\xi} x - 2y\eta_k,$$

which in turn, by (29), gives

$$x_k = \xi_k \eta^2 - 2\eta_k \sum_{l=1}^{3} \xi_l \eta_l \quad (k=1,2,3). \tag{32}$$

Analogously, for the inverse we have

$$\xi_k = x_k y^2 - 2y_k \sum_{l=1}^{3} x_l y_l \quad (k=1,2,3). \tag{33}$$

Thus the transformation obtained is a birational involution. Throughout the derivation it was assumed that the two vectors (x_k), (ξ_k) are real and linearly independent. However, one readily observes that to have (30), (33) as the only solution to (31), (32) it is sufficient to require that $\eta \neq 0$, in which case also $y \neq 0$ and the transformation is canonical.

Before using this transformation to regularize a collision in the three-body problem, we still want to determine the orbits for the two-body problem considered previously, where, we recall, $h=0$. After the canonical transformation (31), (32) the Hamiltonian equations (17) become

$$\xi_k' = 0, \quad \eta_k' = -\lambda_{\xi_k} = -\frac{\xi_k}{\xi} \quad (\xi \neq 0;\ k=1,2,3),$$

with the solutions given by

$$\eta_k = -\frac{\xi_k}{\xi} s + \zeta_k, \tag{34}$$

where ξ_k, ζ_k are real constants. By inserting these solutions into (32), we obtain the x_k as quadratic polynomials in s, and from this the desired orbits. We will show that, in general, these are parabolas. Indeed, if we consider a plane containing the two vectors (ξ_k), (ζ_k), then by (34) it contains the vector (η_k). Hence by (32) the vector (x_k), and with it the whole orbit, lies in this plane. Because of the invariance relative to orthogonal transformations, it is enough to consider the case $\xi_3 = \zeta_3 = 0$, with the plane given by $x_3=0$. The six expressions x_1^2, x_1x_2, x_2^2, x_1, x_2, 1, being polynomials of degree $\leqq 4$ in s, are homogeneous linear functions in the five variables $s^4, s^3, s^2, s, 1$, and therefore there exists a polynomial of degree two in x_1, x_2 that vanishes identically as a function of s. Thus. in any case, the orbit is a conic section. In the subsequent discussion we replace s by $s+c$ for some constant c, whereby (34) becomes

$$\eta_k = -\frac{\xi_k}{\xi}s + \left(\zeta_k - c\frac{\xi_k}{\xi}\right).$$

We choose c so that the two vectors (ξ_k) and $\left(\zeta_k - c\frac{\xi_k}{\xi}\right)$ are orthogonal, i.e. so that

$$c\xi = \sum_{k=1}^{2} \xi_k\zeta_k .$$

If one denotes $\zeta_k - c\frac{\xi_k}{\xi}$ again by ζ_k, then $(\xi_k), (\zeta_k)$ are orthogonal and (34) holds, so that

$$\eta^2 = s^2+\zeta^2, \quad \sum_{k=1}^{2}\xi_k\eta_k = -\xi s,$$

where $\zeta^2=\zeta_1^2+\zeta_2^2$. Hence (32) becomes

$$(35)\quad x_k = \xi_k(s^2+\zeta^2) + 2\left(\zeta_k - \frac{\xi_k}{\xi}s\right)\xi s = 2\xi s\zeta_k + (\zeta^2 - s^2)\xi_k \quad (k=1,2).$$

If $\zeta \neq 0$, this defines a parabola with axis parallel to the vector $(-\xi_k)$. We will show that the focus of this parabola is the origin. For this we appeal to a property characterizing such a parabola, namely, the position vector forms the same angle with the tangent of the curve as the latter does with the axis. Because the direction of the axis is determined by $(-\xi_k)$ while the vector (x_k') gives the direction of the tangent, we need only verify that

$$\sum_{k=1}^{2}\frac{x_k}{x}x_k' = \sum_{k=1}^{2}\frac{-\xi_k}{\xi}x_k' .$$

From (35) we have

$$\frac{x_k'}{2} = \xi\zeta_k - s\xi_k ,$$

so that

$$\frac{1}{2}\sum_{k=1}^{2}\frac{\xi_k}{\xi}x_k' = -s\xi .$$

On the other hand, differentiation of $x^2 = x_1^2 + x_2^2$ and $x = \xi\eta^2$ leads by (34) to

$$\frac{1}{2}\sum_{k=1}^{2}\frac{x_k}{x}x_k' = \frac{1}{2}x' = \xi\sum_{k=1}^{2}\eta_k\eta_k' = \xi\sum_{k=1}^{2}\left(\frac{\xi_k}{\xi}s - \zeta_k\right)\frac{\xi_k}{\xi} = \xi s,$$

as was to be verified. Finally, t and s are related through

$$t' = x = \xi\eta^2 = \xi(s^2 + \zeta^2),$$

which, with a suitable choice of initial time, leads to

$$t = \frac{\xi}{3}s^3 + \xi\zeta^2 s.$$

The case of collision of the two bodies requires that $x = \xi(s^2 + \zeta^2)$ vanish, hence that $\zeta = 0$, $x_k = -s^2\xi_k$ $(k = 1, 2)$, $t = \frac{\xi}{3}s^3$. The parabola then degenerates to a line, the collision occurs at $s = t = 0$, and $t^{\frac{1}{3}}$ serves as a uniformizer for the η_k.

It should be observed that, as a consequence of our earlier analysis, only those parabolas that satisfy the condition $xy^2 = m_1 m_3$, i.e. for which $\xi = m_1 m_3$, are orbits of the original two-body problem with $h = 0$. This accounts for the apparent contradiction that the parabolic orbits obtained contain six parameters, the same number as the general solution to the two-body problem in space when expressed in relative coordinates.

§ 8. Application to the Three-Body Problem

The transformation derived in the previous section will now be applied to the three-body problem. Let $F(x_k, y_k)$ be the function originally introduced through (7; 7), (7; 9), (7; 15), with the twelve variables x_k, y_k $(k = 1, \ldots, 6)$ defined in (7; 4), (7; 5). This leads to the Hamiltonian system (7; 16), in which the independent variable s is given by (7; 12). With s_1 as in (7; 13), the six pairs of functions $x_k(s), y_k(s)$ are all regular for $0 \leqq s < s_1$, while at least one of the three functions $y_1(s), y_2(s), y_3(s)$ has a singularity at $s = s_1$. We now transform the three pairs x_k, y_k $(k = 1, 2, 3)$ according to (7; 30), (7; 33) into the three new pairs ξ_k, η_k $(k = 1, 2, 3)$, retaining the other three pairs as $x_k = \xi_k$, $y_k = \eta_k$ $(k = 4, 5, 6)$. This transformation of the first three pairs was already shown to be canonical, from which it readily follows that the whole transformation of the six pairs x_k, y_k $(k = 1, \ldots, 6)$ is canonical. The transformation being independent of s, the Hamiltonian function is again F, and the transformed equations of motion remain

$$\xi_k' = F_{\eta_k}, \quad \eta_k' = -F_{\xi_k} \quad (k = 1, \ldots, 6), \tag{1}$$

with F now regarded as a function of the ξ_k, η_k. By (7; 7), (7; 9), (7; 29), (7; 31), (7; 32) we have

$$(2)\quad xT = \frac{1}{2}(m_1^{-1}+m_3^{-1})\xi + \frac{1}{2}(m_2^{-1}+m_3^{-1})\xi\eta^2 \sum_{k=1}^{3} \eta_{k+3}^2 + m_3^{-1}\xi \sum_{k=1}^{3} \eta_k\eta_{k+3},$$

$$(3)\quad x = \xi\eta^2, \quad xU = m_1 m_3 + m_2 \xi\eta^2 \left(\frac{m_3}{r_{23}} + \frac{m_1}{r_{12}}\right)$$

with

$$(4)\quad r_{23}^2 = \sum_{k=1}^{3} \xi_{k+3}^2, \quad r_{12}^2 = \sum_{k=1}^{3} (x_k - \xi_{k+3})^2,$$

$$(5)\quad x_k = \xi_k \eta^2 - 2\eta_k \sum_{l=1}^{3} \xi_l \eta_l, \quad y_k = \frac{\eta_k}{\eta^2} \quad (k=1,2,3),$$

$$(6)\quad \xi^2 = \sum_{k=1}^{3} \xi_k^2, \quad \eta^2 = \sum_{k=1}^{3} \eta_k^2,$$

and these expressions are to be inserted into

$$(7)\quad F = xT - xU - hx.$$

It will now be shown, with the aid of Cauchy's existence theorem, that the new coordinates ξ_k, η_k $(k=1, \ldots, 6)$ remain regular as functions of s at $s=s_1$. For this we first use the results of § 6 to investigate the behavior of these coordinates as $s \to s_1$ $(0 \leqq s < s_1)$. We saw that as $t \to t_1$, i.e. as $s \to s_1$, the ξ_k, η_k $(k=4,5,6)$ approach definite limits, and the distances r_{23}, r_{12} approach positive limits. Furthermore, (7; 11) says that

$$(8)\quad \xi = xy^2 \to \frac{2(m_1 m_3)^2}{m_1 + m_3} = c > 0 \quad (s \to s_1),$$

and therefore also ξ has a positive limit. At this point, however, we cannot yet say that each individual ξ_1, ξ_2, ξ_3 has a limit. From $x \to 0$ we have $y \to \infty$ and

$$(9)\quad \eta_k = \frac{y_k}{y^2} \to 0 \quad (s \to s_1; k=1,2,3).$$

Let s_0 in the interval $0 \leqq s < s_1$ be such that

$$(10)\quad \frac{c}{2} \leqq \xi \leqq 2c \quad (s_0 \leqq s < s_1),$$

and consider ξ_1, ξ_2, ξ_3 as independent real variables in the spherical shell S defined by (10). The nine other coordinates ξ_k $(k=4,5,6)$, η_k

($k=1, \ldots, 6$) have limits as $s \to s_1$, and as seen from the definitions (4), (5), (6) and condition (9), the corresponding limit point in nine-dimensional space can be enclosed by a compact real sphere K, so small that in the product domain $P = S \times K$ the functions $\xi, r_{12}^{-1}, r_{23}^{-1}$ are regular in the twelve independent variables ξ_k, η_k ($k=1, \ldots, 6$). Hence by (2), (3), (7) also F is regular there. Now if s_0 is sufficiently near s_1, the solution curve $\xi_k(s), \eta_k(s)$, for $s_0 \leqq s < s_1$, lies entirely within P, and therefore by the result at the end of § 4 the $\xi_k(s), \eta_k(s)$ ($k=1, \ldots, 6$) remain regular at $s = s_1$. In particular, each ξ_1, ξ_2, ξ_3 does indeed have a limit as $s \to s_1$.

For a more precise determination of the behavior of t near s_1 we set $\xi_k(s_1) = \xi_{k1}$ ($k=1,2,3$), $b = \frac{1}{2}(m_1^{-1} + m_3^{-1})$, and use (1), (2), (3), (7) in conjunction with (8), (9) to obtain for $k=1,2,3$ the series expansions

$$\eta_k' = -F_{\xi_k} = -\frac{b}{c}\xi_{k1} + \cdots$$

$$\eta_k = -\frac{b}{c}\xi_{k1}(s-s_1) + \cdots \tag{11}$$

in powers of $s - s_1$. Thus

$$\eta^2 = b^2(s-s_1)^2 + \cdots, \tag{12}$$

while (5) leads to

$$\begin{cases} x_k = \xi_{k1} b^2 (s-s_1)^2 - 2b^2 \xi_{k1}(s-s_1)^2 + \cdots = -b^2 \xi_{k1}(s-s_1)^2 + \cdots \\ \qquad (k=1,2,3), \end{cases} \tag{13}$$

so that

$$x = b^2 c (s-s_1)^2 + \cdots.$$

Since (7; 12) says that $t' = x$, the series expansion of t in a neighborhood of $s = s_1$ then takes the form

$$t = t_1 + \frac{b^2 c}{3}(s-s_1)^3 + \cdots, \tag{14}$$

while inversion of this finally gives

$$s - s_1 = \left\{\frac{3}{b^2 c}(t-t_1)\right\}^{\frac{1}{3}} + \cdots \tag{15}$$

as an expansion in powers of $(t-t_1)^{\frac{1}{3}}$ with real coefficients. Here $(t-t_1)^{\frac{1}{3}}$ is understood to be real also for $t < t_1$. This shows that the original solution $x_k(t), y_k(t)$ of (7; 8), when continued analytically along the interval $\tau \leqq t < t_1$, has a branch point of order two at $t = t_1$. From (13) the x_k are seen to be regular in the local uniformizer $(s - s_1)$, while from

(11), (12) we have

$$y_k = \frac{\eta_k}{\eta^2} = -(bc)^{-1}\xi_{k1}(s-s_1)^{-1} + \cdots \quad (k=1,2,3),$$

so that if $\xi_{k1} \neq 0$ then y_k, as a function of s, has a first order pole at $s=s_1$.

Our result shows that the x_k, y_k can be analytically continued across the singularity $t=t_1$, and this is done by crossing s_1 in the positive direction along the real s-axis. By (14), also t remains real and increases across t_1, while in the process, because the coefficients in all the series expansions are real, the x_k, y_k remain real. From (13) one sees that the two particles P_1, P_3 collide at $t=t_1$ and then are reflected against one another. This assertion is meant, of course, only in a mathematical sense, and has no physical significance. For $t>t_1$ sufficiently near t_1 we again have $x>0$ and y finite, so that we can reintroduce the coordinates x_k, y_k in place of ξ_k, η_k by means of the inverse substitution (7; 31), (7; 32). Being analytic expressions in the variable t, the values of the center of mass integrals, the angular momentum constants, and the energy integral remain unchanged under analytic continuation, as do the differential equations. Consequently (7; 16) remains as is, and from it we revert by (7; 6) to (7; 2).

From now until the end of § 11 it will be assumed that the three angular momentum constants α, β, γ in relative coordinates are not all 0, where we refer to a coordinate system in which the center of mass is at rest.

Selecting a definite value $t>t_1$ to which the solution of (7; 2) has been continued, and denoting it by τ, we can again apply the previous results at the new value of τ. If in continuing the solution analytically along increasing $t>\tau$ a singularity is encountered at a finite $t=t_2$, then again exactly two particles must collide, for by assumption not all the angular momentum constants are 0. The collision need no longer be between P_1, P_3, but in any case, just as at $t=t_1$ a corresponding regularization can be carried out at $t=t_2$. We continue the solution across t_2, and in this way reach further singularities $t=t_n$ $(n=1,2,\ldots)$. Should the number of these singularities be finite, or should t_n tend to infinity as $n\to\infty$, this will result in an analytic continuation for all finite $t\geqq\tau$. We next show that the remaining case, namely, when the t_n accumulate at a finite point t_∞, cannot occur.

For this too we use the previous method. Let t_n $(n=1,2,\ldots)$ be an increasing sequence consisting of all singularities encountered in continuing the solution for $t\geqq\tau$ along the real t-axis, and assume that this sequence has a finite limit t_∞. The potential function U becomes infinite at each $t=t_n$ and is finite elsewhere in the interval $\tau\leqq t<t_\infty$.

We claim that as t tends to t_∞ within this interval, U becomes infinite. If not, there would be a positive number A and an increasing sequence of values of t that converge to t_∞ and at which $U \leqq A$. By the final observation in § 5, the solution would then have to be regular at $t = t_\infty$. But surely, being a limit point of the singularities t_n, the point t_∞ must itself be a singularity. Thus $U \to \infty$ as $t \to \infty$, and the minimum of the three distances r_{12}, r_{23}, r_{31} approaches 0. By Lagrange's formula there is an interval $t_0 \leqq t < t_\infty$ in which $\ddot{I} > 0$, while at all $t = t_n$ the function $\ddot{I}$ becomes infinite. The function $\dot{I}$ was shown to be continuous from the left at $t = t_1$, and similarly at all $t = t_n$, while the same method shows continuity from the right as well. Thus $\dot{I}$ is increasing and continuous in this interval. The reasoning leading to (6; 6) can now be repeated word for word to show that I has a positive lower bound for $t_0 \leqq t < t_\infty$. In analogy to (6; 7), the length of a definite side of the corresponding triangle, say $r_{13} = r$, then tends to 0, while the other two sides stay above a positive bound. Therefore, with t_0 suitably restricted, all collisions at the infinitely many t_n in this interval are between P_1, P_3 and can be regularized by the single transformation (7; 4), (7; 5), (7; 30), (7; 33). With the previous notation, the reasoning leading to (6; 9), together with (7; 11), shows that $\xi = xy^2$ has the limit

$$\lim_{t \to t_\infty} \xi = \frac{2(m_1 m_3)^2}{m_1 + m_3} > 0 .$$

According to (7; 13) the singularities t_n correspond to the values

$$s_n = \int_{t_0}^{t_n} \frac{dt}{x(t)}$$

of the uniformizer s. As in the proof for the existence of the integral (6; 10), the function $\dot{I}$ remains bounded as $t \to t_\infty$, and the integral

$$s_\infty = \int_{t_0}^{t_\infty} \frac{dt}{x(t)}$$

converges. Thus the s_n accumulate at the finite value s_∞. The reasoning used in the beginning of this section to show regularity of $x(s)$ at $s = s_1$ will now show that $x(s)$ is regular also at $s = s_\infty$. On the other hand, $x(s)$ vanishes at the infinitely many s_n that accumulate at s_∞, and since $x(s)$ is analytic and not identically 0, this is a contradiction. The assumption that the singularities t_n have a finite accumulation point was therefore false.

The analytic continuation leading to a solution for all $t \geqq \tau$ can likewise be carried out for $t \leqq \tau$. However, since the equations of motion (7; 2) are invariant under replacement of q_k, p_k, t by $q_k, -p_k, -t$, this case need not be treated separately. Thus, the continuation results in a solution for all finite real values of t. On the other hand, to have a uniform representation of the solution for all time, it would be desirable to have the uniformizer s not depend on which two of the three particles collide at each singularity. Therefore, to realize the uniformization in the large, we will replace the variable s by a new variable ω, so that the coordinates q_k $(k=1,\ldots,9)$ of the three particles, as well as the time t, become regular functions of ω in the unit disk $|\omega|<1$, with the real t-axis corresponding to the interval $-1<\omega<1$.

No additional work is needed to see that such a parameter ω exists. Indeed, up to now we have regularized a singularity by using the substitution (7; 12), in which x denotes the distance between the colliding particles, so that two particles are always distinguished. This asymmetry, however, can be removed by replacing the variable x^{-1} in (7; 12) by U, i.e. by defining

$$s=\int_{\tau}^{t} U\,dt\,. \tag{16}$$

Because at the singularity t_1 the function U behaves asymptotically as $m_1 m_3 x^{-1}$, the new parameter s is easily seen to regularize all collisions. One would still want s to go to $\pm\infty$ with t, and while this is actually true for the s defined by (16), the verification is not entirely trivial and can be avoided if instead we set

$$s=\int_{\tau}^{t}(U+1)\,dt\,. \tag{17}$$

Then clearly s has the desired relation to t, and this parameter also regularizes all collisions. Hence each finite point s_0 on the real s-axis in the complex s-plane is the center of a disk K_0 where t and the nine coordinates q_k have convergent power series expansions in the variable $s-s_0$. The union over s_0 of the disks K_0 is a simply connected domain G, symmetrically enclosing the real s-axis. The Riemann mapping theorem allows this domain to be mapped conformally onto the unit disk in an ω-plane, and indeed, in such a way that the real s-axis goes into the diameter $-1<\omega<1$. The parameter ω introduced in this way has the desired properties.

This proves only the existence of such a parameter, and to carry out the conformal transformation explicitly one must know more about the domain covered by the disks of convergence K_0. It is conceivable, for instance, that as the center s_0 varies, the radii ϱ_0 of K_0 do not stay above a positive lower bound, and therefore G does not contain a parallel strip enclosing the real axis. This actually is not so, and in the next section we will follow Sundman in proving that the radii of convergence ϱ_0 have a positive lower bound δ. The time t and the coordinates q_k $(k=1, \ldots, 9)$, as functions of the parameter $s=\sigma+iv$ defined by (17), will then be regular throughout the strip $-\delta<v<\delta$. The proof will in fact be constructive, enabling δ to be expressed as a function of the initial values and masses in the three-body problem, always under the assumption that the three angular momentum constants are not all 0. To that end we must first derive two important lemmas, also due to Sundman, and of interest in their own right. The first lemma says that the perimeter of the triangle with the particles as vertices remains above a positive bound ϑ for all time. Up to now we have shown only that the perimeter is always positive and triple collision cannot occur; still, it would be conceivable that for large time it does get arbitrarily close to 0. The second lemma says that the velocity of the particle opposite the shortest side remains below a bound $\varkappa$ for all time, whereas up to now we have shown only that it is always finite. The quantities $\vartheta, \varkappa$ will appear as functions of the initial values and masses.

Finally, we still wish to take a general view of the totality of collision orbits as compared to all possible orbits in the three-body problem. For a collision orbit, let $t=t_1$ be a time at which two particles, say P_1, P_3, collide, and let the collision be regularized as before, with ξ_k, η_k $(k=1, \ldots, 6)$ the new coordinates and s, defined by (7; 12), the new independent variable, so that $s=s_1$ corresponds to the collision. The twelve coordinates ξ_k, η_k are regular functions of s in a neighborhood of s_1, while (8), (9) imply that

$$(18) \qquad \xi(s_1)=\frac{2(m_1 m_3)^2}{m_1+m_3}, \quad \eta_k(s_1)=0 \quad (k=1,2,3),$$

where $\xi=(\xi_1^2+\xi_2^2+\xi_3^2)^{\frac{1}{2}}$. Conversely, let twelve real initial values $\xi_k(s_1), \eta_k(s_1)$ $(k=1, \ldots, 6)$ be prescribed at $s=s_1$, subject only to (18). That is, $\eta_1(s_1)=\eta_2(s_1)=\eta_3(s_1)=0$ and $\xi_1(s_1), \xi_2(s_1), \xi_3(s_1)$ are subject only to condition (18) on the sum of their squares, while for $k=4,5,6$ the values can be chosen arbitrarily. The Hamiltonian function F defined by (7) vanishes for these values, and remains identically 0 along the corresponding solution to the system (1). A return to the original coordinates $q_k, \dot{q}_k$ $(k=1, \ldots, 9)$ by the inverse transformation then leads to a collision orbit along which the energy constant has the value h, which appears as a linear parameter in F. Thus the twelve initial values ξ_k, η_k $(k=1, \ldots, 6)$ and the free parameter h are restricted by the four analytic conditions (18), and therefore the collision orbits depend on nine parameters plus the time t, i.e. altogether on ten independent parameters. By the Cauchy existence theorem the solution is easily seen to depend analytically on the parameters. Removal of the original assumption that the center of mass rest at the origin gives six additional parameters, so that the collision orbits form a 16-dimensional analytic manifold in the 18-dimensional space of the $q_k, \dot{q}_k$ $(k=1, \dots, 9)$. Note that there are two other such manifolds coming from collisions of the other two possible pairs of points. Nothing more is known about the structure of these three manifolds in the large. Conceivably they form a dense set in the 18-dimensional space. One can, however, conclude that the Lebesgue measure of this set in $(q, \dot{q})$-space is 0, and in particular, collision orbits have no significance in ergodic theory, where statements are made only about sets of positive measure. In conclusion, we observe that the previously excluded triple-collision orbits, lying in the 15-dimensional algebraic manifold defined by the vanishing of the three angular momentum constants in relative coordinates, also form a set of measure 0.

§ 9. An Estimate of the Perimeter

It is again assumed that the center of mass P_0 rests at the origin. This section deals with the proof of Sundman's first lemma:

If the three angular momentum constants are not all 0, the perimeter of the triangle formed by the three particles remains for all time above a positive constant.

As in § 5, we again denote the coordinates of P_k by x_k, y_k, z_k and use q_k or q when abbreviating. If we denote the distance between P_k and P_0 by ϱ_k, it is enough to show that the quantity

$$I = \sum_q mq^2 = \sum_{k=1}^{3} m_k \varrho_k^2$$

remains above a positive constant. For, because the center of mass is within the triangle, one sees that $\varrho_j + \varrho_k \leqq r_{jl} + r_{kl}$ if j, k, l is a cyclic permutation of 1, 2, 3. Addition of these inequalities gives the lower bound $\varrho_1 + \varrho_2 + \varrho_3 \leqq \sigma$ for the perimeter $r_{12} + r_{23} + r_{31} = \sigma$. Similarly, the triangle inequality $r_{jk} \leqq \varrho_j + \varrho_k$ leads to the estimate $\sigma \leqq 2(\varrho_1 + \varrho_2 + \varrho_3)$ from above. If μ denotes the largest of the three masses m_k, then on the one hand

$$I \leqq \mu \sum_{k=1}^{3} \varrho_k^2 \leqq \mu \sigma^2, \tag{1}$$

while on the other hand, by the Schwarz inequality,

$$\frac{\sigma^2}{4} \leqq \left(\sum_{k=1}^{3} \varrho_k \right)^2 = \left\{ \sum_{k=1}^{3} (m_k^{\frac{1}{2}} \varrho_k) \, m_k^{-\frac{1}{2}} \right\}^2 \leqq I \sum_{k=1}^{3} m_k^{-1}. \tag{2}$$

Thus $I\sigma^{-2}$ lies between two positive bounds that depend only on the masses and, in particular, it is sufficient for our purpose to bound I from below.

As in the earlier proof for the nonvanishing of I, we begin with formulas (6;2), (6;4), which state

$$\tfrac{1}{2}\ddot{I}=T+h=U+2h, \tag{3}$$

$$2IT\geqq\tfrac{1}{4}\dot{I}^2+\eta, \quad \eta=\frac{\alpha^2+\beta^2+\gamma^2}{3}>0, \tag{4}$$

where α, β, γ are the angular momentum constants. Eliminating T between them, we get

$$\ddot{I}-\tfrac{1}{4}\dot{I}^2I^{-1}-\eta I^{-1}-2h\geqq 0. \tag{5}$$

After multiplication by $2\dot{I}I^{-\frac{1}{2}}$ the left side can be readily integrated with respect to t, with

$$L=(\dot{I}^2+4\eta)\,I^{-\frac{1}{2}}-8hI^{\frac{1}{2}} \tag{6}$$

being a corresponding indefinite integral. By (5) both L and I increase or decrease together, and this fact will be used to obtain a lower bound for I. We again use the subscript τ to denote the value of a quantity at time $t=\tau$, and assume that the four inequalities

$$I_\tau<A, \quad U_\tau<A, \quad |h|<A, \quad \eta^{-1}<A \tag{7}$$

hold for some positive number A. It will turn out that there exists a positive number $\Theta=\Theta(A, m)$, depending only on A and the masses m_k, such that $I>\Theta$ for all real t. Our conclusions will also allow an explicit expression of Θ as a function of A and the m_k, although the somewhat unwieldy computations will not be carried out. For convenience we will use $c_l=c_l(A, m)$ with $l=1, \ldots, 58$ to denote certain positive quantities that depend only on A, m_k and are constructively defined; they carry the connotation of upper bounds. To begin with, clearly

$$m_k m_l r_{kl\tau}^{-1}\leqq U_\tau<A \qquad (k<l;\ k, l=1, 2, 3),$$

and hence

$$r_{kl\tau}>c_1^{-1},$$

while by (2) also

$$I_\tau>c_2^{-1}.$$

We first treat the simpler case $h\geqq 0$, wherein according to (3) one has

$$\tfrac{1}{2}\ddot{I}=U+2h>0,$$

and therefore I is a convex function of t. If the initial value $\dot{I}_\tau=0$, then I has an absolute minimum at $t=\tau$, and $I\geqq I_\tau>c_2^{-1}$ is an estimate of the kind desired. Otherwise, replacing t by $-t$ if necessary, we may assume $\dot{I}_\tau < 0$. Consider an interval $\tau \leqq t < t_1$ in which I decreases monotonically. Then also the quantity L, defined by (6), decreases there, and, because $h \geqq 0$, so does $L + 8hI^{\frac{1}{2}}$. Consequently

$$(\dot{I}^2 + 4\eta)\, I^{-\frac{1}{2}} \leqq (\dot{I}_\tau^2 + 4\eta)\, I_\tau^{-\frac{1}{2}} \qquad (\tau \leqq t \leqq t_1),$$

and all the more

$$4\eta I^{-\frac{1}{2}} \leqq (\dot{I}_\tau^2 + 4\eta)\, I_\tau^{-\frac{1}{2}}$$

$$I \geqq I_\tau \left(1 + \frac{1}{4\eta} \dot{I}_\tau^2\right)^{-2}. \tag{8}$$

Moreover, since I is convex, the estimate (8) is valid for all time. In addition, (4) implies that

$$\dot{I}_\tau^2 \leqq 8 I_\tau T_\tau = 8 I_\tau (U_\tau + h) < 16 A^2,$$

which is true even if $h < 0$, and therefore (8) leads to the estimate

$$I \geqq c_2^{-1}(1 + 4A^3)^{-2}, \qquad I > c_3^{-1}.$$

This finishes the case $h \geqq 0$.

For the case $h < 0$ the estimate is not as easy, since I need not be convex and can have infinitely many extrema. We set $k = -2h$ and again restrict ourselves to estimating I for $t \geqq \tau$. If $\dot{I} \geqq 0$ for all $t \geqq \tau$, then also $I \geqq I_\tau > c_2^{-1}$, and therefore we need only consider the case when $\dot{I} < 0$ for some $t > \tau$. Let $\tau \leqq t_0 < t_1$ be such that in the interval $t_0 < t < t_1$ the function I is monotonically decreasing. Then L is also decreasing for these t and, in particular,

$$(\dot{I}^2 + 4\eta)\, I^{-\frac{1}{2}} + 4kI^{\frac{1}{2}} \leqq (\dot{I}_0^2 + 4\eta)\, I_0^{-\frac{1}{2}} + 4kI_0^{\frac{1}{2}} \qquad (t_0 \leqq t \leqq t_1), \tag{9}$$

where the subscript 0 refers to values of the respective functions at $t = t_0$. Thus, since $k > 0$, certainly

$$4\eta I^{-\frac{1}{2}} \leqq (\dot{I}_0^2 + 4\eta)\, I_0^{-\frac{1}{2}} + 4kI_0^{\frac{1}{2}}$$

$$I \geqq I_0 \left(1 + \frac{k}{\eta} I_0 + \frac{1}{4\eta} \dot{I}_0^2\right)^{-2} \qquad (t_0 \leqq t \leqq t_1). \tag{10}$$

With t_1 fixed, let t_0 be the smallest value possible for the left endpoint of such an interval. Then either $t_0 = \tau$ and

$$I \geqq c_2^{-1} \left(1 + \frac{k}{\eta} A + \frac{4}{\eta} A^2\right)^{-2}, \qquad I > c_4^{-1} \qquad (\tau \leqq t \leqq t_1), \tag{11}$$

or $t_0 > \tau$ and I has a maximum at t_0. In the latter case $\dot{I}_0 = 0$ and (9) implies

$$\eta I^{-\frac{1}{2}} + kI^{\frac{1}{2}} \leqq \eta I_0^{-\frac{1}{2}} + kI_0^{\frac{1}{2}} \qquad (t_0 \leqq t \leqq t_1). \tag{12}$$

The assumption, we recall, is that in this interval the function I is decreasing.

The last inequality will give an estimate of I from below, which will readily come from the properties of the function

$$f(x)=\eta x^{-\frac{1}{2}}+kx^{\frac{1}{2}} \quad (x>0).$$

This function is unchanged if x is replaced by $\left(\frac{\eta}{k}\right)^2 x^{-1}$, and for positive x has exactly one extremum, namely, a minimum at $x=\frac{\eta}{k}$. Moreover, in the interval $0<x<\frac{\eta}{k}$ it is monotonically decreasing. When $t_0<t\leqq t_1$, we have $I<I_0$ and, by (12), also $f(I)\leqq f(I_0)$, so that certainly $I_0>\frac{\eta}{k}$. On the other hand, $f(x)=f(I_0)$ at $x=\left(\frac{\eta}{k}\right)^2 I_0^{-1}<\frac{\eta}{k}$, so that $f(x)>f(I_0)$ for $x<\left(\frac{\eta}{k}\right)^2 I_0^{-1}$, and therefore

$$I\geqq\left(\frac{\eta}{k}\right)^2 I_0^{-1} \quad (t_0\leqq t\leqq t_1).$$

If $I_0\leqq k^{-2}$, then clearly

$$I\geqq\eta^2>A^{-2} \quad (t_0\leqq t\leqq t_1). \tag{13}$$

It remains to consider the case $I_0>k^{-2}$. Then either

$$I\geqq k^{-2}=(2h)^{-2}>(2A)^{-2} \quad (t_0\leqq t\leqq t_1), \tag{14}$$

or at some $t=t_2<t_1$ in this interval we have $I=I_2=k^{-2}$. The latter assumption says that

$$I>(2A)^{-2} \quad (t_0\leqq t\leqq t_2), \tag{15}$$

while in the remaining interval $t_2\leqq t\leqq t_1$ we can apply inequality (10) with $t_2, I_2, \dot{I}_2$ in place of $t_0, I_0, \dot{I}_0$ and obtain

$$I\geqq\left(k+\eta^{-1}+\frac{k}{4\eta}\dot{I}_2^2\right)^{-2} \quad (t_2\leqq t\leqq t_1). \tag{16}$$

Subsequently we will derive the estimate

$$\dot{I}_2^2<c_5 k^{-1} \quad (I_2=k^{-2}, \dot{I}_2<0). \tag{17}$$

Assuming (17) has been shown, it can be combined with (13), (14), (15), (16) into one inequality

(18) $$I > c_6^{-1} \quad (t_0 \leqq t \leqq t_1),$$

valid whenever I has a maximum at t_0 and is monotonically decreasing in the interval $t_0 \leqq t \leqq t_1$. There are now three possibilities to consider. If I continues to decrease throughout $t > t_0$, then (18) holds for all $t \geqq t_0$. In the other cases I has a first minimum in $t > t_0$, say at $t = t_1$. If I increases throughout $t > t_1$, then

$$I \geqq I_1 > c_6^{-1} \quad (t \geqq t_1),$$

and again $I > c_6^{-1}$ whenever $t \geqq t_0$. If, on the other hand, I does not increase throughout $t > t_1$, there is a first maximum, say at $t = t_3$, and one has

$$I \geqq I_1 > c_6^{-1} \quad (t_0 \leqq t \leqq t_3)$$

in the interval between successive maxima. Finally, one observes that the times for the individual maxima cannot have a finite accumulation point, since otherwise the zeros of $\dot{I}(t)$ would also accumulate and, in view of our results on analytic continuation of solutions to the three-body problem, $\dot{I}(t)$ would vanish identically. That case, however, was treated earlier, under the assumption $\dot{I} \geqq 0$ $(t \geqq \tau)$. Thus, if t_0 is the first maximum of I for $t > \tau$, the estimate $I > c_6^{-1}$ holds for all $t \geqq t_0$. It remains to look at the interval $\tau \leqq t \leqq t_0$, or if I has no maximum for $t > \tau$, at the half line $\tau \leqq t$. If there is no minimum in the interior of this interval in the first case, or if I is not continually decreasing in the second, then trivially $I \geqq I_\tau > c_2^{-1}$ there. In the other case one can apply (11) and obtain $I > c_4^{-1}$ for $\tau \leqq t \leqq t_0$ or $\tau \leqq t$ respectively. It follows that actually $I > c_7^{-1}$ for all $t \geqq \tau$.

It remains to prove (17), and for this the differential inequality (5) is inadequate; the function I has to be examined more closely. Suppose that at time t the shortest side of the triangle with the particles as vertices is again $r_{13} = r$. If ϱ denotes the distance between P_2 and the center of mass P_0, the latter being also the origin of our coordinate system, the triangle inequality gives

(19) $$\varrho < r + r_{23} \leqq 2r_{23}, \quad \varrho < r + r_{12} \leqq 2r_{12}.$$

Conversely, ϱ can be estimated from below by the sides r_{12}, r_{23} as follows. The center of mass integral

(20) $$m_1 q_1 + m_2 q_2 + m_3 q_3 = 0$$

gives the formula

$$M q_2 = m_1(q_2 - q_1) + m_3(q_2 - q_3),$$

where $M = m_1 + m_2 + m_3$. Since the angle at P_2 is at most $\pi/3$, with its cosine $\geqq \frac{1}{2}$, it follows that

$$M^2 \varrho^2 \geqq (m_1 r_{12})^2 + (m_3 r_{23})^2 + m_1 m_3 r_{12} r_{23} > \frac{1}{2}(m_1 r_{12} + m_3 r_{23})^2$$

$$2M\varrho > m_1 r_{12} + m_3 r_{23}. \tag{21}$$

Because r_{13} is the shortest side of the triangle, there is also the relation $\frac{1}{2} r_{12} \leqq r_{23} \leqq 2r_{12}$, and therefore by (19), (21) the ratios $r_{12}/\varrho, r_{23}/\varrho$ lie between two positive bounds that depend only on the masses.

With this auxiliary consideration aside, we now turn to estimating the value $\dot{I}_2$. The center of mass being at rest gives rise to the equation

$$0 = \sum_{k=1}^{3} m_k(\dot{x}_k x_3 + \dot{y}_k y_3 + \dot{z}_k z_3),$$

which when subtracted from

$$\frac{1}{2}\dot{I} = \sum_q m\dot{q}q = \sum_{k=1}^{3} m_k(\dot{x}_k x_k + \dot{y}_k y_k + \dot{z}_k z_k)$$

leads to

$$\frac{1}{2}\dot{I} = \sum_{k=1}^{2} m_k\{\dot{x}_k(x_k - x_3) + \dot{y}_k(y_k - y_3) + \dot{z}_k(z_k - z_3)\},$$

or more briefly

$$\frac{1}{2}\dot{I} = \sum_q \{m_1\dot{q}_1(q_1 - q_3) + m_2\dot{q}_2(q_2 - q_3)\}. \tag{22}$$

With the aid of (20) the quantity $q_2 - q_3$ can be expressed in terms of $q_1 - q_3$ and q_2; namely,

$$m_1(q_1 - q_3) + (m_1 + m_3)(q_3 - q_2) + (m_1 + m_2 + m_3)q_2 = 0$$

$$q_2 - q_3 = \frac{m_1}{m_1 + m_3}(q_1 - q_3) + \frac{M}{m_1 + m_3} q_2, \tag{23}$$

so that (22) becomes

$$\frac{1}{2}\dot{I} = \sum_q m_1\left(\dot{q}_1 + \frac{m_2}{m_1 + m_3}\dot{q}_2\right)(q_1 - q_3) + \frac{m_2 M}{m_1 + m_3}\sum_q \dot{q}_2 q_2. \tag{24}$$

With v the larger of the velocities of P_1 and P_2, the Schwarz inequality gives the estimate

$$\left|\sum_q m_1\left(\dot{q}_1 + \frac{m_2}{m_1 + m_3}\dot{q}_2\right)(q_1 - q_3)\right| \leqq \frac{m_1 M}{m_1 + m_3} vr. \tag{25}$$

The term on the right can be further estimated, using the fact that $h<0$. For then

(26) $$T=U+h<U,$$

hence clearly

(27) $$rT\leqq rU<c_8,$$

and consequently

$$rv^2<c_9,\quad rv\leqq\sqrt{c_9 r}.$$

On the other hand,

$$0\leqq 2T=2U-k$$

leads to the estimate

(28) $$2U\geqq k$$

from below, whereby

$$r<c_{10}k^{-1}$$

and

(29) $$rv<c_{11}k^{-\frac{1}{2}}.$$

Since

(30) $$\sum_q q_2^2=\varrho^2,\quad \sum_q \dot q_2 q_2=\dot\varrho\varrho,$$

the estimate (24) in conjunction with (25), (29) leads to the differential inequality

(31) $$\left|\dot I-\frac{2m_2M}{m_1+m_3}\varrho\dot\varrho\right|<c_{12}k^{-\frac{1}{2}}.$$

Now at time $t=t_2$ we recall that $I(t_2)=I_2=k^{-2}$, $\dot I(t_2)=\dot I_2<0$. Consequently, if we can derive an estimate of the form

(32) $$-\varrho_2\dot\varrho_2<c_{13}k^{-\frac{1}{2}},$$

where $\varrho_2, \dot\varrho_2$ denote the values of $\varrho, \dot\varrho$ at $t=t_2$, as a consequence of (31) we will have the inequality

$$0<-\dot I_2<c_{14}k^{-\frac{1}{2}},$$

and thereby assertion (17) with $c_5=c_{14}^2$.

It remains to prove (32), and for this, just as we did earlier for I, we set up a differential inequality for ϱ whose integration will result in the desired estimate. Differentiating (30) once more, one obtains

$$\sum_q (\ddot{q}_2 q_2 + \dot{q}_2^2) = \ddot{\varrho}\varrho + \dot{\varrho}^2,$$

and the equations of motion then give

$$\ddot{\varrho}\varrho + \dot{\varrho}^2 = \sum_q q_2 \left(m_1 \frac{q_1 - q_2}{r_{12}^3} + m_3 \frac{q_3 - q_2}{r_{23}^3} \right) + v_2^2, \tag{33}$$

where v_2 is the velocity of P_2. Since by (21) the distance ϱ is positive, the Schwarz inequality says that

$$\dot{\varrho}^2 = \left(\sum_q \dot{q}_2 \frac{q_2}{\varrho} \right)^2 \leqq \sum_q \dot{q}_2^2 = v_2^2. \tag{34}$$

Moreover, (19) leads to the estimate

$$|q_k - q_2| r_{k2}^{-3} \leqq r_{k2}^{-2} < 4\varrho^{-2} \tag{35}$$

for $k = 1, 3$, while certainly $|q_2| \leqq \varrho$, so that (33), (34), (35) combine into the differential inequality

$$\ddot{\varrho} > -c_{15}\varrho^{-2}. \tag{36}$$

For the proof of (32) it is enough to consider the case $\dot{\varrho}_2 < 0$. Let the interval $t_4 \leqq t \leqq t_2$ be chosen so small that in it not only $\dot{\varrho} \leqq 0$, but also r_{13} remains the shortest side of the triangle. By (36) we then have

$$2\dot{\varrho}\ddot{\varrho} \leqq -2c_{15}\dot{\varrho}\varrho^{-2} \qquad (t_4 \leqq t \leqq t_2),$$

which upon integration gives

$$\dot{\varrho}_2^2 - 2c_{15}\varrho_2^{-1} \leqq \dot{\varrho}^2 - 2c_{15}\varrho^{-1} \qquad (t_4 \leqq t \leqq t_2).$$

Therefore, all the more

$$\dot{\varrho}_2^2 < \dot{\varrho}_4^2 + 2c_{15}\varrho_2^{-1},$$

where $\dot{\varrho}_4 = \dot{\varrho}(t_4)$, and, because $\varrho_2 \leqq \varrho_4 = \varrho(t_4)$, this leads to the inequality

$$(\varrho_2\dot{\varrho}_2)^2 < (\varrho_4\dot{\varrho}_4)^2 + 2c_{15}\varrho_2.$$

The expression on the left is exactly what is needed to prove (32). On the right, the second term permits an easy estimate of the desired form, for by definition of I we have

$$I > m_2\varrho^2, \quad \varrho_2 < c_{16} I_2^{\frac{1}{2}} = c_{16} k^{-1}.$$

To estimate the first term, we distinguish several cases. First we assume that the previous value t_0, at which I attains its maximum, is admissible as t_4. Since $\dot{I}_0 = 0$, in that case (31) gives

$$(\varrho_4\dot{\varrho}_4)^2 < c_{17} k^{-1}. \tag{37}$$

Second, suppose that $t_4 = t_0$ is not admissible, and that an admissible t_4 is selected as small as possible. Then $t_0 < t_4 \leqq t_2$ and at t_4 either $\dot{\varrho} = \dot{\varrho}_4 = 0$, or r_{13} ceases to be shortest side as t decreases further. If $\dot{\varrho}_4 = 0$, again (37) holds, this time trivially. In the remaining case, at $t = t_4$ an additional side becomes equal to r and, by the triangle inequality, $r_{k2} \leqq 2r$ $(k = 1, 3)$. Because ϱ/r_{k2} has an upper bound depending only on the masses, the same is true of ϱ/r at $t = t_4$. By (27), at this point $\varrho U < c_{18}$, whereupon (26), (28), (34) combine to give the estimate

$$(\varrho\dot{\varrho})^2 \leqq (\varrho v_2)^2 \leqq c_{19}\varrho^2 T < c_{19}\varrho^2 U < c_{20} U^{-1} \leqq 2c_{20} k^{-1} \qquad (t = t_4).$$

Thus (32) can be deduced in all possible cases, and the proof of Sundman's first lemma is complete.

On account of (1), we have proved the estimate

$$\sigma > c_{21}^{-1}, \tag{38}$$

and clearly the proof allows c_{21} to be expressed explicitly as a function of A and the m_k. Sundman actually carried this out.

§ 10. An Estimate of the Velocity

Assume the center of mass P_0 to lie at the origin. Sundman's second lemma then says:

If the angular momentum constants are not all 0, then the velocity of the particle opposite the shortest side of the triangle remains below a finite constant for all time.

The proof follows rather easily from lemma one. By (9; 38) the perimeter satisfies

$$r_{12} + r_{23} + r_{31} > c_{21}^{-1} > 0 \tag{1}$$

for all time. If at time t the shortest side r satisfies

$$r \geqq \frac{1}{4} c_{21}^{-1},$$

then clearly $T = U + h < c_{22}$ and, indeed, all velocities at this time are smaller than c_{23}. From now on we may therefore restrict ourselves to the case

$$r < \frac{1}{4} c_{21}^{-1}. \tag{2}$$

Let $r_{13} = r$ again be the shortest side of the triangle, and $v_2 = v$ the velocity of P_2. In addition to (9; 36), we need an inequality in the other direction, and for this the same is required for (9; 34). It is seen from (9; 30) that

$$(3)\quad \begin{cases} \varrho^2(v^2-\dot\varrho^2) = \sum_q q_2^2 \sum_q \dot q_2^2 - \left(\sum_q q_2\dot q_2\right)^2 \\ \qquad = (x_2\dot y_2 - y_2\dot x_2)^2 + (y_2\dot z_2 - z_2\dot y_2)^2 + (z_2\dot x_2 - x_2\dot z_2)^2, \end{cases}$$

and to estimate this expression we use the angular momentum integrals. Transforming

$$(4)\qquad \gamma = \sum_{k=1}^{3} m_k(x_k\dot y_k - y_k\dot x_k)$$

with the aid of the center of mass integrals into

$$\gamma = \sum_{k=1}^{2} m_k\{(x_k - x_3)\dot y_k - (y_k - y_3)\dot x_k\},$$

we use (9; 23) to eliminate $x_2 - x_3$, $y_2 - y_3$, and obtain

$$(5)\quad \begin{cases} \gamma = m_1\left\{(x_1-x_3)\left(\dot y_1 + \dfrac{m_2}{m_1+m_3}\dot y_2\right) - (y_1-y_3)\left(\dot x_1 + \dfrac{m_2}{m_1+m_3}\dot x_2\right)\right\} \\ \qquad + \dfrac{m_2 M}{m_1+m_3}(x_2\dot y_2 - y_2\dot x_2). \end{cases}$$

The first expression on the right is in absolute value less than

$$c_{24} r T^{\frac12} \leqq c_{24} r(U+|h|)^{\frac12} \leqq c_{24} r U^{\frac12} + c_{24} r |h|^{\frac12} < c_{24} r U^{\frac12} + c_{25},$$

while from $rU < c_8$ also

$$r^2 U < c_{26}.$$

Inserting the initial values into the right side of (4), we find by the Schwarz inequality that

$$\gamma^2 \leqq 2I_\tau T_\tau \leqq 2I_\tau(U_\tau + |h|) < 4A^2.$$

Thus (5) leads to the estimate

$$|x_2\dot y_2 - y_2\dot x_2| < c_{27},$$

with two corresponding inequalities for cyclic permutations of x, y, z, and consequently (3) gives

$$0 \leqq v^2 - \dot\varrho^2 < c_{28}\varrho^{-2}.$$

Recall that the quotients r_{12}/ϱ, r_{23}/ϱ are bounded, while by (1), (2) and the triangle inequality both $r_{12}^{-1} < 4c_{21}$ and $r_{23}^{-1} < 4c_{21}$. Thus also

$$(6)\qquad \varrho^{-1} < c_{29},$$

and therefore

(7) $$0 \leqq v^2 - \dot{\varrho}^2 < c_{30}\varrho^{-1}.$$

The first expression on the right side of the differential equation (9; 33) can, as before, be estimated from above by $c_{31}\varrho^{-1}$, and together with (7) this gives the inequality

(8) $$|\ddot{\varrho}| < c_{32}\varrho^{-2},$$

which for the case we are considering replaces (9; 36) by a two-sided estimate of $\ddot{\varrho}$.

Integrating (8), we now obtain the desired result. Again it is sufficient to consider the case $t \geqq \tau$. If at time t the derivative $\dot{\varrho} = 0$, then (6), (7) immediately give $v^2 < c_{30}c_{29}$. Consider therefore a point in time at which $\dot{\varrho} \neq 0$, and enclose it by an interval $t_1 < t < t_2$ throughout which (2) holds and $\dot{\varrho}$ does not vanish. Then $r_{13} = r$ remains the shortest side throughout this interval. One sees from (8) that

$$|2\dot{\varrho}\ddot{\varrho}| < 2c_{32}|\dot{\varrho}|\varrho^{-2} \qquad (t_1 < t < t_2),$$

and, since $\dot{\varrho}$ has a fixed sign throughout the interval, it follows that

$$|\dot{\varrho}^2 - \dot{\varrho}_1^2| < 2c_{32}|\varrho^{-1} - \varrho_1^{-1}|,$$

where $\varrho_1 = \varrho(t_1)$, $\dot{\varrho}_1 = \dot{\varrho}(t_1)$. By (6) this implies that

(9) $$\dot{\varrho}^2 < \dot{\varrho}_1^2 + 2c_{32}c_{29}.$$

Let t_1 be as small as possible, subject to the additional condition $t_1 \geqq \tau$. If $t_1 = \tau$, then (7) gives

(10) $$\dot{\varrho}_1^2 \leqq v_\tau^2 \leqq 2m_2^{-1}T_\tau < c_{33}.$$

If on the other hand $t_1 > \tau$, then either $\dot{\varrho}_1 = 0$ or

$$r = \frac{1}{4}c_{21}^{-1} \qquad (t = t_1).$$

In the first case (10) still holds, this time trivially, while in the second case we have

$$U < c_{34}, \quad T = U + h < c_{35} \qquad (t = t_1),$$

and again

(11) $$\dot{\varrho}_1^2 < c_{36}.$$

Thus (6), (7), (9), (10), (11) combine to give an inequality $v^2 < c_{37}$ valid in all cases, which completes the proof of the lemma. Moreover, the end result is an estimate

(12) $$v < c_{38}$$

in which c_{38} can be expressed explicitly in terms of A and the masses.

§ 11. Sundman's Theorem

With the aid of Sundman's two lemmas we now investigate the coordinates q_k $(k=1,\ldots,9)$ in the three-body problem as functions of the new independent variable

$$(1) \qquad s=\int_{\tau}^{t}(U+1)dt$$

introduced already in (8; 17). Deviating from the definition given in (7; 15), we set

$$(2) \qquad F=\frac{T-U-h}{U+1}=\frac{T-h+1}{U+1}-1,$$

where h is the energy constant of the solution we are considering. Just as (7; 6) led to (7; 16), the equations of motion (7; 2) lead to the Hamiltonian system

$$(3) \qquad q_k'=F_{p_k}, \quad p_k'=-F_{q_k} \quad (k=1,\ldots,9),$$

with the prime denoting differentiation with respect to s. In addition, we still have

$$(4) \qquad t'=(U+1)^{-1},$$

while the function F is identically 0 along the solution considered. As a third lemma, we now wish to prove the following statement:

If the initial values satisfy the inequalities in (9;7), then there exists a positive quantity $\delta=\delta(A,m)$, depending only on A and the masses, such that the coordinates q, the distances between the three bodies, and the time t are all regular analytic functions of $s=\sigma+i\nu$ in the strip $-\delta<\nu<\delta$ of the complex s-plane.

For the proof we will again appeal to Cauchy's existence theorem. The solution has already been continued analytically for all real time, and we mentioned that according to (1) the real t-axis corresponds to the real s-axis. It remains to show that for an arbitrary real value $s=s_1$ the q_k $(k=1,\ldots,9)$, $r_{\varkappa\lambda}$ $(1\leqq\varkappa<\lambda\leqq 3)$, and t are all regular in a disk $|s-s_1|<\delta$, where δ depends only on A and the masses but not on s_1. The subscript 1 will be used in this section to denote values of various functions at $s=s_1$. Introducing yet another real number $B\geqq A+1$, to be determined later, we distinguish between two cases.

First suppose that at $s=s_1$ the value of $U=U_1\leqq B$. For the application of the existence theorem to (3), (4) it is sufficient to find a positive number b that depends only on B and the masses, such that F and $(U+1)^{-1}$ as functions of the q_k, p_k are regular in the complex region

$$|q_k-q_{k1}|<b, \quad |p_k-p_{k1}|<b \quad (k=1,\ldots,9)$$

and in absolute value remain below a bound that also depends only on B and the masses. We see from (2) that this has to be shown only for T and $(U+1)^{-1}$. In what follows, $b_1, \ldots, b_5$ are understood to be suitable positive numbers that again depend only on B and the masses. At $s=s_1$ certainly

$$T = T_1 = U_1 + h < B + A < 2B,$$

and a glance at (7;1) will show that in the complex region

$$(5) \qquad |p_k - p_{k1}| < 1 \quad (k = 1, \ldots, 9)$$

there is an estimate of the form $|T| < b_1$, while regularity of T is trivial. Turning to a corresponding investigation of $(U+1)^{-1}$, we first wish to determine a neighborhood of the q_{k1} in which $|U+1| > \frac{1}{4}$. Indeed, it is enough to have $|U - U_1| < \frac{3}{4}$ there, since then

$$|U+1| \geqq (U_1 + 1) - |U_1 - U| > \tfrac{1}{4}.$$

With μ the smallest and m the largest of the three masses m_k $(k = 1, 2, 3)$, we see from (5;2) that it is sufficient to have a neighborhood of the q_{k1} in which the three conditions

$$(6) \qquad \left| \frac{r_{\varkappa\lambda 1}}{r_{\varkappa\lambda}} - 1 \right| < \frac{\mu^2}{4m^2 B} = b_2 \quad (1 \leqq \varkappa < \lambda \leqq 3)$$

are met; for then, because

$$(7) \qquad \frac{\mu^2}{r_{\varkappa\lambda 1}} < U_1 \leqq B,$$

it will follow that

$$(8) \qquad \left| \frac{1}{r_{\varkappa\lambda}} - \frac{1}{r_{\varkappa\lambda 1}} \right| = \frac{1}{r_{\varkappa\lambda 1}} \left| \frac{r_{\varkappa\lambda 1}}{r_{\varkappa\lambda}} - 1 \right| < B\mu^{-2} \frac{\mu^2}{4m^2 B} = \frac{1}{4m^2}$$

$$|U - U_1| = \left| \sum_{\varkappa<\lambda} m_\varkappa m_\lambda \left(\frac{1}{r_{\varkappa\lambda}} - \frac{1}{r_{\varkappa\lambda 1}} \right) \right| < 3m^2 \frac{1}{4m^2} = \frac{3}{4}.$$

Consider now the expression $(u^2 + v^2 + w^2)^{-\frac{1}{2}} - 1$ as a function of the three complex variables u, v, w. At all points on the real sphere $u_1^2 + v_1^2 + w_1^2 = 1$ it is regular and assumes the value 0, so that in a suitable complex region

$$|u - u_1| < b_3, \quad |v - v_1| < b_3, \quad |w - w_1| < b_3$$

it is less than b_2. Setting

$$u = \frac{x_\varkappa - x_\lambda}{r_{\varkappa\lambda 1}}, \quad v = \frac{y_\varkappa - y_\lambda}{r_{\varkappa\lambda 1}}, \quad w = \frac{z_\varkappa - z_\lambda}{r_{\varkappa\lambda 1}}$$

with x, y, z having the original meaning as Cartesian coordinates, we then see that (6) holds whenever

$$\frac{|q_k - q_{k1}|}{r_{\varkappa\lambda 1}} < \frac{b_3}{2} \quad (k=1, \ldots, 9),$$

and therefore, on account of (7), certainly whenever

$$(9) \qquad |q_k - q_{k1}| < \frac{b_3 \mu^2}{2B} = b_4 \quad (k=1, \ldots, 9).$$

Hence, in connection with (8), it follows that in this region the function $(U+1)^{-1}$ has absolute value less than 4 and is regular. Thus, according to (5), (9), the choice $b = \min(1, b_4)$ serves our purpose, and the Cauchy existence theorem implies that the q_k, p_k, and t are regular for $|s - s_1| < b_5$.

Next we discuss the case $U_1 > B$. In particular, this covers the case of collision when U_1 is infinite. Let r_{13} again be the shortest side at $s = s_1$. Using the canonical transformations (7;4), (7;5) and (7;30), (7;33) we introduce the new variables ξ_k, η_k $(k=1, \ldots, 6)$ in place of q_k, p_k. The Hamiltonian system (3) then becomes

$$(10) \qquad \xi_k' = F_{\eta_k}, \quad \eta_k' = -F_{\xi_k} \quad (k=1, \ldots, 6),$$

where F, defined by (2), is to be expressed as a function of the ξ_k, η_k. For the side $r_{13} = x$, we again have $x = \xi\eta^2$ by (7;29), where, as before,

$$\xi^2 = \xi_1^2 + \xi_2^2 + \xi_3^2, \quad \eta^2 = \eta_1^2 + \eta_2^2 + \eta_3^2.$$

Introducing x as a multiplicative factor, we also have

$$(11) \qquad F = \frac{xT + (1-h)x}{xU + x} - 1, \quad \frac{1}{U+1} = \frac{x}{xU + x}.$$

To apply the existence theorem to (10), (4) we must still take a closer look at the three functions x, xT, $(xU+x)^{-1}$ of the twelve independent variables ξ_k, η_k $(k=1, \ldots, 6)$ in a sufficiently small complex neighborhood of ξ_{k1}, η_{k1}. We first look at x and xT.

Because $U_1 > B$, certainly $3m^2 x^{-1} > B$ at $s = s_1$, and therefore

$$(12) \qquad x < \frac{3m^2}{B} \quad (s = s_1).$$

Since the initial values are assumed to satisfy inequalities (9;7), also estimate (9;38) holds. From now on let

$$(13) \qquad B \geqq 12 m^2 c_{21}.$$

Then all the more

$$(x)_1 < \tfrac{1}{4} c_{21}^{-1},$$

and therefore the other two sides satisfy

(14) $$r_{121}^{-1} < 4c_{21}, \quad r_{231}^{-1} < 4c_{21}.$$

From

$$xT = x(U+h) = m_1m_3 + m_1m_2\frac{x}{r_{12}} + m_2m_3\frac{x}{r_{23}} + hx$$

it now follows that

(15) $$|(xT)_1 - m_1m_3| < \frac{c_{39}}{B} < c_{39},$$

whereupon (8;2) gives

(16) $$(\xi)_1 < c_{40}.$$

In view of (7;5) the velocity v of P_2 satisfies

$$(m_2v)^2 = y_4^2 + y_5^2 + y_6^2 = \eta_4^2 + \eta_5^2 + \eta_6^2,$$

which by (10;12) leads to an estimate

(17) $$(\eta_4^2 + \eta_5^2 + \eta_6^2)_1 < c_{41}.$$

Furthermore,

$$\xi\eta = (\xi\eta^2)^{\frac{1}{2}}\xi^{\frac{1}{2}} = x^{\frac{1}{2}}\xi^{\frac{1}{2}},$$

so that (12), (16) give

(18) $$(\xi\eta)_1 < c_{42}B^{-\frac{1}{2}}.$$

Thus (8;2), (17), (18) combine to yield

$$|(xT)_1 - \tfrac{1}{2}(m_1^{-1} + m_3^{-1})(\xi)_1| < c_{43}B^{-\frac{1}{2}},$$

and, by (15), finally

$$\left|(\xi)_1 - \frac{2(m_1m_3)^2}{m_1+m_3}\right| < c_{44}B^{-\frac{1}{2}}.$$

Again let

$$\frac{2(m_1m_3)^2}{m_1+m_3} = c,$$

and impose on B the additional condition

(19) $$4c_{44}B^{-\frac{1}{2}} \leqq c.$$

In that case

(20) $$\tfrac{3}{4}c < (\xi)_1 < \tfrac{5}{4}c,$$

and because $\xi\eta^2 = x$, also

$$(21)\qquad (\eta_1^2+\eta_2^2+\eta_3^2)_1 = (\eta^2)_1 < \frac{4m^2}{cB} < c_{45}\,.$$

In the complex region

$$(22)\qquad |\xi_k - \xi_{k1}| < \frac{c}{10} \qquad (k=1,2,3)$$

we have

$$|\xi^2 - (\xi^2)_1| < \frac{3}{100}c^2 + \frac{\sqrt{3}}{4}c^2 < \frac{1}{2}c^2$$

$$(23)\qquad \tfrac{1}{4}c < |\xi| < 2c\,,$$

so that, in particular, $\xi = (\xi_1^2+\xi_2^2+\xi_3^2)^{\frac{1}{2}}$ is regular there. From (8;2) one sees that the function $xT\xi^{-1}$ is a fourth degree polynomial in the η_k $(k=1,\ldots,6)$. As a consequence of (17), (20), (21), (23), the functions xT and $x=\xi\eta^2$ are regular in the complex domain defined by (22) together with

$$(24)\qquad |\eta_k - \eta_{k1}| < \frac{c}{10} \qquad (k=1,\ldots,6)\,,$$

and there satisfy estimates of the form

$$(25)\qquad |xT| < c_{46}\,, \qquad |x| < c_{47}\,.$$

We now turn to a corresponding investigation of $(xU+x)^{-1}$. According to (12), (14) one has

$$0 \leqq xU + x - m_1m_3 = \frac{m_2m_3}{r_{23}}x + \frac{m_1m_2}{r_{12}}x + x < c_{48}B^{-1} \qquad (s=s_1)\,.$$

In addition, let

$$(26)\qquad c_{48}B^{-1} \leqq \tfrac{1}{2}m_1m_3\,;$$

whereupon

$$(27)\qquad m_1m_3 \leqq (xU+x)_1 < \tfrac{3}{2}m_1m_3\,.$$

The expression xU, given as a function of ξ_k $(k=1,\ldots,6)$ and η_k $(k=1,2,3)$ by (8;3), (8;4), (8;5), (8;6), surely is regular as long as ξ, r_{12}, r_{23} are all nonzero. We next determine a constant c_{49} so as to have in the complex domain

$$(28)\quad |\xi_k - \xi_{k1}| < c_{49}^{-1} \quad (k=1,\ldots,6)\,, \quad |\eta_k - \eta_{k1}| < c_{49}^{-1} \quad (k=1,2,3)$$

the inequality

$$(29)\qquad |(xU+x)-(xU+x)_1| < \tfrac{1}{2}m_1m_3\,.$$

With r being either of the two lengths r_{12}, r_{23}, the inequality (29) will certainly hold if

$$\left|\frac{x}{r}-\left(\frac{x}{r}\right)_1\right| < \frac{\mu}{8m} \quad (r=r_{12}, r_{23}), \quad |x-(x)_1| < \frac{1}{8}m_1m_3\,.$$

Because the second estimate of (25) gives $|x| < c_{47}$ in the region defined by (22), (24), it is enough to find $c_{49} > 10c^{-1}$ so as to meet in the domain (28) the conditions

$$(30)\qquad \begin{cases} \left|\dfrac{(r)_1}{r}-1\right| < \dfrac{c_{21}\mu}{64c_{47}m} & (r=r_{12}, r_{23}), \\ |x-(x)_1| < \min\left(\dfrac{m_1m_3}{8}, \dfrac{c_{21}\mu}{64m}\right); \end{cases}$$

indeed, then

$$\left|\frac{x}{r}-\left(\frac{x}{r}\right)_1\right| \le \frac{|x|}{(r)_1}\left|\frac{(r)_1}{r}-1\right| + \frac{|x-(x)_1|}{(r)_1} < \frac{4c_{47}}{c_{21}}\frac{c_{21}\mu}{64c_{47}m} + \frac{4}{c_{21}}\frac{c_{21}\mu}{64m} = \frac{\mu}{8m}\,.$$

As in (6), the first condition in (30) can be met by restricting the variables x_k, ξ_{k+3} $(k=1,2,3)$, which enter into r^2 in the form given by the expressions (8; 4), to suitable regions

$$(31)\quad |x_k-x_{k1}| < c_{50}^{-1} \quad (k=1,2,3), \quad |\xi_k-\xi_{k1}| < c_{50}^{-1} \quad (k=4,5,6),$$

and this can be done so that the second condition in (30) is met simultaneously. Finally, according to (8; 5) the variables x_1, x_2, x_3 are cubic polynomials in ξ_k, η_k $(k=1,2,3)$, and thus the quantity c_{49} can be determined as desired. In particular, the expressions ξ, r_{12}, r_{23} are all different from 0 in the region (28), and therefore $xU+x$ is regular there, while in addition (27), (29) give the inequalities

$$|xU+x| > \frac{1}{2}m_1m_3\,, \quad |xU+x|^{-1} < \frac{2}{m_1m_3}\,.$$

This shows that the functions F and $(U+1)^{-1}$ are both regular in the region

$$(32)\qquad |\xi_k-\xi_{k1}| < c_{49}^{-1}\,, \quad |\eta_k-\eta_{k1}| < c_{49}^{-1} \quad (k=1,\dots,6)$$

where they remain in absolute value below a constant c_{51}. The Cauchy existence theorem, applied to (4), (10), then shows that ξ_k, η_k $(k=1,\dots,6)$, and t are regular for $|s-s_1|<c_{52}^{-1}$, as are the original coordinates q_k $(k=1,\ldots,9)$, as is seen from (7; 4), (8; 5), and the known behavior of the center of mass.

Let B now be chosen as the smallest number $\geqq A+1$ that meets the conditions (13), (19), (26) we imposed along the way. Then $B=c_{53}$, whereupon $b_5=c_{54}^{-1}$. Setting $\delta=\min(c_{52}^{-1}, c_{54}^{-1})$, we contend that δ has the property stated in the third lemma. Indeed, it remains only to verify that the three distances r_{kl} are regular functions of $s=\sigma+iv$ in the strip $-\delta<v<\delta$. In the case $U_1\leqq B$ we saw that assumption (9) leads to the inequalities in (6), whereby certainly $r_{kl}\neq 0$; on the other hand, Cauchy's existence theorem shows that for $|s-s_1|<c_{54}^{-1}$ the $q_k(s)$ remain within the range defined by (9). For the case $U_1>B$, nonvanishing of r_{12}, r_{23} follows from (30), while by the existence theorem the expression $\xi=(\xi_1^2+\xi_2^2+\xi_3^2)^{\frac{1}{2}}$ as a function of s remains in the region defined by (23), in which certainly $\xi\neq 0$, so that also $r_{13}=x=\xi(\eta_1^2+\eta_2^2+\eta_3^2)$ is regular. This completes the proof of the third lemma.

The strip $-\delta<v<\delta$ in the s-plane is now mapped by the conformal transformation

$$\omega=\frac{e^{\frac{\pi s}{2\delta}}-1}{e^{\frac{\pi s}{2\delta}}+1}$$

onto the unit disk $|\omega|<1$, with the origin $s=0$ going into $\omega=0$ and the real s-axis onto the real diameter. In this way the rectangular Cartesian coordinates x_k, y_k, z_k, the distances r_{kl}, and the time t become regular functions of the variable ω and have power series expansions that converge for $|\omega|<1$. These power series give a complete description of the motion for all real time, with the interval $0<\omega<1$ corresponding to future time $(t>\tau)$ and $-1<\omega<0$ to past time $(t<\tau)$. Finally, a uniform translation of the coordinate system in a fixed direction removes the assumption that the center of mass rest at the origin, and, this being a linear transformation of x, y, z, t, the coordinates are again regular in ω. At this point we formulate once more our main result, Sundman's theorem:

If the three angular momentum constants in relative coordinates are not all 0, then the Cartesian coordinates, the distances between the three particles, and the time t all have power series expansions in ω that converge for $|\omega|<1$ and describe the motion for all real time; here ω is defined by means of the substitutions

$$s=\int_{\tau}^{t}(U+1)\,dt\,, \qquad \omega=\frac{e^{\frac{\pi s}{2\delta}}-1}{e^{\frac{\pi s}{2\delta}}+1}\,,$$

with δ a positive number determined by the masses, the initial values of the coordinates, and the velocity components at time $t=\tau$.

In Sundman's work the result appears in a somewhat different form, in that he used a different auxiliary variable in place of s. It can, however, be readily ascertained that qualitatively both forms express the same meaning. Sundman also gave explicit expressions for the constants in the estimates, which for the sake of brevity we chose to forgo.

If the equations of motion are rewritten directly in terms of the variable ω, the coordinates q_k $(k=1, \ldots, 9)$ can be computed by means of power series with undetermined coefficients. To this end, ω is introduced through

$$\frac{dt}{d\omega}=\frac{dt}{ds}\frac{ds}{d\omega}=4\pi^{-1}\delta(1-\omega^2)^{-1}(1+U)^{-1},$$

where δ, we know, can at the outset be expressed and estimated in terms of the initial values. From the various estimates one can also obtain information about the accuracy of the approximation given by the partial sums of these series. The process, however, does not lead to a practical method for determining orbits.

The possible collisions can be obtained from these expansions by computing the zeros of the derivative $\frac{dt}{d\omega}$ in the interval $-1<\omega<1$. Because the zeros of a function analytic in $|\omega|<1$, and not identically 0, cannot accumulate in the interior of the unit circle, we again see that the collision times do not accumulate at a finite point. It is, however, quite possible that these zeros accumulate at $\omega=1$ or $\omega=-1$, and such examples can be exhibited. To be useful in statements about the motion as $t\to\pm\infty$, these power series would have to converge uniformly in the entire open interval $-1<\omega<1$; yet nothing is known about their convergence behavior at $\omega=\pm 1$. In conclusion we will show that the lengths of the time intervals between successive collisions, should there be any, lie above a positive bound. Assume for this that a collision occurs at $s=s_1$. At that point $\eta_{k1}=0$ $(k=1,2,3)$, $(\xi)_1=c>0$, $(x)_1=0$, $(xU)_1=m_1m_3$, whereupon (8; 2), (8; 3), (8; 4), (8; 5), (11) combine to give

$$(F_{\xi_k})_1=c^{-1}\left(\frac{\xi_k}{\xi}\right)_1 \quad (k=1,2,3).$$

Rotating the coordinate axes, if necessary, one may also assume that $(\xi_1)_1=(\xi)_1$. The function F in the complex domain (32) is regular and in absolute value less than c_{51}. By Cauchy's integral formula

$$|F_{\xi_1}-(F_{\xi_1})_1|<\frac{1}{2c},$$

provided only that the conditions

(33) $$|\xi_k - \xi_{k1}| < c_{55}^{-1} < c_{49}^{-1}, \quad |\eta_k - \eta_{k1}| < c_{55}^{-1} \quad (k = 1, \ldots, 6),$$

with a suitable c_{55}, are satisfied. In particular, for ξ_k, η_k real, then

$$F_{\xi_1} > \frac{1}{2c}.$$

If s is real and lies in an interval throughout which (33) remains valid, which by the existence theorem will be true if $|s - s_1| \leqq c_{56}^{-1} < c_{52}^{-1}$, then by (10) we have, along the solution in question,

(34) $$|\eta_1| = |\eta_1 - (\eta_1)_1| = \left| \int_{s_1}^{s} F_{\xi_1} ds \right| \geqq \frac{1}{2c} |s - s_1|.$$

In addition (23), (27), (29) still imply that

$$\xi > \frac{1}{4} c, \quad xU + x < 2m_1 m_3,$$

which in conjunction with (34) leads to

(35) $$x = \xi\eta^2 \geqq \frac{1}{16c}(s - s_1)^2, \quad \frac{1}{U+1} = \frac{x}{xU + x} \geqq c_{57}^{-1}(s - s_1)^2,$$

and by (4) finally to

(36) $$|t - t_1| = \left| \int_{s_1}^{s} \frac{ds}{U+1} \right| \geqq \frac{1}{3} c_{57}^{-1} |s - s_1|^3.$$

The last inequality in (35) shows that in the interval $s_1 - c_{56}^{-1} \leqq s \leqq s_1 + c_{56}^{-1}$ the function U becomes infinite only at $s = s_1$, and therefore a second collision at $s \neq s_1$ cannot occur. Thus, according to (36), the time between two collisions has a lower bound $\frac{1}{3} c_{57}^{-1} c_{56}^{-3} = c_{58}^{-1}$ that depends only on A and the masses. This contains the statement expressed above.

To apply these results to a system comprised of the earth, sun, and moon, we take those bodies to be particles for which Newton's law holds exactly; moreover, we neglect the effect of all other celestial bodies and forces of nature. It is well known through observation that these three bodies do not stay in a fixed plane, and one could possibly determine a numerical value for A at a certain fixed time. Under the above assumptions, a direct computation will then give two positive numbers ϱ and ε such that the moon, at a potential collision of the earth and sun, must be at least a distance ϱ from the earth and require at least an additional time ε before possibly colliding with it. Thus, Sundman's theory allows us to view the future with confidence.

§ 12. Triple Collision

In this and the next section we will investigate behavior of coordinates in the three-body problem under the assumption that at the singularity $t=t_1$ all three particles collide. Although in general this gives rise to essential singularities in the coordinates, it will nevertheless be possible to describe all triple-collision orbits in terms of suitable series expansions. For what follows it is convenient to replace t by the variable t_1-t and denote the latter again by t, with $t_1-\tau$ accordingly denoted by τ. The original equations of motion are thereby left intact. Thus we assume the nine coordinates $q=q(t)$ of the points P_1, P_2, P_3 to be regular in the interval $0<t\leqq\tau$, open on the left, and we wish to determine their behavior as $t\to 0$. Just as in § 6 and § 9, the expression

$$(1)\qquad I=\sum_q mq^2=\sum_{k=1}^{3} m_k(x_k^2+y_k^2+z_k^2)=\sum_{k=1}^{3} m_k\varrho_k^2$$

plays an important role in this investigation. The center of mass of the three particles is again assumed to rest at the origin of the coordinate system.

Because $t=0$ corresponds to a triple collision, as t monotonically decreases to 0, the positive function $I=I(t)$ approaches 0 while $U=U(t)$ tends to $+\infty$. In view of (6; 2), this means that in a sufficiently small interval $0<t\leqq t_0\leqq\tau$ the function $I(t)$ decreases monotonically with t, and its derivative satisfies $\dot I>0$. It further follows from (6; 2) that

$$\ddot I I^{-\frac{1}{4}}-\frac{1}{4}\dot I^2 I^{-\frac{5}{4}}=\frac{1}{4}(8IT-\dot I^2)I^{-\frac{5}{4}}+2hI^{-\frac{1}{4}}$$

$$(2)\quad \dot I_0 I_0^{-\frac{1}{4}}-\dot I I^{-\frac{1}{4}}=\frac{1}{4}\int_t^{t_0}(8IT-\dot I^2)I^{-\frac{5}{4}}dt+2h\int_t^{t_0} I^{-\frac{1}{4}}dt\qquad (0<t\leqq t_0),$$

where $I_0, \dot I_0$ denote values of $I, \dot I$ at $t=t_0$. Because $\dot I I^{-\frac{1}{4}}>0$, the left side of (2) is bounded from above, while according to (6; 3) the integrand in the first term on the right is nonnegative. The constant h in the second term, however, may be negative. In any case, to establish convergence of the first integral up to $t=0$ it is enough to prove that the second integral converges when $t=0$, and for this we derive an estimate of I from below.

With μ_1 denoting the smallest of the three masses m_1, m_2, m_3, the Schwarz inequality implies that

$$I\geqq\mu_1\sum_{k=1}^{3}\varrho_k^2\geqq\frac{\mu_1}{3}(\varrho_1+\varrho_2+\varrho_3)^2\geqq\frac{\mu_1}{3}r_{12}^2,$$

and therefore

$$U > m_1 m_2 r_{12}^{-1} \geqq \mu_2 I^{-\frac{1}{2}}, \quad \mu_2 = m_1 m_2 \sqrt{\frac{\mu_1}{3}}$$

$$\ddot{I} = 2U + 4h > \mu_2 I^{-\frac{1}{2}}$$

for all sufficiently small $t > 0$. Consequently

$$(\dot{I}^2)^{\cdot} > 4\mu_2 (I^{\frac{1}{2}})^{\cdot},$$

whereupon, integrating twice between 0 and t, we obtain

$$\dot{I}^2 \geqq 4\mu_2 I^{\frac{1}{2}}, \quad (I^{\frac{3}{4}})^{\cdot} = \frac{3}{4} I^{-\frac{1}{4}} \dot{I} \geqq \mu_3 = \frac{3}{2}\sqrt{\mu_2}$$

(3) $$I^{\frac{3}{4}} \geqq \mu_3 t .$$

From this the convergence of the aforementioned integral becomes evident.

In addition, (2) now shows that the function $\dot{I} I^{-\frac{1}{4}}$ approaches a finite limit $\delta \geqq 0$ as $t \to 0$, and therefore

$$(I^{\frac{3}{4}})^{\cdot} = \frac{3}{4} I^{-\frac{1}{4}} \dot{I} \to \frac{3}{4}\delta, \quad I^{\frac{3}{4}} = \frac{3}{4}\delta t + o(t)$$

(4) $$I \sim \varkappa t^{\frac{4}{3}} \quad (t \to 0),$$

with (3) implying that

(5) $$\varkappa = \left(\frac{3}{4}\delta\right)^{\frac{4}{3}} > 0 .$$

Moreover,

$$\dot{I} \sim \delta I^{\frac{1}{4}} \sim \delta \left(\frac{3}{4}\delta t\right)^{\frac{1}{3}}$$

(6) $$\dot{I} \sim \frac{4}{3}\varkappa t^{\frac{1}{3}} \quad (t \to 0) .$$

The above asymptotic formula for $\dot{I}$ refines the one for I given in (4) and can formally be derived from it by differentiation. It will be shown next that even

$$\ddot{I} \sim \frac{4}{9}\varkappa t^{-\frac{2}{3}} \quad (t \to 0)$$

is true. This, according to (6; 2), is equivalent to

(7) $$U \sim \frac{2}{9}\varkappa t^{-\frac{2}{3}} .$$

If we introduce the function

(8) $$(8IT - \dot{I}^2) t^{-\frac{2}{3}} = g(t) = g$$

and make use of (5; 9), (4) and (6) in conjunction with (5), our statement (7) reduces to showing that

$$\lim_{t\to 0} g(t) = 0 . \tag{9}$$

In view of (6; 3), certainly $g(t) \geqq 0$. To prove (9), we appeal to the already established convergence of the first integral in (2), up to $t=0$, which together with (4) and (5) implies also convergence of

$$\int_0^\tau g(t)\frac{dt}{t} = G .$$

The nonnegative function g must therefore satisfy

$$\varliminf_{t\to 0} g(t) = 0 .$$

Now if (9) were false, because $g(t)$ is continuous in the interval $0 < t \leqq \tau$, there would exist a sufficiently small positive number ε and a sequence $\tau, \tau_1, \tau_2, \ldots$ monotonically decreasing to 0 such that

$$g(\tau_{2j}) = \varepsilon, \quad g(\tau_{2j-1}) = 2\varepsilon, \quad \varepsilon \leqq g(t) \leqq 2\varepsilon \quad (\tau_{2j} \leqq t \leqq \tau_{2j-1}; j = 1, 2, \ldots). \tag{10}$$

Hence over each of these intervals the function g would increase by an amount ε. To reach a contradiction, we estimate the derivative

$$\dot{g} = (8\dot{I}T + 8I\dot{T} - 2\dot{I}\ddot{I})t^{-\frac{2}{3}} - \frac{2}{3}gt^{-1} \tag{11}$$

from above.

First of all (4), (5), and (6) say that

$$I = O(t^{\frac{4}{3}}), \quad I^{-1} = O(t^{-\frac{4}{3}}), \quad \dot{I} = O(t^{\frac{1}{3}}) \quad (t \to 0) .$$

Therefore (8) and (10) combine to give

$$T = \frac{1}{8} I^{-1}(gt^{\frac{2}{3}} + \dot{I}^2) = O(t^{-\frac{2}{3}}) \quad (\tau_{2j} \leqq t \leqq \tau_{2j-1}; j \to \infty),$$

so that in view of (5; 10) each coordinate satisfies

$$\dot{q} = O(t^{-\frac{1}{3}}),$$

while also

$$U = T - h = O(t^{-\frac{2}{3}}), \quad r_{kl}^{-1} = O(t^{-\frac{2}{3}})$$

$$\begin{aligned}(r_{kl}^{-1})^{\cdot} &= r_{kl}^{-3}\{(x_k - x_l)(\dot{x}_l - \dot{x}_k) + (y_k - y_l)(\dot{y}_l - \dot{y}_k) + (z_k - z_l)(\dot{z}_l - \dot{z}_k)\} \\ &= O(t^{-\frac{4}{3}} \cdot t^{-\frac{1}{3}}) = O(t^{-\frac{5}{3}})\end{aligned}$$

$$\dot{T} = \dot{U} = O(t^{-\frac{5}{3}})$$

and

$$\ddot{I} = 2T + 2h = O(t^{-\frac{2}{3}}),$$

all this being true for $\tau_{2j} \leqq t \leqq \tau_{2j-1}, j \to \infty$. These estimates together with (10) and (11) show that

(12) $$\dot{g} < bt^{-1} \quad (\tau_{2j} \leqq t \leqq \tau_{2j-1}; j = 1, 2, \ldots)$$

for some positive constant b.

From (10) and (12) we now have

$$\varepsilon = g(\tau_{2j-1}) - g(\tau_{2j}) = \int_{\tau_{2j}}^{\tau_{2j-1}} \dot{g}(t)dt < b \int_{\tau_{2j}}^{\tau_{2j-1}} \frac{dt}{t}$$

$$\int_{\tau_{2j}}^{\tau_{2j-1}} g(t)\frac{dt}{t} > \varepsilon \int_{\tau_{2j}}^{\tau_{2j-1}} \frac{dt}{t} > \varepsilon^2 b^{-1},$$

which upon summation over j contradicts convergence of the integral G. This proves statements (9) and (7) and in turn leads to the estimates

(13) $$r_{kl}^{-1} = O(t^{-\frac{2}{3}}), \quad \dot{q} = O(t^{-\frac{1}{3}}) \quad (t \to 0)$$

which up to now were known only in the intervals $\tau_{2j} \leqq t \leqq \tau_{2j-1}$ for $j \to \infty$.

An important consequence of (8), (9) and (6; 3) is the asymptotic formula

(14) $$p\dot{q} - q\dot{p} = o(t^{\frac{1}{3}}) \quad (t \to 0)$$

valid for any two coordinates p and q. This, in particular, shows once more that in the case of triple collision all 3 angular momentum constants are 0.

Since according to (1) and (4) we have

$$q = O(t^{\frac{2}{3}}),$$

it is convenient to introduce

$$\bar{q} = qt^{-\frac{2}{3}}, \quad \bar{p} = pt^{-\frac{2}{3}} \quad (t > 0).$$

Then

(15) $$\bar{q} = O(1), \quad \bar{p} = O(1) \quad (t \to 0),$$

and (14) becomes

(16) $$\bar{p}\dot{\bar{q}} - \bar{q}\dot{\bar{p}} = (p\dot{q} - q\dot{p})t^{-\frac{4}{3}} = o(t^{-1}) \quad (t \to 0).$$

In general, for f a homogeneous function of degree ν in the coordinates q, let $\bar{f}$ denote the corresponding function in the variables $\bar{q}$,

so that

$$\bar{f} = f t^{\frac{-2\nu}{3}}.$$

By (4) and (6) then

$$\text{(17)} \qquad \bar{I} = I t^{-\frac{4}{3}} \to \varkappa, \quad \dot{\bar{I}} = \dot{I} t^{-\frac{4}{3}} - \frac{4}{3} I t^{-\frac{7}{3}} = o(t^{-1}) \quad (t \to 0),$$

while on the other hand

$$\bar{I} = \sum_{\bar{q}} m \bar{q}^2, \quad \frac{1}{2} \dot{\bar{I}} = \sum_{\bar{q}} m \bar{q} \dot{\bar{q}},$$

so that (15), (16), and (17) give

$$\frac{1}{2} \bar{p} \dot{\bar{I}} - \bar{I} \dot{\bar{p}} = \sum_{\bar{q}} m \bar{q} (\bar{p} \dot{\bar{q}} - \bar{q} \dot{\bar{p}}) = o(t^{-1})$$

$$\bar{I} \dot{\bar{p}} = o(t^{-1}), \quad \dot{\bar{p}} = o(t^{-1}).$$

Thus for each coordinate q we have

$$\text{(18)} \qquad \dot{\bar{q}} = o(t^{-1}) \quad (t \to 0).$$

In addition to the triangle Δ formed by the three particles P_1, P_2, P_3, we consider the triangle $\bar{\Delta}$ whose vertices $\bar{P}_1, \bar{P}_2, \bar{P}_3$ have the corresponding coordinates $\bar{q}$. Thus, the new triangle is obtained from the original one by a dilatation in the ratio of 1 to $t^{\frac{2}{3}}$. At triple collision Δ collapses to the origin of the coordinate system, while in the end it will be shown that also the large triangle $\bar{\Delta}$ has a definite limiting position as $t \to 0$. First we will show that the lengths of the three sides of $\bar{\Delta}$ have positive limits, which we will also determine. For this we consider the equations of motion

$$m \ddot{q} = U_q$$

and express these in terms of $\bar{q}$ in place of q.

The relations

$$\bar{U} = U t^{\frac{2}{3}}, \quad \bar{U}_{\bar{q}} = U_q t^{\frac{4}{3}},$$

$$\ddot{q} = (\bar{q} t^{\frac{2}{3}})^{\cdot\cdot} = \ddot{\bar{q}} t^{\frac{2}{3}} + \frac{4}{3} \dot{\bar{q}} t^{-\frac{1}{3}} - \frac{2}{9} \bar{q} t^{-\frac{4}{3}} = (\dot{\bar{q}} t^{\frac{4}{3}})^{\cdot} t^{-\frac{2}{3}} - \frac{2}{9} \bar{q} t^{-\frac{4}{3}}$$

combine to give

$$\text{(19)} \qquad (\dot{\bar{q}} t^{\frac{4}{3}})^{\cdot} t^{\frac{2}{3}} - \frac{2}{9} \bar{q} = m^{-1} \bar{U}_{\bar{q}}.$$

We now take the mean value of both sides of this differential equation over the interval from t to $2t$, where $0 < 2t \leqq \tau$. In view of (18), integration by parts leads to the estimate

$$(20) \qquad \int_t^{2t} (\dot{\bar{q}} t^{\frac{4}{3}})^{\cdot} t^{\frac{2}{3}} dt = [\dot{\bar{q}} t^{\frac{4}{3}} t^{\frac{2}{3}}]_t^{2t} - \frac{2}{3} \int_t^{2t} \dot{\bar{q}} t^{\frac{4}{3}} t^{-\frac{1}{3}} dt = o(t) \qquad (t \to 0),$$

while for $t \leqq t^* \leqq 2t$ also

$$(21) \qquad \bar{q}(t^*) - \bar{q}(t) = \int_t^{t^*} \dot{\bar{q}} dt = o(1).$$

To estimate the mean value of the function $\bar{U}_{\bar{q}}$ we use the relations

$$\bar{U}_{\bar{q}}(t^*) - \bar{U}_{\bar{q}}(t) = \int_t^{t^*} \dot{\bar{U}}_{\bar{q}} dt, \quad \dot{\bar{U}}_{\bar{q}} = \sum_{\bar{p}} \bar{U}_{\bar{q}\bar{p}} \dot{\bar{p}},$$

and note that according to (13) and (18) we then have

$$(22) \qquad \bar{r}_{kl}^{-1} = O(1)$$

$$\bar{U}_{\bar{q}\bar{p}} = O(1), \quad \dot{\bar{U}}_{\bar{q}} = o(t^{-1}), \quad \bar{U}_{\bar{q}}(t^*) - \bar{U}_{\bar{q}}(t) = o(1) \qquad (t \to 0).$$

Thus (19) in conjunction with (20) and (21) leads to the formula

$$(23) \qquad -\frac{2}{9} \bar{q} = m^{-1} \bar{U}_{\bar{q}} + o(1) \qquad (t \to 0).$$

This, of course, is no longer a differential equation but rather an algebraic relation asymptotically satisfied by the coordinates $\bar{q}$.

It was shown already in § 6 that in case of a triple collision the three particles move in a fixed plane, and this may be taken as the $z = 0$ plane, so that z_1, z_2, z_3 are identically 0 and only the 6 coordinates x_k, y_k $(k = 1, 2, 3)$ have to be considered. From their derivation, or from a direct calculation, the corresponding 6 equations (23) are seen to be invariant under arbitrary rotations of the coordinate axes in the (x, y)-plane. Let us introduce new coordinates X_k, Y_k in place of $\bar{x}_k, \bar{y}_k$ $(k = 1, 2, 3)$ with the same origin as before, but with the new abscissa axis always parallel to the direction of the vector $P_3 P_1$. In this new moving coordinate system $Y_1 = Y_3$, while (23) leads to the 3 equations

$$(24) \qquad \begin{cases} \frac{2}{9} Y_1 = m_2 (Y_1 - Y_2) \bar{r}_{12}^{-3} + m_3 (Y_1 - Y_3) \bar{r}_{13}^{-3} + o(1) \\ \frac{2}{9} Y_2 = m_1 (Y_2 - Y_1) \bar{r}_{12}^{-3} + m_3 (Y_2 - Y_3) \bar{r}_{23}^{-3} + o(1) \\ \frac{2}{9} Y_3 = m_1 (Y_3 - Y_1) \bar{r}_{13}^{-3} + m_2 (Y_3 - Y_2) \bar{r}_{23}^{-3} + o(1) \end{cases}$$

and 3 analogous equations for X_1, X_2, X_3. According to (15) all 6 coordinates $X_k(t), Y_k(t)$ $(k = 1, 2, 3)$ remain bounded as $t \to 0$. Consider an arbitrary sequence of values for $t \to 0$ along which these coordinates tend to definite limits $\hat{X}_k, \hat{Y}_k$. The distances $\bar{r}_{kl} = r_{kl} t^{-\frac{2}{3}}$ then also have limits $\hat{r}_{kl}$ which in view of (22) are all positive. Because $Y_1 = Y_3$, subtraction of the third equation in (24) from the first leads to the relation

$$m_2(\hat{Y}_1 - \hat{Y}_2)(\hat{r}_{12}^{-3} - \hat{r}_{23}^{-3}) = 0 .$$

Consequently either $\hat{r}_{12} = \hat{r}_{23}$ or $\hat{Y}_1 = \hat{Y}_2$.

If $\hat{r}_{12} \neq \hat{r}_{23}$, then $\hat{Y}_1 = \hat{Y}_2$ and, since also $\hat{Y}_1 = \hat{Y}_3$ and $m_1\hat{Y}_1 + m_2\hat{Y}_2 + m_3\hat{Y}_3 = 0$, it follows that $\hat{Y}_k = 0$ for all three ordinates. Thus in this case, as $t \to 0$ along the sequence considered, the 3 points $\bar{P}_1, \bar{P}_2, \bar{P}_3$ tend to limiting positions on the abscissa axis of the new coordinate system. In the other case $\hat{r}_{12} = \hat{r}_{13}$, while the previous analysis repeated relative to a coordinate system with abscissa axis parallel to P_1P_2 then shows that also $\hat{r}_{13} = \hat{r}_{23}$. The limiting configuration in the second case is therefore an equilateral triangle. Thus we see that as $t \to 0$ along the chosen sequence, either all three angles of the triangle $\bar{\Delta}$ go to $\frac{\pi}{3}$; or one angle goes to π and the other two to 0. Moreover, because for $t > 0$ the angles are continuous functions of t, their limit values do not depend on the choice of the sequence $t \to 0$. The two possible configurations will in the future be referred to as the equilateral case and the collinear case. In addition, set

$$m_1 + m_2 + m_3 = M .$$

For the equilateral case, let $\hat{r}_{12} = \hat{r}_{13} = \hat{r}_{23} = r$. From (24) we then have

$$\tfrac{2}{9}\hat{Y}_k r^3 = M\hat{Y}_k \qquad (k = 1, 2, 3),$$

and since not all the $\hat{Y}_k$ are 0 it follows that

(25) $$r^3 = \tfrac{9}{2}M .$$

Thus r has a well determined value, while from (7) we also have

(26) $$\varkappa = \tfrac{9}{2}(m_1m_2 + m_1m_3 + m_2m_3)\, r^{-1} .$$

If, in addition, the orientation of the coordinate system is so chosen that upon traversal of the triangle $\bar{\Delta}$ in the positive direction the vertices $\bar{P}_1, \bar{P}_2, \bar{P}_3$ follow in that order, then

$$\hat{X}_1 - \hat{X}_3 = r, \quad \hat{X}_2 - \hat{X}_3 = \frac{1}{2}r, \quad \hat{Y}_1 - \hat{Y}_3 = 0, \quad \hat{Y}_2 - \hat{Y}_3 = \frac{\sqrt{3}}{2}r$$

$$0 = \sum_{k=1}^{3} m_k\hat{X}_k = M\hat{X}_1 - \frac{1}{2}m_2 r - m_3 r, \; 0 = \sum_{k=1}^{3} m_k\hat{Y}_k = M\hat{Y}_1 + \frac{\sqrt{3}}{2}m_2 r$$

$$\hat{X}_1 = \frac{\frac{1}{2}m_2 + m_3}{M}r, \quad \hat{X}_2 = \frac{m_3 - m_1}{2M}r, \quad \hat{X}_3 = -\frac{m_1 + \frac{1}{2}m_2}{M}r,$$

$$\hat{Y}_1 = \hat{Y}_3 = -\frac{\sqrt{3}m_2}{2M}r, \quad \hat{Y}_2 = \sqrt{3}\frac{m_1 + m_3}{2M}r .$$

For the collinear case let $\hat{r}_{13}=\hat{X}_1-\hat{X}_3=a$ be the longest side, so that $\hat{r}_{12}=\hat{X}_1-\hat{X}_2=\varrho a$, $\hat{r}_{23}=\hat{X}_2-\hat{X}_3=\sigma a$ with $\varrho+\sigma=1$, $0<\varrho<1$. The corresponding equations (24) for the abscissas then become

$$\tfrac{2}{9}\hat{X}_1 a^2=m_2\varrho^{-2}+m_3,\ \tfrac{2}{9}\hat{X}_2 a^2=-m_1\varrho^{-2}+m_3\sigma^{-2},\ \tfrac{2}{9}\hat{X}_3 a^2=-m_1-m_2\sigma^{-2},$$

and upon subtraction these give

$$(27)\quad \tfrac{2}{9}a^3=m_1+m_2(\varrho^{-2}+\sigma^{-2})+m_3,\ \tfrac{2}{9}\sigma a^3=m_1(1-\varrho^{-2})+m_2\sigma^{-2}+m_3\sigma^{-2}.$$

Elimination of a now leads to the fifth degree equation

$$(28)\quad m_1\sigma^2(\varrho^3-1)+m_2(\varrho^3-\sigma^3)+m_3\varrho^2(1-\sigma^3)=0 \qquad (\sigma=1-\varrho)$$

for ϱ. If this equation is written as

$$\frac{m_1+m_2\sigma}{m_1+m_2\sigma^{-2}}=\frac{m_3+m_2\varrho}{m_3+m_2\varrho^{-2}},$$

it can be seen that in the interval $0\leqq\varrho\leqq1$ the left side, as a function of ϱ, decreases monotonically from 1 to 0, while the right side monotonically increases from 0 to 1. Hence (28) has exactly one positive solution $\varrho<1$, and (27) then also uniquely determines the length a. Equation (7) now implies that

$$(29)\qquad \varkappa=\tfrac{9}{2}(m_1m_2\varrho^{-1}+m_1m_3+m_2m_3\sigma^{-1})a^{-1},$$

while the coordinates are given by

$$0=\sum_{k=1}^{3}m_k\hat{X}_k=M\hat{X}_1-m_2\varrho a-m_3a,\qquad \hat{X}_1=\frac{m_2\varrho+m_3}{M}a,$$

$$\hat{X}_2=\frac{m_3\sigma-m_1\varrho}{M}a,\qquad \hat{X}_3=-\frac{m_1+m_2\sigma}{M}a,\qquad \hat{Y}_1=\hat{Y}_2=\hat{Y}_3=0.$$

These values correspond to the assumption that at $t=0$ the point $\bar{P}_2$ lies between $\bar{P}_1$ and $\bar{P}_3$. The two other possibilities in the collinear case are obtained from this by cyclic permutation of m_1, m_2, m_3.

Thus we have proved that in both the equilateral and the collinear cases the coordinates X_k, Y_k of $\bar{P}_k$ $(k=1,2,3)$ have well determined limit values that do not depend on the initially considered sequence $t\to0$. Consequently in reference to the moving coordinate system the triangle $\bar{\Delta}$ tends to a limiting position that in both cases depends only on the values m_1, m_2, m_3.

In this section we have derived the results of Sundman [1] insofar as they relate to triple collisions; in particular, the theorem that the sides of the triangle, when expanded in the ratio of 1 to $t^{\frac{2}{3}}$, have definite limits that correspond to either the equilateral or the collinear case. It does not follow, however, that the old coordinates $\bar{x}_k, \bar{y}_k$ of the points $\bar{P}_k$ $(k=1,2,3)$ themselves have limit values. For this it remains to show that as $t\to 0$ the angle between the old and the new coordinate systems also tends to a limit. Moreover, in complete analogy to the case of a simple collision, it will be possible also in this case to obtain suitable series expansions for the coordinates x_k, y_k, and thereby collectively determine the possible triple-collision orbits. This will be carried out in the next section.

§ 13. Triple-Collision Orbits

First we will construct a special solution to the planar three-body problem which will show that both the equilateral case of a triple collision as well as the collinear case with its 3 possible permutations actually do occur. For this, let the 6 coordinates $q=x_k, y_k$ $(k=1,2,3)$ have the form

$$q=q(t)=\hat{q}w\,,\qquad w=w(t)\,, \tag{1}$$

where w is a twice differentiable positive function in the interval $0<t\leqq\tau$ and the 6 constants $\hat{q}$ are such that for $t>0$ the 3 points are distinct. Also, to have triple collision at $t=0$, assume that $w(t)\to 0$ as $t\to 0$. In view of (1), the function

$$\hat{U}=Uw$$

depends only on the parameters $\hat{q}$ and is a homogeneous function of degree -1 in these values. Consequently

$$\sum_{\hat{q}}\hat{q}\hat{U}_{\hat{q}}=-\hat{U}<0\,,$$

and therefore $\hat{q}\hat{U}_{\hat{q}}<0$ for at least one coordinate. With assumption (1) the equations of motion (5;3) become

$$m\hat{q}\ddot{w}=\hat{U}_{\hat{q}}w^{-2}\,,$$

implying that the expression $\ddot{w}w^2$ must be a negative constant b_1. If w is multiplied by a positive factor to achieve the normalization $b_1=-\frac{2}{9}$, we are left with the equations

$$\ddot{w}=-\tfrac{2}{9}w^{-2}\,,\qquad -\tfrac{2}{9}\hat{q}=m^{-1}\hat{U}_{\hat{q}}\,. \tag{2}$$

Integration of the differential equation in (2) gives

$$\dot{w}^2=\tfrac{4}{9}(w^{-1}+b_2)\,, \tag{3}$$

where b_2 is an arbitrary constant. If the constant is taken to be 0, it follows that

$$\tfrac{3}{2}w^{\frac{1}{2}}\dot{w}=1\,,\qquad w^{\frac{3}{2}}=t\,,\qquad w=t^{\frac{2}{3}}\,.$$

The second equation in (2), on the other hand, coincides with (12;23) if the functions $\bar{q}=\bar{q}(t)$ are replaced by the constants $\hat{q}$. From our earlier result the solutions are given by three fixed points $\hat{P}_1, \hat{P}_2, \hat{P}_3$ with coordinates $\hat{q}$, corresponding either to the equilateral or the collinear case. If the abscissa axis is chosen in the direction of the vector $\hat{P}_3\hat{P}_1$, the values of the $\hat{q}$ are precisely those found for $\hat{X}_k$, $\hat{Y}_k$ $(k=1,2,3)$. This shows that each of the individual cases for a triple collision does indeed appear as a special solution to the three-body problem.

One can still introduce an arbitrary real parameter into the above solutions by not assuming that $b_2=0$ in (3) and using the series expansion

$$(1+b_2w)^{-\frac{1}{2}}=1-\tfrac{1}{2}b_2w+\cdots \qquad (|b_2w|<1).$$

Then

$$\tfrac{3}{2}(w^{\frac{1}{2}}-\tfrac{1}{2}b_2w^{\frac{3}{2}}+\cdots)\,\dot{w}=1,\; w^{\frac{3}{2}}-\tfrac{3}{10}b_2w^{\frac{5}{2}}+\cdots=t,\; w-\tfrac{1}{5}b_2w^2+\cdots=t^{\frac{2}{3}}$$

$$w=t^{\frac{2}{3}}\bar{w}, \quad \bar{w}=1+\tfrac{1}{5}b_2t^{\frac{2}{3}}+\cdots, \quad q=\bar{q}t^{\frac{2}{3}}, \quad \bar{q}=\hat{q}\bar{w}, \tag{4}$$

with $\bar{w}$ a power series in $b_2t^{\frac{2}{3}}$ that converges for sufficiently small positive values of t. Moreover, the quantity b_2 can readily be expressed in terms of the energy constant h.

The following investigation will show that the particular solutions to the three-body problem appearing in (4) are still not the most general triple-collision orbits [1, 2, 3]. In determining all of them one may assume that for the collinear case, with the notation of the previous section, the point $\bar{P}_2$ at $t=0$ lies between $\bar{P}_3$ and $\bar{P}_1$, since the other two possibilities can be reduced to this by a cyclic permutation of the indices.

In accordance with § 7 we introduce the relative coordinates

$$\xi_1=x_1-x_3, \quad \xi_2=x_2-x_3, \quad \xi_3=y_1-y_3, \quad \xi_4=y_2-y_3 \tag{5}$$

of P_1 and P_2 with respect to P_3, whereupon

$$r_{13}^2=\xi_1^2+\xi_3^2, \quad r_{23}^2=\xi_2^2+\xi_4^2, \quad r_{12}^2=(\xi_1-\xi_2)^2+(\xi_3-\xi_4)^2,$$

and because the center of mass was taken to rest at the origin, we also have

$$\begin{cases} Mx_1=(m_2+m_3)\,\xi_1-m_2\xi_2, \quad Mx_2=-m_1\xi_1+(m_1+m_3)\,\xi_2, \\ \qquad Mx_3=-m_1\xi_1-m_2\xi_2 \\ My_1=(m_2+m_3)\,\xi_3-m_2\xi_4, \quad My_2=-m_1\xi_3+(m_1+m_3)\,\xi_4, \\ \qquad My_3=-m_1\xi_3-m_4\xi_4, \end{cases} \tag{6}$$

where again $M=m_1+m_2+m_3$. If in addition

$$\eta_1=m_1\dot{x}_1, \quad \eta_2=m_2\dot{x}_2, \quad \eta_3=m_1\dot{y}_1, \quad \eta_4=m_2\dot{y}_2, \tag{7}$$

then

$$m_3\dot{x}_3 = -\eta_1 - \eta_2, \quad m_3\dot{y}_3 = -\eta_3 - \eta_4$$

$$T = \frac{1}{2m_1}(\eta_1^2 + \eta_3^2) + \frac{1}{2m_2}(\eta_2^2 + \eta_4^2) + \frac{1}{2m_3}\{(\eta_1 + \eta_2)^2 + (\eta_3 + \eta_4)^2\}.$$

As was already established in § 7, this leads for the variables ξ_k, η_k to the Hamiltonian system

$$\dot{\xi}_k = E_{\eta_k}, \quad \dot{\eta}_k = -E_{\xi_k} \quad (k = 1, \ldots, 4)$$

with $E = T - U$, or more explicitly

$$(8)\begin{cases} \dot{\xi}_k = \left(\dfrac{1}{m_1} + \dfrac{1}{m_3}\right)\eta_k + \dfrac{1}{m_3}\eta_{k+1}, \quad \dot{\xi}_{k+1} = \left(\dfrac{1}{m_2} + \dfrac{1}{m_3}\right)\eta_{k+1} + \dfrac{1}{m_3}\eta_k \\ \qquad\qquad\qquad\qquad\qquad\qquad\qquad\qquad\qquad\qquad (k = 1, 3) \\ \dot{\eta}_k = U_{\xi_k} \quad (k = 1, 2, 3, 4). \end{cases}$$

For a further discussion of triple collision it is necessary to introduce a new rectangular coordinate system whose origin coincides with P_3 and whose abscissa axis always passes through P_1. Let the new abscissa axis form an angle p_4 relative to the old one, with $p_1, 0$ and p_2, p_3 the new coordinates of P_1 and P_2. Abbreviating

$$\cos p_4 = c, \quad \sin p_4 = s,$$

we then have

$$(9) \qquad \xi_1 = p_1 c, \quad \xi_2 = p_2 c - p_3 s, \quad \xi_3 = p_1 s, \quad \xi_4 = p_2 s + p_3 c,$$

$$(10) \qquad r_{13} = p_1, \quad r_{23}^2 = p_2^2 + p_3^2, \quad r_{12}^2 = (p_1 - p_2)^2 + p_3^2.$$

To extend the transformation (9) of the variables $\xi_1, \ldots, \xi_4$ into $p_1, \ldots, p_4$ to a canonical transformation of the 8 independent variables ξ_k, η_k $(k = 1, \ldots, 4)$, one considers the generating function

$$W = W(p, \eta) = \eta_1 p_1 c + \eta_2(p_2 c - p_3 s) + \eta_3 p_1 s + \eta_4(p_2 s + p_3 c).$$

Then the four-by-four determinant

$$|W_{\eta_k p_l}| = -p_1 \neq 0.$$

In accordance with § 3, the expression

$$W_{p_k} = q_k, \quad W_{\eta_k} = \xi_k \quad (k = 1, \ldots, 4)$$

then gives a canonical transformation that clearly satisfies (9) and contributes the 4 additional equations

$$(11) \qquad \begin{cases} q_1 = \eta_1 c + \eta_3 s, \quad q_2 = \eta_2 c + \eta_4 s, \quad q_3 = -\eta_2 s + \eta_4 c, \\ q_4 = p_1(-\eta_1 s + \eta_3 c) + p_2(-\eta_2 s + \eta_4 c) - p_3(\eta_2 c + \eta_4 s). \end{cases}$$

With the aid of the auxiliary variable

(12) $$q_0 = -\eta_1 s + \eta_3 c$$

these lead to

(13) $$q_4 = p_1 q_0 + p_2 q_3 - p_3 q_2, \quad q_0 = (q_4 - p_2 q_3 + p_3 q_2)\, p_1^{-1}$$

$$\eta_1 = q_1 c - q_0 s, \quad \eta_2 = q_2 c - q_3 s, \quad \eta_3 = q_1 s + q_0 c, \quad \eta_4 = q_2 s + q_3 c$$

and

(14) $$T = \frac{1}{2m_1}(q_1^2 + q_0^2) + \frac{1}{2m_2}(q_2^2 + q_3^2) + \frac{1}{2m_3}\{(q_1 + q_2)^2 + (q_0 + q_3)^2\}.$$

It should be observed that here and henceforth the q_k have a different meaning than the symbol q used earlier.

The real advantage in the canonical transformation defined by (9) and (11) is that as a consequence of (10), (13), and (14) the function E in its dependence on the 8 new variables p_k, q_k $(k = 1, \ldots, 4)$ no longer contains the angle p_4. As seen from § 2, the new Hamiltonian system becomes

(15) $$\dot{p}_k = E_{q_k}, \quad \dot{q}_k = -E_{p_k} \quad (k = 1, \ldots, 4)$$

and, because $E_{p_4} = 0$, it follows that q_4 is constant. On the other hand, (5; 8) in conjunction with (5), (7), (9), and (11) says that

$$\begin{aligned} q_4 = \xi_1\eta_3 + \xi_2\eta_4 - \xi_3\eta_1 - \xi_4\eta_2 &= m_1(x_1 - x_3)\,\dot{y}_1 + m_2(x_2 - x_3)\,\dot{y}_2 \\ &\quad - m_1(y_1 - y_3)\,\dot{x}_1 - m_2(y_2 - y_3)\,\dot{x}_2 \\ &= m_1(x_1\dot{y}_1 - y_1\dot{x}_1) + m_2(x_2\dot{y}_2 - y_2\dot{x}_2) + m_3(x_3\dot{y}_3 - y_3\dot{x}_3) = \gamma = 0, \end{aligned}$$

since for a triple collision the angular momentum constants are 0. Thus (15) reduces to the system

$$\dot{p}_k = (E_{q_k})_{q_4=0}, \quad \dot{q}_k = -(E_{p_k})_{q_4=0} \quad (k = 1, 2, 3)$$

of only 6 equations, free from p_4 and q_4, plus the supplementary equation

$$\dot{p}_4 = (E_{q_4})_{q_4=0}$$

to be solved by quadrature. For the purpose of investigating these differential equations we now determine the asymptotic behavior of p_k and q_k $(k = 1, 2, 3)$ as $t \to 0$.

The quantities

(16) $$p_k t^{-\frac{2}{3}} = \bar{p}_k \quad (k = 1, 2, 3)$$

correspond in the notation of the previous section to

$$\bar{p}_1 = X_1 - X_3, \quad \bar{p}_2 = X_2 - X_3, \quad \bar{p}_3 = Y_2 - Y_3$$

and therefore have at $t = 0$ the limit values

$$\hat{p}_1 = r, \quad \hat{p}_2 = \frac{1}{2} r, \quad \hat{p}_3 = \frac{\sqrt{3}}{2} r$$

in the equilateral case, and

$$\hat{p}_1 = a, \quad \hat{p}_2 = \sigma a, \quad \hat{p}_3 = 0$$

in the collinear case, where r, a, σ have the old connotation. Next, let

$$(17) \qquad q_k t^{\frac{1}{3}} = \bar{q}_k \quad (k = 0, 1, 2, 3).$$

It is very significant that also the $\bar{q}_k$ have limits at $t = 0$, and these will now be determined.

First, from (12; 1) and (5), (7), (9), (11) we have

$$\tfrac{1}{2}\dot{I} = \sum_{k=1}^{3} m_k(x_k \dot{x}_k + y_k \dot{y}_k) = \sum_{k=1}^{2} m_k((x_k - x_3)\,\dot{x}_k + (y_k - y_3)\,\dot{y}_k)$$

$$= \sum_{k=1}^{4} \xi_k \eta_k = \sum_{k=1}^{3} p_k q_k = t^{\frac{1}{3}} \sum_{k=1}^{3} \bar{p}_k \bar{q}_k,$$

which in view of (12; 6) shows that

$$(18) \qquad \sum_{k=1}^{3} \bar{p}_k \bar{q}_k \to \tfrac{2}{3}\varkappa \quad (t \to 0).$$

Here the positive constant $\varkappa$ is given by (12; 26) in the equilateral case, and (12; 29) in the collinear case. Because $q_4 = 0$, it follows from (13) that also

$$(19) \qquad \bar{p}_1 \bar{q}_0 + \bar{p}_2 \bar{q}_3 - \bar{p}_3 \bar{q}_2 = 0.$$

Additional asymptotic relations can be obtained from (12; 14), which in particular says that

$$x_1 \dot{y}_1 - y_1 \dot{x}_1 = o(t^{\frac{1}{3}}), \quad x_2 \dot{y}_2 - y_2 \dot{x}_2 = o(t^{\frac{1}{3}}),$$

$$x_1 \dot{x}_2 - x_2 \dot{x}_1 + y_1 \dot{y}_2 - y_2 \dot{y}_1 = o(t^{\frac{1}{3}}).$$

With the aid of (6), (9), (11), (12), (16) and (17) this leads to

$$(m_2+m_3)(\xi_1\eta_3-\xi_3\eta_1)-m_2(\xi_2\eta_3-\xi_4\eta_1)=o(t^{\frac{4}{3}})$$
$$(m_1+m_3)(\xi_2\eta_4-\xi_4\eta_2)-m_1(\xi_1\eta_4-\xi_3\eta_2)=o(t^{\frac{4}{3}})$$
$$m_1(m_2+m_3)(\xi_1\eta_2+\xi_3\eta_4)-m_2(m_1+m_3)(\xi_2\eta_1+\xi_4\eta_3) + m_1m_2(\xi_1\eta_1+\xi_3\eta_3-\xi_2\eta_2-\xi_4\eta_4)=o(t^{\frac{4}{3}})$$

(20) $$(m_2+m_3)\bar{p}_1\bar{q}_0-m_2(\bar{p}_2\bar{q}_0-\bar{p}_3\bar{q}_1)\to 0$$

(21) $$(m_1+m_3)(\bar{p}_2\bar{q}_3-\bar{p}_3\bar{q}_2)-m_1\bar{p}_1\bar{q}_3\to 0$$

(22) $$m_1(m_2+m_3)\bar{p}_1\bar{q}_2-m_2(m_1+m_3)(\bar{p}_2\bar{q}_1+\bar{p}_3\bar{q}_0) + m_1m_2(\bar{p}_1\bar{q}_1-\bar{p}_2\bar{q}_2-\bar{p}_3\bar{q}_3)\to 0\,,$$

all for $t\to 0$. On the basis of (12; 13) and (7), (11), (12), (17) we also have

(23) $$\bar{q}_k=O(1)\qquad(k=0,1,2,3)\,.$$

In the equilateral case (18), (19), (20), (21) and (23) combine to give the relations

$$2\bar{q}_1+\bar{q}_2+\sqrt{3}\,\bar{q}_3\to\tfrac{4}{3}\varkappa r^{-1}\,,\qquad 2\bar{q}_0-\sqrt{3}\,\bar{q}_2+\bar{q}_3\to 0$$
$$2\sqrt{3}m_2\bar{q}_1+(m_2+2m_3)(\sqrt{3}\bar{q}_2-\bar{q}_3)\to 0,\ \sqrt{3}(m_1+m_3)\bar{q}_2+(m_1-m_3)\bar{q}_3\to 0.$$

If we then set

$$\bar{q}_3=\frac{m_2}{\sqrt{3}}(m_1+m_3)\,\bar{q}\,,$$

it follows that

$$\bar{q}_2=\frac{m_2}{3}(m_3-m_1)\,\bar{q}+o(1)\,,\qquad \bar{q}_1=\frac{m_1}{3}(m_2+2m_3)\,\bar{q}+o(1)\,,$$

$$\tfrac{4}{3}(m_1m_2+m_1m_3+m_2m_3)\,\bar{q}\to\tfrac{4}{3}\varkappa r^{-1}\,,$$

so that by (12; 25) and (12; 26) we have

$$\bar{q}\to\frac{r}{M}\,.$$

Consequently in the equilateral case $\bar{q}_1, \bar{q}_2, \bar{q}_3$ have the limit values

$$\hat{q}_1=\frac{m_1}{3M}(m_2+2m_3)\,r,\quad \hat{q}_2=\frac{m_2}{3M}(m_3-m_1)\,r,\quad \hat{q}_3=\frac{m_2}{\sqrt{3}M}(m_1+m_3)\,r\,,$$

while

$$\bar{q}_0\to\hat{q}_0=\frac{\sqrt{3}}{2}\hat{q}_2-\frac{1}{2}\hat{q}_3=-\frac{m_1m_2}{\sqrt{3}M}\,r\,.$$

Just like $\hat{p}_1, \hat{p}_2, \hat{p}_3$, these values depend only on the masses m_1, m_2, m_3.

In the collinear case one obtains from (19), (20), (22), and (23) the relations

$$\bar{q}_0+\sigma\bar{q}_3\to 0,\ (m_2\varrho+m_3)\bar{q}_0\to 0,\ m_2(m_1\varrho-m_3\sigma)\bar{q}_1+m_1(m_2\varrho+m_3)\bar{q}_2\to 0,$$

and therefore

$$\bar{q}_0\to 0, \quad \bar{q}_3\to 0.$$

If this time we set

$$\bar{q}_1=m_1(m_2\varrho+m_3)\,\bar{q},$$

then

$$\bar{q}_2=m_2(m_3\sigma-m_1\varrho)\,\bar{q}+o(1),$$

and it follows from (18) that

$$\{m_1(m_2\varrho+m_3)+m_2\sigma(m_3\sigma-m_1\varrho)\}\,\bar{q}\to\tfrac{2}{3}\varkappa a^{-1}.$$

On the other hand, in view of (12; 27) and (12; 29) we have

$$\tfrac{2}{9}a^3m_1(m_2\varrho+m_3)+\tfrac{2}{9}\sigma a^3m_2(m_3\sigma-m_1\varrho)$$
$$=M(m_1m_2\varrho^{-1}+m_1m_3+m_2m_3\sigma^{-1})=\tfrac{2}{9}M\varkappa a,$$

so that finally

$$\bar{q}\to\frac{2a}{3M}.$$

Thus in the collinear case $\bar{q}_0, \bar{q}_1, \bar{q}_2, \bar{q}_3$ have the limits

$$\hat{q}_0=0,\quad \hat{q}_1=\frac{2m_1}{3M}(m_2\varrho+m_3)\,a,\quad \hat{q}_2=\frac{2m_2}{3M}(m_3\sigma-m_1\,\varrho)\,a,\quad \hat{q}_3=0,$$

which likewise depend only on the three masses.

Next, with q_4 again arbitrary, we introduce into the differential equations (15) the substitutions

$$p_k=\bar{p}_k t^{\frac{1}{3}}\ (k=1,2,3),\quad q_k=\bar{q}_k t^{-\frac{1}{3}}\ (k=0,1,2,3),\quad q_4=\bar{q}_4 t^{\frac{1}{3}},\quad p_4=\bar{p}_4$$

as well as

$$t=e^{-u},\qquad dt=-t\,du.$$

The function $\bar{E}$, obtained from E through direct replacement of the 7 variables p_k $(k=1,2,3)$ and q_k $(k=1,2,3,4)$ by $\bar{p}_k$ and $\bar{q}_k$, then satisfies

$$E=\bar{E}t^{-\frac{2}{3}},$$

and the differential equations transform into the system

$$(24)\quad \begin{cases} \dfrac{d\bar{p}_k}{du}=\dfrac{2}{3}\bar{p}_k-\bar{E}_{\bar{q}_k}, & \dfrac{d\bar{q}_k}{du}=-\dfrac{1}{3}\bar{q}_k+\bar{E}_{\bar{p}_k}\quad (k=1,2,3) \\[2ex] \dfrac{d\bar{p}_4}{du}=-\bar{E}_{\bar{q}_4}, & \dfrac{d\bar{q}_4}{du}=\dfrac{1}{3}\bar{q}_4. \end{cases}$$

Now let

$$(25)\qquad \begin{cases} \bar{p}_k = \hat{p}_k + \delta_k, & \bar{q}_k = \hat{q}_k + \delta_{k+2} \quad (k=1,2) \\ \bar{p}_3 = \hat{p}_3 + \delta_5, & \bar{q}_3 = \hat{q}_3 + \delta_6, \quad \bar{q}_4 = \delta_7, \quad \bar{p}_4 = \delta_8, \end{cases}$$

where $\hat{p}_k, \hat{q}_k$ $(k=1,2,3)$ are the previously derived limits for either the equilateral or the collinear case. The values $\delta_k = 0$ $(k=1,\ldots,8)$ correspond precisely to the coordinates of the special triple-collision orbits found in the beginning of this section, and because these too lead to solutions of (24), it follows that the right sides of the 8 differential equations, as functions of the δ_k, all vanish at the point $\delta_k=0$ $(k=1,\ldots,8)$. On the other hand, in a sufficiently small complex neighborhood of this point the right sides are regular and therefore have convergent series expansions in powers of the δ_k. The system (24) thus takes on the form

$$(26)\qquad \frac{d\,\delta_k}{du} = \sum_{l=1}^{8} a_{kl}\,\delta_l + \phi_k \quad (k=1,\ldots,8),$$

where the ϕ_k are power series in $\delta_1,\ldots,\delta_7$ beginning with quadratic terms, and the a_{kl} are real constants. Here

$$a_{k8}=0 \ (k=1,\ldots,8), \quad a_{77}=\tfrac{1}{3}, \quad a_{7l}=0 \ (l \neq 7),$$

giving the eight-by-eight matrix (a_{kl}) the structure

$$(27)\qquad \mathfrak{A} = \begin{pmatrix} \mathfrak{B} & * & 0 \\ 0 & \frac{1}{3} & 0 \\ * & * & 0 \end{pmatrix}$$

with $\mathfrak{B}$ a six-by-six matrix. Moreover, $\phi_7=0$.

Since $u=\log t^{-1}$, as t decreases to 0 the variable u increases to ∞. Moreover, $\delta_7 = q_4 t^{-\frac{1}{3}} = 0$ for every triple-collision orbit. Thus, to obtain all triple-collision orbits, one solves the 6 differential equations

$$(28)\qquad \frac{d\,\delta_k}{du} = \sum_{l=1}^{6} a_{kl}\,\delta_l + \phi_k \quad (k=1,\ldots,6),$$

with $\delta_7=0$, to find all solutions $\delta_k=\delta_k(u)$ $(k=1,\ldots,6)$ that are defined for u sufficiently large and asymptotically approach the solution $\delta_k=0$ as $u\to\infty$.

To solve this problem we appeal to a result that will first be derived in § 28 independently of the present considerations. According to (27), the linear part of the right side in (28) has the coefficient matrix $\mathfrak{B}$. Let $\lambda_1,\ldots,\lambda_6$ be the eigenvalues of $\mathfrak{B}$, that is, the roots of the polynomial $|\lambda\mathfrak{E}-\mathfrak{B}|$, with $\lambda_1,\ldots,\lambda_f$ the ones having a negative real part. In view of the restrictions imposed in § 28, it is further assumed that the eigenvalues

$\lambda_1, \ldots, \lambda_6$ are all simple, none is purely imaginary or 0, and for all systems of nonnegative integers $n_1, \ldots, n_f$ with $n_1 + \cdots + n_f > 1$ we have

$$\text{(29)} \qquad \sum_{l=1}^{f} n_l \lambda_l \neq \lambda_k \quad (k=1, \ldots, f).$$

In accordance with § 28, the desired solutions of (28) are then given by

$$\text{(30)} \qquad \delta_k = \psi_k(v_1, \ldots, v_f) \ (k=1, \ldots, 6), \quad v_l = c_l e^{\lambda_l u} \ (l=1, \ldots, f),$$

where the ψ_k are convergent power series in the variables $v_1, \ldots, v_f$ without a constant term, and the c_l are arbitrary constants sufficiently small in absolute value. The coefficients in the series ψ_k are uniquely determined by a recursive process. Once the desired solutions $\delta_1, \ldots, \delta_6$ of (28) have been found from (30), the angle $p_4 = \bar{p}_4 = \delta_8$ is obtained by quadrature from the last equation in (26), which with $\delta_7 = 0$ reads

$$\text{(31)} \qquad \frac{dp_4}{du} = -\left(\frac{1}{m_1} + \frac{1}{m_3}\right)\frac{\bar{q}_0}{\bar{p}_1} - \frac{1}{m_3}\frac{\bar{q}_3}{\bar{p}_1}, \quad \bar{q}_0 = \frac{\bar{q}_2\bar{p}_3 - \bar{p}_2\bar{q}_3}{\bar{p}_1}.$$

In light of (30) the right side of this differential equation is again a power series in $v_1, \ldots, v_f$ without a constant term, and because $\lambda_1, \ldots, \lambda_f$ have negative real parts, the integral converges up to $u = \infty$. Consequently for each solution resulting in a triple collision also p_4 has a finite limit $\hat{p}_4$ as $t \to 0$. This finally proves that collision of the three particles takes place in definite directions.

The calculation of the 6 eigenvalues $\lambda_1, \ldots, \lambda_6$ is facilitated by a return to the earlier variables. From (27) we have

$$\lambda \mathfrak{E}_8 - \mathfrak{A} = \begin{pmatrix} \lambda\mathfrak{E}_6 - \mathfrak{B} & * & 0 \\ 0 & \lambda - \frac{1}{3} & 0 \\ * & * & \lambda \end{pmatrix}$$

$$\text{(32)} \qquad |\lambda\mathfrak{E}_8 - \mathfrak{A}| = \lambda(\lambda - \tfrac{1}{3})\, |\lambda\mathfrak{E}_6 - \mathfrak{B}|.$$

We will transform the 8 variables δ_k $(k=1, \ldots, 8)$ via a substitution

$$\omega_k = \sum_{l=1}^{8} c_{kl}\, \delta_l + \chi_k \quad (k=1, \ldots, 8),$$

where the coefficients c_{kl} form an invertible constant matrix $\mathfrak{C}$ and the χ_k are convergent power series in the δ_l $(l=1, \ldots, 8)$ beginning with quadratic terms. This results in (26) being replaced by the new differential equations

$$\frac{d\omega_k}{du} = \sum_{l=1}^{8} g_{kl}\, \omega_l + \cdots \quad (k=1, \ldots, 8)$$

whose coefficient matrix $\mathfrak{G}=(g_{kl})$ is determined from the relation

$$\mathfrak{A}=\mathfrak{C}^{-1}\mathfrak{G}\mathfrak{C}. \tag{33}$$

This assertion is purely algebraic in nature, and has nothing to do with properties of solutions to the differential equations. In defining ω_k we refer back to the previously used canonical transformation

$$\left\{\begin{array}{l} \xi_1=p_1c, \quad \xi_2=p_2c-p_3s, \quad \xi_3=p_1s, \quad \xi_4=p_2s+p_3c \\ \eta_1=q_1c-q_0s, \quad \eta_2=q_2c-q_3s, \quad \eta_3=q_1s+q_0c, \quad \eta_4=q_2s+q_3c \\ q_0=(q_4-p_2q_3+p_3q_2)\,p_1^{-1}, \quad c=\cos p_4, \quad s=\sin p_4 \end{array}\right. \tag{34}$$

and carry out the additional invertible homogeneous linear substitution

$$\left\{\begin{array}{l} -\left(\dfrac{1}{m_1}+\dfrac{1}{m_3}\right)\eta_k-\dfrac{1}{m_3}\eta_{k+1}=\xi_{k+4}, \\ -\dfrac{1}{m_3}\eta_k-\left(\dfrac{1}{m_2}+\dfrac{1}{m_3}\right)\eta_{k+1}=\xi_{k+5} \qquad (k=1,3). \end{array}\right. \tag{35}$$

In terms of

$$\bar{\xi}_k=\xi_k t^{-\frac{2}{3}}, \qquad \bar{\xi}_{k+4}=\xi_{k+4}t^{\frac{1}{3}} \qquad (k=1,\ldots,4),$$

with $t=e^{-u}$, the system (8) then becomes

$$\frac{d\bar{\xi}_k}{du}=\frac{2}{3}\bar{\xi}_k+\bar{\xi}_{k+4}, \qquad \frac{d\bar{\xi}_{k+4}}{du}=-\frac{1}{3}\bar{\xi}_{k+4}-F_k \qquad (k=1,\ldots,4), \tag{36}$$

where

$$F_k=\left\{\begin{array}{ll} (m_1+m_3)\,R_2\bar{\xi}_k+m_2R_1\bar{\xi}_{k+1}+m_2R_3(\bar{\xi}_k-\bar{\xi}_{k+1}) & (k=1,3) \\ m_1R_2\bar{\xi}_{k-1}+(m_2+m_3)\,R_1\bar{\xi}_k+m_1R_3(\bar{\xi}_k-\bar{\xi}_{k-1}) & (k=2,4) \end{array}\right.$$

$$R_1=(\bar{\xi}_2^2+\bar{\xi}_4^2)^{-\frac{3}{2}}, \quad R_2=(\bar{\xi}_1^2+\bar{\xi}_3^2)^{-\frac{3}{2}}, \quad R_3=\{(\bar{\xi}_1-\bar{\xi}_2)^2+(\bar{\xi}_3-\bar{\xi}_4)^2\}^{-\frac{3}{2}}.$$

Finally, let $\hat{\xi}_k$ $(k=1,\ldots,8)$ be the value at $\delta_l=0$ of $\bar{\xi}_k$ as a function of the eight variables δ_l $(l=1,\ldots,8)$, and define

$$\omega_k=\bar{\xi}_k-\hat{\xi}_k \qquad (k=1,\ldots,8). \tag{37}$$

The right sides of (36) then lead to convergent power series in the ω_k without a constant term, with the corresponding coefficient matrix

$$\mathfrak{G}=\begin{pmatrix} \frac{2}{3}\mathfrak{E}_4 & \mathfrak{E}_4 \\ -\mathfrak{S} & -\frac{1}{3}\mathfrak{E}_4 \end{pmatrix}, \tag{38}$$

where the elements s_{kl} $(k,l=1,\ldots,4)$ in the four-by-four matrix $\mathfrak{S}$ are the partial derivatives $\dfrac{\partial F_k}{\partial \bar{\xi}_l}$ evaluated at $\bar{\xi}_r=\hat{\xi}_r$ $(r=1,\ldots,4)$.

From (34) we have $\hat{\xi}_1 = \hat{p}_1$, $\hat{\xi}_2 = \hat{p}_2$, $\hat{\xi}_3 = 0$, $\hat{\xi}_4 = \hat{p}_3$, and therefore in the equilateral case

$$r^3\mathfrak{S} = \begin{pmatrix} \frac{1}{4}m_2 - 2(m_1+m_3) & 0 & \frac{3\sqrt{3}}{4}m_2 & -\frac{3\sqrt{3}}{2}m_2 \\ -\frac{9}{4}m_1 & \frac{1}{4}(m_1+m_2+m_3) & -\frac{3\sqrt{3}}{4}m_1 & \frac{3\sqrt{3}}{4}(m_1-m_2-m_3) \\ \frac{3\sqrt{3}}{4}m_2 & -\frac{3\sqrt{3}}{2}m_2 & m_1+m_3-\frac{5}{4}m_2 & 0 \\ -\frac{3\sqrt{3}}{4}m_1 & \frac{3\sqrt{3}}{4}(m_1-m_2-m_3) & \frac{9}{4}m_1 & -\frac{5}{4}(m_1+m_2+m_3) \end{pmatrix},$$

while in the collinear case

$$a^3\mathfrak{S} = \begin{pmatrix} -2\mathfrak{Q} & 0 \\ 0 & \mathfrak{Q} \end{pmatrix}, \quad \mathfrak{Q} = \begin{pmatrix} m_1+m_3+m_2\varrho^{-3} & m_2(\sigma^{-3}-\varrho^{-3}) \\ m_1(1-\varrho^{-3}) & m_1\varrho^{-3}+(m_2+m_3)\sigma^{-3} \end{pmatrix}.$$

With the abbreviation

$$(\lambda+\tfrac{1}{3})(\lambda-\tfrac{2}{3}) = \zeta,$$

one sees from (32), (33), (38) that

$$|\lambda\mathfrak{E}_8 - \mathfrak{A}| = |\lambda\mathfrak{E}_8 - \mathfrak{G}| = \begin{vmatrix} (\lambda-\frac{2}{3})\mathfrak{E}_4 & -\mathfrak{E}_4 \\ \mathfrak{S} & (\lambda+\frac{1}{3})\mathfrak{E}_4 \end{vmatrix}$$

$$(\zeta+\tfrac{2}{9})|\lambda\mathfrak{E}_6 - \mathfrak{B}| = |\zeta\mathfrak{E}_4 + \mathfrak{S}|,$$

whereupon a direct calculation of the four-by-four determinant gives

$$|\zeta\mathfrak{E}_4 + \mathfrak{S}| = (\zeta+\tfrac{2}{9})(\zeta-\tfrac{4}{9})(\zeta^2-\tfrac{2}{9}\zeta-\tfrac{8}{81}+\tfrac{1}{3}\mu)$$

with

$$\mu = \frac{m_1m_2+m_1m_3+m_2m_3}{(m_1+m_2+m_3)^2}$$

in the equilateral case, and

$$|\zeta\mathfrak{E}_4 + \mathfrak{S}| = (\zeta+\tfrac{2}{9})(\zeta-\tfrac{4}{9})(\zeta+\tfrac{2}{9}+\tfrac{2}{9}\nu)(\zeta-\tfrac{4}{9}-\tfrac{4}{9}\nu)$$

with

$$\nu = \frac{m_1(1+\varrho^{-1}+\varrho^{-2})+m_3(1+\sigma^{-1}+\sigma^{-2})}{m_1+m_2(\varrho^{-2}+\sigma^{-2})+m_3}$$

in the collinear case. The determinant $|\lambda\mathfrak{E}_6 - \mathfrak{B}|$ is then obtained from these two polynomials by discarding the factor $(\zeta+\frac{2}{9})$. Thus in the equilateral case the 6 eigenvalues of $\mathfrak{B}$ are

$$\lambda_1=-\tfrac{2}{3},\ \lambda_2=\tfrac{1}{6}(1-\sqrt{13+12\sqrt{1-3\mu}}),\ \lambda_3=\tfrac{1}{6}(1-\sqrt{13-12\sqrt{1-3\mu}}),$$
$$\lambda_4=\tfrac{1}{6}(1+\sqrt{13-12\sqrt{1-3\mu}}),\ \lambda_5=\tfrac{1}{6}(1+\sqrt{13+12\sqrt{1-3\mu}}),\ \lambda_6=1,$$

while in the collinear case they are

$$\lambda_1=-\tfrac{2}{3},\quad \lambda_2=\tfrac{1}{6}(1-\sqrt{25+16\nu}),\quad \lambda_3=\tfrac{1}{6}(1-\sqrt{1-8\nu}),$$
$$\lambda_4=\tfrac{1}{6}(1+\sqrt{1-8\nu}),\quad \lambda_5=\tfrac{1}{6}(1+\sqrt{25+16\nu}),\quad \lambda_6=1.$$

The identity

$$2M^2(1-3\mu)=(m_1-m_2)^2+(m_1-m_3)^2+(m_2-m_3)^2$$

shows that in the equilateral case the 6 eigenvalues are all real, with $\lambda_1, \lambda_2, \lambda_3$ negative and $\lambda_4, \lambda_5, \lambda_6$ positive. Moreover, unless $m_1=m_2=m_3$, the eigenvalues are all distinct, while condition (29) is easily seen to coincide here with the stipulation that neither λ_1/λ_3 nor λ_2/λ_3 be an integer. With these exceptions excluded, the previously named result from § 28 then gives the coordinates in the equilateral case of a triple collision as power series in the three variables

$$v_1=c_1t^{\frac{2}{3}},\quad v_2=c_2t^{-\lambda_2},\quad v_3=c_3t^{-\lambda_3} \tag{39}$$

containing three arbitrary real constants c_1, c_2, c_3. Another constant, which fixes the direction of collision for the 3 points in the orbital plane, comes from determining p_4 by quadrature. Furthermore, one can undertake an arbitrary orientation of the (x, y)-plane in space and impart to it a translation with constant velocity, obtaining in this way 8 additional constants. The equilateral case of a triple collision thus contains altogether 12 independent real parameters. Finally, observe that for an arbitrary constant $c_0>0$ the transformation $t\to c_0t$, $c_k\to c_kc_0^{\lambda_k}$ leaves the variables v_k $(k=1,2,3)$ in (39) unchanged, so that also the functions $\bar{p}_k, \bar{q}_k$ $(k=1,2,3)$ constructed from these through (25) and (30) remain unaltered, whereas the $p_k=\bar{p}_kt^{\frac{2}{3}}$ themselves do not. On account of the independent variable t, the triple-collision orbits in the equilateral case then form a 13-dimensional manifold in an 18-dimensional space.

In the collinear case the two eigenvalues λ_1, λ_2 are real and negative, λ_5, λ_6 real and positive, while λ_3, λ_4 are either real and positive or complex conjugate with the positive real part $\frac{1}{6}$. To have the eigenvalues all distinct, it is necessary to stipulate that $\nu\neq\frac{1}{8}$, whereas (29) means here that λ_2/λ_1 must not be an integer. The coordinates are then given as power series in the two variables

$$v_1=c_1t^{\frac{2}{3}},\quad v_2=c_2t^{-\lambda_2}, \tag{40}$$

with two arbitrary real constants c_1, c_2.

In conclusion we show that in the collinear case of a triple collision the angle p_4 is constant, so that the 3 particles move along a fixed line. For this, in addition to equation (31) for p_4, we isolate the two differential equations for $\bar{p}_3$ and $\bar{q}_3$, which in view of (10), (13), (14) and (24) are explicitly given by

$$(41)\begin{cases} \dfrac{d\bar{p}_3}{du}=\dfrac{2}{3}\bar{p}_3+\left\{\left(\dfrac{1}{m_1}+\dfrac{1}{m_3}\right)\dfrac{\bar{p}_2}{\bar{p}_1}-\dfrac{1}{m_3}\right\}\bar{q}_0+\left(\dfrac{1}{m_3}\dfrac{\bar{p}_2}{\bar{p}_1}-\dfrac{1}{m_2}-\dfrac{1}{m_3}\right)\bar{q}_3 \\ \dfrac{d\bar{q}_3}{du}=-\dfrac{1}{3}\bar{q}_3+m_2m_3(\bar{p}_2^2+\bar{p}_3^2)^{-\frac{3}{2}}\bar{p}_3+m_1m_2\{(\bar{p}_1-\bar{p}_2)^2+\bar{p}_3^2\}^{-\frac{3}{2}}\bar{p}_3 \\ \qquad +\left(\dfrac{1}{m_1}+\dfrac{1}{m_3}\right)\dfrac{\bar{q}_2}{\bar{p}_1}\bar{q}_0+\dfrac{1}{m_3}\dfrac{\bar{q}_2}{\bar{p}_1}\bar{q}_3 \end{cases}$$

with

$$\bar{q}_0=(\bar{q}_4-\bar{p}_2\bar{q}_3+\bar{q}_2\bar{p}_3)\,\bar{p}_1^{-1}.$$

Because in the collinear case $\hat{p}_3=0$, $\hat{q}_3=0$, and by (25) therefore $\bar{p}_3=\delta_5$, $\bar{q}_3=\delta_6$, the right sides of both equations in (41) for $\delta_7=0$ are of the form $\Phi\delta_5+\Psi\delta_6$, with Φ and Ψ convergent power series in $\delta_1,\ldots,\delta_6$. If now the two eigenvalues corresponding to the linear part in the right side of (41) both have a positive real part, the considerations in § 28 imply that the solutions $\delta_5=\delta_5(u)$ and $\delta_6=\delta_6(u)$, subject to the conditions $\delta_k\to 0$ $(k=1,\ldots,6)$ as $u\to\infty$, must vanish identically in u. The condition on the two eigenvalues, on the other hand, will follow from the contention that these are just λ_3 and λ_4. Since a direct calculation of the eigenvalues using (41) appears too laborious, it is better to proceed as we did in determining the 6 eigenvalues of $\mathfrak{B}$. One verifies from (24) and (25) that in the collinear case the coefficients a_{kl} in (26) are 0 if $k=1,\ldots,4$ and $l=5,\ldots,8$. Now if in the definition of the ω_k in (37) one interchanges the indices 3 and 5 as well as 4 and 6 on the left side, then according to (34) and (35) the correspondingly constructed new matrix $\mathfrak{C}$ decomposes like $\mathfrak{A}$, while in place of (38) one now has

$$\mathfrak{G}=\begin{pmatrix} \frac{2}{3}\mathfrak{E}_2 & \mathfrak{E}_2 & 0 & 0 \\ 2a^{-3}\mathfrak{Q} & -\frac{1}{3}\mathfrak{E}_2 & 0 & 0 \\ 0 & 0 & \frac{2}{3}\mathfrak{E}_2 & \mathfrak{E}_2 \\ 0 & 0 & -a^{-3}\mathfrak{Q} & -\frac{1}{3}\mathfrak{E}_2 \end{pmatrix}.$$

From this one verifies the contention about the two eigenvalues in question, whereupon it follows that in the collinear case $\bar{p}_3=0$, $\bar{q}_3=0$. Since then also $\bar{q}_0=0$, equation (31) now does indeed imply that $p_4=\hat{p}_4$ is constant.

To obtain the general solution in the collinear case one has to replace the orbital line by any straight line in the old coordinate system moving parallel to itself with constant velocity. This gives 7 new constants, in addition to c_1, c_2, p_4, altogether 10 independent real parameters, whereas there are 12 parameters in the equilateral case.

The exponents $-\lambda_2, -\lambda_3$ in (39) for the equilateral case, and also the exponent $-\lambda_2$ in (40) for the collinear case, are in general irrational, namely, whenever the quantities μ, ν, which depend on the masses m_1, m_2, m_3, avoid a certain countable set of exceptional values. Thus, generally, a triple collision gives rise to an essential singularity in the coordinates x_k, y_k $(k=1,2,3)$. On the other hand, by setting $c_2=0$, $c_3=0$ in (39) and $c_2=0$ in (40) one obtains special triple-collision solutions with the same type of singularity as in simple collisions, for the coordinates are then power series in the variable $t^{\frac{2}{3}}$. A closer inspection shows these special solutions to be the same as those given in (4), with the constant b_2 there bearing a simple relation to c_1.

The considerations in § 28 can be extended to give analogous results for the case when the 6 eigenvalues are not all distinct or do not satisfy condition (29), a case that up to now has been excluded. This, however, will not be pursued here any further.

A final point to be discussed is the possibility of a singularity in the three-body problem due to simple collisions at a sequence of times $t=t_1, t_2, \ldots$ decreasing to 0. It follows from the reasoning in § 8 that a triple collision would then have to occur at $t=0$. But the reasoning of § 12, appropriately translated to this case, would now show (12; 7) again to be true, contradicting that $U=U(t)$, on the other hand, would have to become infinite at all t_k $(k=1,2,\ldots)$. Thus, extending the result obtained in § 8, we have proved that simple collisions in the three-body problem do not accumulate at a finite time, even if the angular momentum constants with respect to the fixed center of mass are all 0.

This concludes our investigation into the singularities of the three-body problem. The restriction of t to real values was essential here, and a satisfactory extension of this analysis to the complex case appears hopeless.

Chapter Two

Periodic Solutions

§ 14. The Solutions of Lagrange

The theorem of Sundman discussed in chapter one represents the most far reaching result yet known about general solutions to the three-body problem. Unfortunately his ingenious method cannot be extended to the case $n > 3$. Although the results in § 6 can be used to exclude at the outset the simultaneous collision of all n bodies, the investigations of § 12, § 13 show that already a triple collision may give rise to essential singularities.

In this chapter we will develop methods that go beyond the n-body problem and apply to many general questions in mechanics as well. The methods deal with finding periodic solutions for a given problem. A solution of period τ has the special property that it is completely determined by knowledge over a finite time interval of length τ, and therefore difficulties arising from the unboundedness of time, as for example in Sundman's first lemma, are no longer present. Because planetary motion is approximately periodic, periodic solutions to the n-body problem are important also in astronomy.

The simplest periodic solutions are ones whose coordinates do not depend on time. These are known as equilibrium solutions. We observe first that there are no such solutions to the n-body problem $(n > 1)$, for if each coordinate q were constant in time, then from $\ddot{q} = 0$ the equations of motion (5; 3) would imply that $U_q = 0$, and it would follow from Euler's theorem on homogeneous functions that

$$\sum_q q U_q = -U = 0,$$

whereas from its definition U must certainly be positive. Since there are no equilibrium solutions to the n-body problem, we look for the next simplest periodic solutions. Restricting ourselves again to three particles P_1, P_2, P_3, we ask: Does the three-body problem have solutions for which the particles move uniformly along circular orbits in a fixed plane? In answering this question affirmatively, we will obtain a special class of solutions to the three-body problem that were given already by Lagrange [1] in the year 1772.

In the plane in question we introduce a rectangular Cartesian coordinate system and denote the coordinates of P_k by q_{2k-1}, q_{2k} $(k=1,2,3)$. Again let

$$(1)\quad T=\frac{1}{2}\sum_{k=1}^{3} m_k^{-1}(p_{2k-1}^2+p_{2k}^2),\quad U=\sum_{k<l} m_k m_l r_{kl}^{-1},\qquad E=T-U,$$

with r_{kl} the distance between P_k and P_l, so that the equations of motion take the Hamiltonian form

$$(2)\qquad \dot{q}_k=E_{p_k},\quad \dot{p}_k=-E_{q_k}\qquad (k=1,\ldots,6).$$

Retaining the same origin, we introduce a new rectangular Cartesian coordinate system that rotates uniformly about the origin in the given plane. The rotational velocity is to be determined so that in the new coordinate system the particles are at rest. The new coordinates x_{2k-1}, x_{2k} of P_k are given by

$$(3)\quad \begin{cases} x_{2k-1}=q_{2k-1}c+q_{2k}s, \quad x_{2k}=-q_{2k-1}s+q_{2k}c \quad (k=1,2,3), \\ \qquad\qquad c=\cos\lambda,\quad s=\sin\lambda, \end{cases}$$

where the angle of rotation λ takes the form $\lambda=\omega t$, with $\omega\neq 0$ a real constant still to be determined. To introduce these coordinates into the equations of motion (2), we complete (3) to a canonical transformation by choosing suitable additional variables $y_1,\ldots,y_6$. For this purpose we base a canonical transformation on the representation given by (3; 4), and readily obtain

$$w=w(q,y)=\sum_{k=1}^{3}\{(q_{2k-1}c+q_{2k}s)y_{2k-1}+(-q_{2k-1}s+q_{2k}c)y_{2k}\}$$

as an appropriate generating function. The matrix of the binary bilinear form in the curved bracket being orthogonal, the 6-by-6 determinant $|w_{q_k y_l}|$ has the value $1\neq 0$. In accordance with (3; 4), the corresponding canonical transformation then reads

$$p_k=w_{q_k},\quad x_k=w_{y_k}\qquad (k=1,\ldots,6),$$

so that (3) is augmented by

$$(4)\qquad p_{2k-1}=y_{2k-1}c-y_{2k}s,\quad p_{2k}=y_{2k-1}s+y_{2k}c\qquad (k=1,2,3).$$

It follows that

$$p_{2k-1}^2+p_{2k}^2=y_{2k-1}^2+y_{2k}^2,$$

and to introduce the new coordinates into T one has merely to replace p_k by y_k in expression (1). Similarly, since U depends only on the distances between the particles, q_k has merely to be replaced by x_k in the corresponding expression for U. Moreover, the derivatives of c, s are $c_t = -\omega s$, $s_t = \omega c$, so that by (3; 4) the new Hamiltonian function is

$$F = E + w_t = E + \omega \sum_{k=1}^{3} (x_{2k} y_{2k-1} - x_{2k-1} y_{2k})$$

and the equations of motion become

$$\dot{x}_k = F_{y_k}, \quad \dot{y}_k = -F_{x_k} \quad (k = 1, \ldots, 6),$$

or

$$(5) \quad \begin{cases} \dot{x}_{2k-1} = E_{y_{2k-1}} + \omega x_{2k}, & \dot{y}_{2k-1} = -E_{x_{2k-1}} + \omega y_{2k}, \\ \dot{x}_{2k} = E_{y_{2k}} - \omega x_{2k-1}, & \dot{y}_{2k} = -E_{x_{2k}} - \omega y_{2k-1} \quad (k = 1, 2, 3). \end{cases}$$

Here

$$E_{x_k} = -U_{x_k} \ (k = 1, \ldots, 6), \ E_{y_{2k-1}} = m_k^{-1} y_{2k-1}, \ E_{y_{2k}} = m_k^{-1} y_{2k} \ (k = 1, 2, 3).$$

Note that the expression

$$\sum_{k=1}^{3} (x_{2k} y_{2k-1} - x_{2k-1} y_{2k}) = Q$$

appearing in the definition of F is just the angular momentum integral, for in accordance with (3), (4), this expression is unaltered if x and y are replaced by q and $p = m\dot{q}$. That $\dot{Q} = 0$ can of course also be verified directly from (5).

The equilibrium solutions to (5) arise from the twelve conditions

$$m_k^{-1} y_{2k-1} + \omega x_{2k} = 0, \quad U_{x_{2k-1}} + \omega y_{2k} = 0, \quad m_k^{-1} y_{2k} - \omega x_{2k-1} = 0,$$
$$U_{x_{2k}} - \omega y_{2k-1} = 0 \quad (k = 1, 2, 3),$$

which upon elimination of $y_k \ (k = 1, \ldots, 6)$ become

$$(6) \quad m_k \omega^2 x_{2k-1} = -U_{x_{2k-1}}, \quad m_k \omega^2 x_{2k} = -U_{x_{2k}} \quad (k = 1, 2, 3).$$

Conversely, each solution $x_k \ (k = 1, \ldots, 6)$ of the above system of equations leads to an equilibrium solution, while addition over k in (6) in conjunction with (5; 6) shows that the center of mass of the three bodies lies at the origin. The solving of equations (6), which we now undertake, was carried out in a different setting already in § 12.

Let us first assume that the triangle $P_1 P_2 P_3$ is not equilateral, and select the indices so that $r_{13} \neq r_{23}$. If in addition the abscissa axis is taken to pass through P_3, then $x_6 = 0$ and the second equation of (6) for $k = 3$ gives rise to the condition

$$m_1 x_2 r_{13}^{-3} + m_2 x_4 r_{23}^{-3} = 0 .$$

Since

$$m_1x_2+m_2x_4+m_3x_6=0, \quad x_6=0,$$

it follows that

$$m_1x_2(r_{13}^{-3}-r_{23}^{-3})=0,$$

hence $x_2=0$ and also $x_4=0$. Therefore, if the three particles do not form an equilateral triangle they are collinear.

For the case of an equilateral triangle, let $r_{12}=r_{23}=r_{31}=r$, so that from (6) and the condition on the center of mass we have

$$\omega^2x_{2k-1}=r^{-3}\sum_{l=1}^{3}m_l(x_{2k-1}-x_{2l-1})=Mr^{-3}x_{2k-1},$$
$$\omega^2x_{2k}=Mr^{-3}x_{2k} \quad (k=1,2,3),$$

where $M=m_1+m_2+m_3$. Because the x_k $(k=1,\dots,6)$ are not all 0, it follows that

$$M=\omega^2r^3, \quad \omega=\pm\sqrt{Mr^{-3}}. \tag{7}$$

Conversely, (7) leads back to (6), so that we have indeed found a solution to the three-body problem for which $P_1P_2P_3$ is an equilateral triangle with side r, while ω is determined from (7). This is known as the equilateral solution.

It remains to investigate the case when the three particles lie on a straight line. Selecting this line as the abscissa axis, we have $x_{2k}=0$ $(k=1,2,3)$, whereby the second equation of (6) is automatically satisfied. Assume now that P_2 lies between P_1 and P_3, while the positive direction along the abscissa axis is from P_1 to P_3. If $r_{13}=a$, $r_{12}=\varrho a$, $r_{23}=\sigma a$, then $\varrho+\sigma=1$, $0<\varrho<1$, and the first equation in (6) for $k=1,2,3$ gives rise to the formulas

$$\begin{cases}-m_2(\varrho a)^{-2}-m_3a^{-2}=\omega^2x_1, \quad m_1(\varrho a)^{-2}-m_3(\sigma a)^{-2}=\omega^2x_3,\\ m_1a^{-2}+m_2(\sigma a)^{-2}=\omega^2x_5.\end{cases} \tag{8}$$

One of these, say the middle one, can be replaced by the condition $m_1x_1+m_2x_3+m_3x_5=0$ on the center of mass, which also says that

$$Mx_1+m_2\varrho a+m_3a=0, \quad Mx_5-m_2\sigma a-m_1a=0,$$

where again $M=m_1+m_2+m_3$. Using this to eliminate x_1 and x_5 in the first and third equation of (8), we obtain

$$\begin{cases}m_2\varrho^{-2}+m_3=M^{-1}\omega^2a^3(m_2\varrho+m_3),\\ m_2\sigma^{-2}+m_1=M^{-1}\omega^2a^3(m_2\sigma+m_1),\end{cases} \tag{9}$$

and because $\sigma = 1 - \varrho$, finally

$$(10) \quad M^{-1}\omega^2 a^3 = \frac{m_2\varrho^{-2} + m_3}{m_2\varrho + m_3} = \frac{m_2(1-\varrho)^{-2} + m_1}{m_2(1-\varrho) + m_1}, \quad 0 < \varrho < 1.$$

Since in the interval $0 < \varrho < 1$ the difference

$$\frac{m_2\varrho^{-2} + m_3}{m_2\varrho + m_3} - \frac{m_2(1-\varrho)^{-2} + m_1}{m_2(1-\varrho) + m_1} = f(\varrho)$$

is a monotonically decreasing function of ϱ which tends to $+\infty$ as $\varrho \to 0$ and to $-\infty$ as $\varrho \to 1$, the equation $f(\varrho) = 0$ has exactly one root ϱ in this interval. Setting $M^{-1}\omega^2 a^3$ equal to the value given by (10) and determining x_1, x_3, x_5 from (8), we obtain a solution to (6). Thus again this leads to an actual solution of the three-body problem, with ω determined up to sign by the value of a. This is known as the collinear solution. To determine ϱ one has to solve a fifth degree equation with coefficients depending on the masses. We saw that there is only one solution in the interval $0 < \varrho < 1$. It was assumed here that P_2 lies between P_1 and P_3, and two additional solutions can be obtained by a cyclic permutation of the indices.

In the opinion of Lagrange the solutions he found had no significance in astronomy. However, more recently it was discovered that the sun, Jupiter, and the small planets of the Trojan group form an approximately equilateral triangle, and it is therefore of interest to study solutions of the three-body problem near the Lagrange solutions. This will be done in § 18.

Lagrange generalized the above solutions further by asking for, and constructing, all additional solutions for which the triangles formed by the particles remain similar throughout the motion. We will consider the analogous question for the planar n-body problem, assuming, in view of the center of mass integral, that the center of mass is at rest. The rectangular Cartesian coordinates x_k, y_k of P_k then take the form

$$(11) \qquad x_k + iy_k = z_k = \zeta_k q \qquad (k = 1, \ldots, n),$$

where $\zeta_1, \ldots, \zeta_n$ are distinct complex constants and q is an unknown complex function of the real variable t. The distances are then given by

$$r_{kl} = |z_k - z_l| = |\zeta_k - \zeta_l|\,|q|,$$

and the complex form of the equations of motion

$$\ddot{z}_k = \sum_{l \neq k} m_l \frac{z_l - z_k}{r_{kl}^3} \qquad (k = 1, \ldots, n)$$

leads for $q \neq 0$ to the equations

$$\zeta_k \ddot{q} = q|q|^{-3} \sum_{l \neq k} m_l \frac{\zeta_l - \zeta_k}{|\zeta_l - \zeta_k|^3} \qquad (k=1, \ldots, n).$$

Because the ζ_k are not all 0, the expression

$$\ddot{q} q^{\frac{1}{2}} \bar{q}^{\frac{3}{2}} = c \tag{12}$$

is independent of t, while

$$\zeta_k c = \sum_{l \neq k} m_l \frac{\zeta_l - \zeta_k}{|\zeta_l - \zeta_k|^3} \qquad (k=1, \ldots, n). \tag{13}$$

The problem is thereby reduced to solving the differential equation (12), which is independent of n, together with the system of algebraic equations (13). For the case $n=2$ this leads to the general solution of the two-body problem, since the segments $P_1 P_2$ trivially remain similar throughout. In addition, we have $m_1 \zeta_1 + m_2 \zeta_2 = 0$, hence $\zeta_1 = m_2 \zeta, \zeta_2 = -m_1 \zeta$ with a complex $\zeta \neq 0$, and it follows from (13) that in this case the quantity $c = -(m_1 + m_2)^{-2} |\zeta|^{-3}$ is real and negative.

The solution of the two-body problem is now assumed as known. If $q = q(t)$ is any solution of the differential equation (12) with c a negative constant, and if $\zeta \neq 0$ is a complex constant, then

$$x + iy = z = \zeta q$$

is a parametric representation of a conic section in the (x, y)-plane, and, indeed, of one for which the origin is a focus. Let n again be arbitrary and suppose that $\zeta_1, \ldots, \zeta_n$ are n distinct complex numbers satisfying the system of equations (13) with $c < 0$. Each solution of (12) then gives, by means of (11), a solution to the n-body problem, whereby each particle describes a conic section and the polygons formed by the n points remain similar throughout. In particular, if the conic section is taken to be an ellipse, $q(t)$ is periodic, and one obtains in this way a periodic solution.

For the case $n=3$ all solutions of (13) with a negative constant c are already known to us, for this system of equations corresponds to (6) with $\omega^2 = -c$ and $x_{2k-1} + i x_{2k} = \zeta_k$, and therefore leads to solutions of the three-body problem in which $\zeta_1, \zeta_2, \zeta_3$ are either vertices of an equilateral triangle or three collinear points with the ratio ϱ of their distances determined by (10). These are then the generalized solutions of Lagrange, with the special case of circular orbits obtained by setting $q = e^{i\omega t}$. Another special case arises from taking the solution $q(t)$ of (12) to be real, and this was already considered in § 13. The collinear case of the latter was treated already by Euler [2], who thus became the first to find a particular solution to the three-body problem. In this connection the fifth degree equation $f(\varrho) = 0$ also appears in his work.

§ 15. Eigenvalues

We consider a system of m first order differential equations expressed in vector form as

(1) $$\dot{x} = f(x),$$

where x and $f(x)$ are columns with entries x_k and $f_k(x)$ $(k = 1, \ldots, m)$ respectively. Let $x = x^{(0)}$ be an equilibrium solution to (1), and assume that the f_k are regular functions of $x_1, \ldots, x_m$ in a neighborhood of $x^{(0)}$ which do not depend explicitly on t. It may be assumed without loss of generality that $x^{(0)} = 0$, so that the Taylor series of $f_k(x)$ at $x = 0$ takes the form

$$f_k(x) = \sum_{l=1}^{m} a_{kl} x_l + \cdots,$$

or in vector notation

$$f(x) = \mathfrak{A}x + \cdots, \qquad \mathfrak{A} = (a_{kl}).$$

At first we shall neglect the higher order terms and, instead of (1), solve the linear system

(2) $$\dot{x} = \mathfrak{A}x.$$

From these solutions we will try to arrive at solutions to the nonlinear system (1), although this will not always be possible. It will be shown, however, that if (1) is a Hamiltonian system, then one can, in general, construct a periodic solution of (1) from a periodic solution of (2).

To solve (2), we use a known algebraic result about the normal form of a square matrix. The eigenvalues of $\mathfrak{A}$ are the m solutions $\lambda_1, \ldots, \lambda_m$ of the m-th degree characteristic equation

(3) $$|\lambda\mathfrak{E} - \mathfrak{A}| = 0,$$

with $\mathfrak{E}$ the identity matrix of m rows, and henceforth it will be assumed that these m roots are distinct from one another. According to the result mentioned, there exists an invertible complex matrix $\mathfrak{C}$ of m rows such that

(4) $$\mathfrak{C}^{-1}\mathfrak{A}\mathfrak{C} = \mathfrak{L} = [\lambda_1, \ldots, \lambda_m]$$

is the diagonal matrix with the entries $\lambda_1, \ldots, \lambda_m$. We still want to investigate the extent to which $\mathfrak{C}$ is determined by $\mathfrak{A}$. If $c^{(1)}, \ldots, c^{(m)}$ are the individual columns of $\mathfrak{C}$, the matrix equation $\mathfrak{A}\mathfrak{C} = \mathfrak{C}\mathfrak{L}$ is equivalent to the m vector equations

$$\mathfrak{A}c^{(k)} = c^{(k)}\lambda_k \quad (k = 1, \ldots, m),$$

hence also to

$$(5) \qquad (\lambda_k \mathfrak{E} - \mathfrak{A})\, c^{(k)} = 0 .$$

Because the λ_k are distinct, the matrix

$$\lambda_k \mathfrak{E} - \mathfrak{A} = \mathfrak{C}(\lambda_k \mathfrak{E} - \mathfrak{L})\, \mathfrak{C}^{-1}$$

has rank $m-1$, and consequently (5) determines $c^{(k)}$ up to an arbitrary scalar factor p_k. If, on the other hand,

$$\mathfrak{P} = [p_1, \dots, p_m], \qquad \mathfrak{B} = \mathfrak{C}\mathfrak{P},$$

with the $p_k \neq 0$ arbitrary, then also $|\mathfrak{B}| \neq 0$ and $\mathfrak{B}^{-1}\mathfrak{A}\mathfrak{B} = \mathfrak{L}$. Thus the $c^{(k)}$ are determined precisely up to nonzero scalar factors. This algebraic reasoning applies to any complex $\mathfrak{A}$.

If $\mathfrak{A}$ is real, the characteristic equation (3) has real coefficients and hence for each root λ_k also the complex conjugate $\bar{\lambda}_k$ appears as one of the roots λ_l. In this way, to each $k = 1, \dots, m$ there corresponds a unique $l = l_k$ in the range $1, \dots, m$ such that $\bar{\lambda}_k = \lambda_l$; in particular, $k = l$ if λ_k is real. The λ_k being distinct, this correspondence between k and l_k is one-to-one. Now, according to (5), we have

$$(\lambda_l \mathfrak{E} - \mathfrak{A})\, \overline{c^{(k)}} = 0 \qquad (l = l_k),$$

and consequently

$$(6) \qquad c^{(l)} = \overline{c^{(k)}}\, \varrho_k \qquad (l = l_k)$$

with ϱ_k a scalar. Similarly $c^{(k)} = \overline{c^{(l)}}\, \varrho_l$, and since $|\mathfrak{C}| \neq 0$ implies that $c^{(k)} \neq 0$, it follows that

$$(7) \qquad \bar{\varrho}_k \varrho_l = 1 .$$

By replacing $c^{(k)}$ with $c^{(k)} \varrho_l^{\frac{1}{2}}$ one may achieve the normalization $\varrho_k = 1$ and therefore have $c^{(l)} = \overline{c^{(k)}}$, a possibility that can also be inferred directly from (5). For subsequent application to Hamiltonian systems, however, it is more convenient not to assume as yet that $\varrho_k = 1$.

The linear substitution

$$x = \mathfrak{C} y, \qquad y = \mathfrak{C}^{-1} x$$

transforms (2) into the system

$$(8) \qquad \dot{y} = \mathfrak{L} y ,$$

whose complete solution is given by

$$y_k = \alpha_k e^{\lambda_k t} \qquad (k = 1, \dots, m),$$

where $y_1, \ldots, y_m$ are elements of the column y, and $\alpha_1, \ldots, \alpha_m$ are m integration constants. In particular, if λ_k is purely imaginary, then y_k is a periodic function of the real variable t. To discuss reality conditions, i.e. conditions under which the solutions are real, one has to consider the relation (6). For

$$x = \sum_{k=1}^{m} c^{(k)} y_k$$

to be real, it must be true that

$$x = \sum_{k=1}^{m} \overline{c^{(k)}}\, \overline{y_k},$$

and because $|\mathfrak{C}| \neq 0$, if $\mathfrak{A}$ is real we need $\overline{y}_k = \varrho_k y_{l_k}$. The integration constants must therefore be chosen so that $\overline{\alpha}_k = \varrho_k \alpha_{l_k}$.

Let us again consider the general nonlinear system

(9) $$\dot{x} = \mathfrak{A}x + \cdots$$

and introduce new variables y by a transformation

$$x_k = \phi_k(y) \qquad (k = 1, \ldots, m)$$

analytic in a neighborhood of $y_1 = 0, \ldots, y_m = 0$ and taking $y = 0$ into $x = 0$. If the corresponding Taylor series is

(10) $$x = \mathfrak{B}y + \cdots,$$

with $|\mathfrak{B}| \neq 0$, then in a neighborhood of $x = 0$ there is an inverse transformation $y = \mathfrak{B}^{-1}x + \cdots$, also analytic, and the system (9) becomes

$$\dot{y} = \mathfrak{B}^{-1}\mathfrak{A}\mathfrak{B}y + \cdots.$$

This shows that under analytic transformation of the differential equation (1) the eigenvalues $\lambda_1, \ldots, \lambda_m$ remain invariant.

We now specialize (1) to a Hamiltonian system

(11) $$\dot{u}_k = H_{v_k}, \quad \dot{v}_k = -H_{u_k} \qquad (k = 1, \ldots, n).$$

To express this in vector form, let w be the column with the $2n$ entries $u_1, \ldots, u_n, v_1, \ldots, v_n$ and H_w the column having the corresponding derivatives of H as entries. If $\mathfrak{E}$ is again the $n \times n$ identity matrix and

$$\mathfrak{J} = \begin{pmatrix} 0 & \mathfrak{E} \\ -\mathfrak{E} & 0 \end{pmatrix},$$

then (11) can be written in the abbreviated form

(12) $$\dot{w} = \mathfrak{J}H_w.$$

The Hamiltonian function $H = H(w)$ is assumed to be regular in a neighborhood of $w = 0$. Because the constant term of the Taylor expansion of H at $w = 0$ plays no role in the differential equation (11), it may as well be set equal to 0. Moreover, if $w = 0$ is to be an equilibrium solution of (11), then all the first derivatives of H must vanish there. The power series of H therefore begins with quadratic terms and can be expressed in the form

$$H = \tfrac{1}{2} w' \mathfrak{S} w + \cdots, \tag{13}$$

where $\mathfrak{S}$ is a symmetric matrix of $2n$ rows and w' is the transpose of w. Consequently

$$H_w = \mathfrak{S} w + \cdots,$$

so that (12) becomes

$$\dot{w} = \mathfrak{A} w + \cdots, \qquad \mathfrak{A} = \mathfrak{J}\mathfrak{S},$$

and the characteristic polynomial to be considered is $p(\lambda) = |\lambda\mathfrak{E} - \mathfrak{J}\mathfrak{S}|$.

Since $\mathfrak{J}' = \mathfrak{J}^{-1} = -\mathfrak{J}$, $|\mathfrak{J}| = 1$, $\mathfrak{S}' = \mathfrak{S}$, we have

$$(\lambda\mathfrak{E} - \mathfrak{J}\mathfrak{S})' = \lambda\mathfrak{E} + \mathfrak{S}\mathfrak{J} = \mathfrak{J}(-\lambda\mathfrak{E} - \mathfrak{J}\mathfrak{S})\mathfrak{J}$$

$$p(\lambda) = |\lambda\mathfrak{E} - \mathfrak{J}\mathfrak{S}| = |-\lambda\mathfrak{E} - \mathfrak{J}\mathfrak{S}| = p(-\lambda),$$

which shows that $p(\lambda)$ is an even function. Thus if λ is a root of $p(\lambda)$, so also is $-\lambda$, with both having the same multiplicity, while if 0 is a root its multiplicity must be even. If all the eigenvalues are again assumed to be simple, and therefore $\neq 0$, after a suitable reordering they can be denoted by $\lambda_k, \lambda_{k+n} = -\lambda_k$ $(k = 1, \ldots, n)$. Setting $\mathfrak{L}_0 = [\lambda_1, \ldots, \lambda_n]$, we can then find an invertible complex matrix $\mathfrak{C}$, of $2n$ rows, such that the normal form of $\mathfrak{J}\mathfrak{S}$ becomes

$$\mathfrak{C}^{-1}\mathfrak{J}\mathfrak{S}\mathfrak{C} = \mathfrak{L} = \begin{pmatrix} \mathfrak{L}_0 & 0 \\ 0 & -\mathfrak{L}_0 \end{pmatrix}, \tag{14}$$

whereby upon transposition we have

$$\mathfrak{C}'\mathfrak{S}\mathfrak{J} = -\mathfrak{L}\mathfrak{C}'. \tag{15}$$

On the other hand, since the matrix

$$-\mathfrak{L}\mathfrak{J}^{-1} = \begin{pmatrix} 0 & \mathfrak{L}_0 \\ \mathfrak{L}_0 & 0 \end{pmatrix}$$

is symmetric, it follows that

$$\mathfrak{L}\mathfrak{J}^{-1} = (\mathfrak{L}\mathfrak{J}^{-1})' = (\mathfrak{J}^{-1})'\,\mathfrak{L} = \mathfrak{J}\mathfrak{L}, \tag{16}$$

and together with (15) this gives

$$(\mathfrak{J}^{-1}\mathfrak{C}'\mathfrak{J})\,\mathfrak{J}\mathfrak{S} = \mathfrak{J}\mathfrak{C}'\mathfrak{S} = -\mathfrak{J}\mathfrak{L}\mathfrak{C}'\mathfrak{J}^{-1} = \mathfrak{L}(\mathfrak{J}^{-1}\mathfrak{C}'\mathfrak{J}).$$

Thus, setting $\mathfrak{B}=(\mathfrak{J}^{-1}\mathfrak{C}'\mathfrak{J})^{-1}$, we have $|\mathfrak{B}|\neq 0$ and

$$\mathfrak{B}^{-1}\mathfrak{J}\mathfrak{S}\mathfrak{B}=\mathfrak{L}.$$

Our earlier result about the extent to which $\mathfrak{C}$ is determined implies that $\mathfrak{C}=\mathfrak{B}\mathfrak{P}$, where $\mathfrak{P}$ is an invertible diagonal matrix which we will express in the form

$$\mathfrak{P}=\begin{pmatrix}\mathfrak{P}_1 & 0\\ 0 & \mathfrak{P}_2\end{pmatrix},$$

with $\mathfrak{P}_1$, $\mathfrak{P}_2$ two $n\times n$ diagonal matrices. It follows that

(17) $$\mathfrak{C}'\mathfrak{J}\mathfrak{C}=\mathfrak{J}\mathfrak{B}^{-1}\mathfrak{C}=\mathfrak{J}\mathfrak{P}=\begin{pmatrix}0 & \mathfrak{P}_2\\ -\mathfrak{P}_1 & 0\end{pmatrix},$$

and since $\mathfrak{J}$, and therefore also $\mathfrak{C}'\mathfrak{J}\mathfrak{C}$, is alternating, we have $\mathfrak{P}_1=\mathfrak{P}_2$. Consequently, if

$$\mathfrak{Q}=\begin{pmatrix}\mathfrak{P}_1 & 0\\ 0 & \mathfrak{E}\end{pmatrix},$$

then $\mathfrak{Q}'\mathfrak{J}\mathfrak{Q}=\mathfrak{J}\mathfrak{P}$. Finally, with $\mathfrak{C}\mathfrak{Q}^{-1}$ again denoted by $\mathfrak{C}$, equation (14) remains as is, while now

$$\mathfrak{C}'\mathfrak{J}\mathfrak{C}=\mathfrak{J},$$

and therefore $\mathfrak{C}$ is symplectic. Furthermore,

(18) $$\mathfrak{C}'\mathfrak{S}\mathfrak{C}=-(\mathfrak{C}'\mathfrak{J}\mathfrak{C})\mathfrak{C}^{-1}\mathfrak{J}\mathfrak{S}\mathfrak{C}=-\mathfrak{J}\mathfrak{L}=\begin{pmatrix}0 & \mathfrak{L}_0\\ \mathfrak{L}_0 & 0\end{pmatrix},$$

so that $\mathfrak{S}$ is brought into the above normal form, and indeed through a symplectic transformation.

Let z now be a column with the $2n$ entries $x_1,\dots,x_n$, $y_1,\dots,y_n$, and consider the linear substitution

(19) $$w=\mathfrak{C}z\,.$$

Because the Jacobian matrix $\mathfrak{C}$ is symplectic, (2; 20) implies that the transformation (19) is canonical. The Hamiltonian system (12) then becomes

(20) $$\dot{z}=\mathfrak{J}H_z\,,$$

with (13), (18) combining to give

(21) $$H=\tfrac{1}{2}z'\mathfrak{C}'\mathfrak{S}\mathfrak{C}z+\cdots=\sum_{k=1}^{n}\lambda_k x_k y_k+\cdots.$$

Thus the quadratic terms of H take the above normal form.

If the power series $H(w)$ has only real coefficients, the matrices $\mathfrak{S}$ and $\mathfrak{A}=\mathfrak{J}\mathfrak{S}$ are real, whereas to have also w real, z must satisfy the condition $\mathfrak{C}z=\overline{\mathfrak{C}}\bar{z}$. Now $\mathfrak{C}$ is not yet uniquely determined by equations (14), (15), for it may still be multiplied on the right by an arbitrary symplectic diagonal matrix

$$\mathfrak{R}=\begin{pmatrix}\mathfrak{R}_0 & 0\\ 0 & \mathfrak{R}_0^{-1}\end{pmatrix}, \qquad \mathfrak{R}_0=[r_1,\dots,r_n]$$

of $2n$ rows. By a suitable choice of $\mathfrak{R}$ the reality condition on z can be simplified as follows. First, suppose that an eigenvalue λ_k $(k\leqq n)$ is not purely imaginary and that $\lambda_l=\bar{\lambda}_k$ $(l=l_k)$. If $l>n$, it follows that $\lambda_k \neq -\lambda_l = \lambda_{l-n}$ and $\lambda_k \neq \lambda_l$, so that an interchange of λ_{l-n}, λ_l leads to the case $l\leqq n$. It may therefore be assumed at the outset that $l\leqq n$, and thus also that $\lambda_{l+n}=\bar{\lambda}_{k+n}$. On the other hand, for λ_k purely imaginary we have $\bar{\lambda}_k=-\lambda_k=\lambda_{k+n}$. With the columns of $\mathfrak{C}$ again denoted by $c^{(1)},\dots,c^{(2n)}$, the identity $\mathfrak{C}'\mathfrak{J}\mathfrak{C}=\mathfrak{J}$ gives

$$c^{(k)\prime}\mathfrak{J}c^{(k+n)}=1 \qquad (k=1,\dots,n).$$

If λ_k is not purely imaginary, (6) now implies that

$$1=c^{(l)\prime}\mathfrak{J}c^{(l+n)}=\varrho_k\overline{c^{(k)\prime}}\mathfrak{J}\overline{c^{(k+n)}}\varrho_{k+n}=\varrho_k\varrho_{k+n}, \tag{22}$$

while if λ_k is purely imaginary, then $l_k=k+n$ and

$$-1=-(c^{(k)\prime}\mathfrak{J}c^{(k+n)})'=c^{(k+n)\prime}\mathfrak{J}c^{(k)}=\varrho_k\overline{c^{(k)\prime}}\mathfrak{J}\overline{c^{(k+n)}}\varrho_{k+n}=\varrho_k\varrho_{k+n}. \tag{23}$$

Let $\mathfrak{C}\mathfrak{R}$ again be replaced by $\mathfrak{C}$, so that the original columns $c^{(k)}, c^{(k+n)}$ are now multiplied by the scalar factors r_k, r_k^{-1}. For λ_k not purely imaginary, (6) shows that ϱ_k then gets multiplied by the factor $r_l\bar{r}_k^{-1}$. If in addition λ_k is not real and $k<l_k$, the choice $r_k=\bar{\varrho}_k$, $r_l=1$ leads to the normalization $\varrho_k=1$, and by (7), (22) then also $\varrho_{k+n}=1$, $\varrho_l=1$, $\varrho_{l+n}=1$; whereas if λ_k is real, then $k=l_k$ and by (7) the absolute value of ϱ_k is 1, so that the same normalization is achieved by taking $r_k=\varrho_k^{-\frac{1}{2}}$. Finally, for λ_k purely imaginary, ϱ_k takes on the positive factor $(r_k\bar{r}_k)^{-1}$, and the identity $\bar{\varrho}_k=-\varrho_k$ coming from (7),(23) remains valid, showing that ϱ_k is purely imaginary. Consequently, in this case one can normalize $\varrho_k=\varrho_{k+n}=\pm i$, and since an interchange of λ_k and λ_{k+n} replaces $c^{(k)}, c^{(k+n)}$ by $c^{(k+n)}, -c^{(k)}$ with $\mathfrak{C}$ remaining symplectic and $-\varrho_{k+n}$ replacing ϱ_k, it is possible to have $\varrho_k=-i$. The above process thus leads to the factor $\varrho_k=-i$ for each purely imaginary eigenvalue λ_k, and $\varrho_k=1$ otherwise. The reality condition $\mathfrak{C}z=\overline{\mathfrak{C}}\bar{z}$, which is equivalent to $\bar{z}_k=\varrho_k z_{l_k}$, then becomes $x_l=\bar{x}_k$, $y_l=\bar{y}_k$ if $\lambda_l=\bar{\lambda}_k \neq -\lambda_k$ $(l=l_k;\ k=1,\dots,n)$ and $y_k=i\bar{x}_k$ if $\bar{\lambda}_k=-\lambda_k$.

If on the right side of (20) only the linear term is retained, one has the linear system $\dot{z} = \mathfrak{Q}z$, or, more exactly, the system

$$\dot{x}_k = \lambda_k x_k, \quad \dot{y}_k = -\lambda_k y_k \qquad (k = 1, \dots, n),$$

whose complete solution is given by

$$x_k = \xi_k e^{\lambda_k t}, \quad y_k = \eta_k e^{-\lambda_k t}$$

with ξ_k, η_k constants of integration. Should one of the eigenvalues, say λ_1, be purely imaginary, then

$$x_k = y_k = 0 \ (k = 2, \dots, n), \quad x_1 = \xi_1 e^{\lambda_1 t}, \quad y_1 = \eta_1 e^{-\lambda_1 t}$$

gives a periodic solution to the linear system, with the condition $\eta_1 = i\bar{\xi}_1$ making w real. Our aim in the next two sections will be to find near this solution a periodic solution to the general nonlinear Hamiltonian system (11).

Incidentally, if the system of differential equations (1) is not Hamiltonian, it may well happen that apart from the equilibrium solution there are no periodic solutions, even if the corresponding linear system does have nonconstant periodic solutions. This, for example, can be seen from the system

$$\text{(24)} \qquad \dot{x} = -y + x(x^2 + y^2)^g, \quad \dot{y} = x + y(x^2 + y^2)^g$$

for two unknown functions x and y, wherein g is a fixed natural number. The corresponding linear system $\dot{x} = -y, \dot{y} = x$ has only periodic solutions, namely, $x = \alpha \cos t + \beta \sin t$, $y = \alpha \sin t - \beta \cos t$ with α, β constants of integration. The corresponding orbits in the (x, y)-plane are concentric circles centered at the origin and traversed with constant angular velocity 1. On the other hand, in polar coordinates r, ϕ with $x = r\cos\phi$, $y = r\sin\phi$ the nonlinear system (24) becomes

$$r\dot{r} = x\dot{x} + y\dot{y} = (x^2 + y^2)^{g+1} = r^{2g+2}, \quad r^2\dot{\phi} = x\dot{y} - y\dot{x} = x^2 + y^2 = r^2,$$

and apart from the trivial solution $r = 0$, its solutions are given by

$$\dot{r} = r^{2g+1}, \quad r = (a - 2gt)^{-\frac{1}{2g}}, \quad \dot{\phi} = 1, \quad \phi = b + t$$

with two integration constants a, b. These solutions are spirals, and therefore certainly not periodic. Since the parameter $g = 1, 2, \dots$ is still free, this example shows also that the property of the system $\dot{x} = -y$, $\dot{y} = x$ having periodic solutions can be destroyed by adding suitable terms of arbitrarily high order to the right sides. The system in this example, however, is not canonical. In what follows we will investigate how to obtain a periodic solution to a Hamiltonian system, under the assumption that the corresponding linear system has such a solution and satisfies a certain additional condition.

§ 16. An Existence Theorem

Consider a Hamiltonian system $\dot{w}=\mathfrak{J}H_w$ satisfying the same conditions as in the previous section. Thus, $H=\frac{1}{2}w'\mathfrak{S}w+\cdots$ is a real power series that begins with quadratic terms and converges in some neighborhood of $w=0$, while the $2n$ eigenvalues $\lambda_1,\ldots,\lambda_n,-\lambda_1,\ldots,-\lambda_n$ of $\mathfrak{J}\mathfrak{S}$ are distinct. Our aim is to prove the following existence theorem:

Let λ_1 be purely imaginary, and assume that none of the $n-1$ quotients $\frac{\lambda_2}{\lambda_1},\ldots,\frac{\lambda_n}{\lambda_1}$ is an integer. Then there exists a family of real periodic solutions to the Hamiltonian system which depend analytically on one real parameter ϱ, with $\varrho=0$ corresponding to the equilibrium solution. The period $\tau(\varrho)$, likewise, is analytic in ϱ and, moreover, $\tau(0)=\frac{2\pi}{|\lambda_1|}$.

To prove this theorem we will assume the desired solutions in the form of power series with undetermined coefficients and work as in § 4 with formal power series, proving their convergence only in the next section. For now the variables are to be viewed as finitely many indeterminates $z_1,\ldots,z_m$ having no definite numerical values, and the coefficients as certain complex numbers. With equality, sum, and product defined as between convergent series, the formal power series form a ring $R(z)$ without zero divisors. If new variables $\zeta_1,\ldots,\zeta_q$ are introduced as power series in $z_1,\ldots,z_m$ without constant terms, each series in the new variables ζ can be rearranged to become a series in the old variables z, the process thus giving rise to an isomorphism between $R(\zeta)$ and a subring of $R(z)$. In addition, we define partial differentiation with respect to $z_1,\ldots,z_m$ in $R(z)$ by applying this operation termwise to each series, observing that for polynomials differentiation can be defined purely algebraically. The usual rules for differentiating sums and products then remain valid, as does the chain rule.

First, instead of a Hamiltonian system, let us consider again the general nonlinear system (15; 1), this time under the assumption that the matrix $\mathfrak{A}$ has two eigenvalues $\lambda_1,\lambda_2=-\lambda_1$ differing in sign only. We look for a particular solution where $x_1,\ldots,x_m$ are power series in two unknown functions $\xi=\xi(t),\eta=\eta(t)$, whereby (15; 1) takes the form

$$x_\xi\dot{\xi}+x_\eta\dot{\eta}=f(x)\,.$$

Under the additional assumption that the functions ξ,η satisfy the two differential equations

$$\dot{\xi}=\alpha\xi,\quad \dot{\eta}=\beta\eta \tag{1}$$

with α and β power series in ξ,η, our search for the particular solution is divided into two steps. Namely, suitably adjusting $\alpha(\xi,\eta),\beta(\xi,\eta)$, we first consider the linear partial differential equation

$$x_\xi \xi\alpha + x_\eta \eta\beta = f(x) \tag{2}$$

for $x(\xi, \eta)$, and after that integrate the system (1). One advantage to this approach is that (2) can be treated within the ring of formal power series, without initial concern about convergence. The linear substitution $x = \mathfrak{C}y$ transforms (2) into

$$y_\xi \xi\alpha + y_\eta \eta\beta - \mathfrak{L}y = g(y), \tag{3}$$

where the power series

$$g(y) = \mathfrak{C}^{-1} f(\mathfrak{C}y) - \mathfrak{L}y \tag{4}$$

begins with quadratic terms and, as before, $\mathfrak{L} = [\lambda_1, \ldots, \lambda_m]$. In order that the $m+2$ power series $y_k(\xi, \eta)$ $(k = 1, \ldots, m)$, $\alpha(\xi, \eta), \beta(\xi, \eta)$ be uniquely determined by comparison of coefficients in (3), we impose the following three conditions: The series $y_1 - \xi, y_2 - \eta$, and y_k $(k = 3, \ldots, m)$ shall all begin with quadratic terms; there shall be no term of the form $\xi(\xi\eta)^l$ in $y_1 - \xi$, and no term of the form $\eta(\xi\eta)^l$ in $y_2 - \eta$; the series for α and β shall be in powers of the product $\xi\eta = \omega$ only. The m eigenvalues $\lambda_1, \lambda_2 = -\lambda_1, \lambda_3, \ldots, \lambda_m$ have already been assumed to be distinct, and in addition we stipulate that none of the $m-2$ ratios $\dfrac{\lambda_3}{\lambda_1}, \ldots, \dfrac{\lambda_m}{\lambda_1}$ be an integer.

Inserting the $m+2$ power series y_k, α, β with undetermined coefficients into equation (3) and comparing the linear parts, we see that the constant terms in α, β must be $\lambda_1, -\lambda_1$. For s a natural number, suppose that by comparing the coefficients of order $\leq s$ in (3) we have already uniquely determined the terms up to order s in y and order $s-1$ in α, β. Comparing the coefficients in terms of the form $\xi^p \eta^q$ $(p+q = s+1)$ in (3), we observe that, since $g(y)$ begins with quadratic terms, for $g(y)$ the coefficient in question is a polynomial in the already known coefficients of $y_1, \ldots, y_m$. Let γ be the desired coefficient of $\xi^p \eta^q$ in y_k. This term contributes $\{\lambda_1(p-q) - \lambda_k\}\gamma$ to the corresponding coefficient on the left side of (3). If it is not the case that both $k = 1$ and $p = q+1$, nor that both $k = 2$ and $q = p+1$, then only a polynomial in the already known coefficients of y, α, β must still be added to the left side, and because the factor in front of γ is different from 0, the coefficient γ is uniquely determined. On the other hand, if either $k = 1, p = q+1$ or $k = 2, q = p+1$, then the second of our three conditions says that $\gamma = 0$. Now, however, one must still add to the left side the unknown coefficient of ω^q in α if $k = 1$, or the coefficient of ω^p in β if $k = 2$, while the other terms there are already known. Thus the inductive assumption holds also for $s+1$ in place of s, and because the assumption is valid for $s=1$, equation (3) can be uniquely solved under the above conditions.

Let us now assume that the series $f(x)$ has only real coefficients and, furthermore, let λ_1 be purely imaginary, so that $\lambda_2=\bar{\lambda}_1$. We wish to investigate the effect of this on the series $y(\xi,\eta), \alpha(\xi,\eta), \beta(\xi,\eta)$. Setting $\mathfrak{C}^{-1}\bar{\mathfrak{C}}=\mathfrak{T}$ and $\mathfrak{T}^{-1}y=y^*$, we observe that by (15; 6), (15; 7) the latter substitution has the simple form

$$y_l=\bar{\varrho}_l y_k^* \quad (l=l_k;\, k=1,\dots,m), \tag{5}$$

so that, in particular, $y_1=\bar{\varrho}_1 y_2^*, y_2=\bar{\varrho}_2 y_1^*, \bar{\varrho}_1\varrho_2=1$. Let us consider identity (3) transformed by complex conjugation of the coefficients, with the indeterminates ξ,η remaining fixed. Since

$$\bar{\mathfrak{C}}^{-1}\bar{f}(\bar{\mathfrak{C}}\bar{y})=\mathfrak{T}^{-1}\mathfrak{C}^{-1}f(\mathfrak{C}\mathfrak{T}\bar{y}),$$

where $\bar{y}=\bar{y}(\xi,\eta)$, equation (3) in context with (4) remains valid if y,α,β are replaced by $\mathfrak{T}\bar{y},\bar{\alpha},\bar{\beta}$. The first two components of the vector $\mathfrak{T}\bar{y}$ are $\bar{\varrho}_1\bar{y}_2=\bar{\varrho}_1\eta+\cdots, \bar{\varrho}_2\bar{y}_1=\bar{\varrho}_2\xi+\cdots$, while the other components again begin with quadratic terms. If ξ,η are now replaced by $\varrho_1\eta,\varrho_2\xi$, it is apparent that also $\mathfrak{T}\bar{y}(\varrho_1\eta,\varrho_2\xi), \bar{\beta}(\varrho_1\eta,\varrho_2\xi), \bar{\alpha}(\varrho_1\eta,\varrho_2\xi)$ represent a solution of (3) fulfilling our three conditions, and it follows from the established uniqueness that

$$\mathfrak{C}y(\xi,\eta)=\bar{\mathfrak{C}}\bar{y}(\varrho_1\eta,\varrho_2\xi), \quad \alpha(\xi,\eta)=\bar{\beta}(\varrho_1\eta,\varrho_2\xi).$$

Therefore, if $\bar{\xi}=\varrho_1\eta$, then in case of convergence the values $\alpha(\xi,\eta), \beta(\xi,\eta)$ are complex conjugates and $\mathfrak{C}y(\xi,\eta)=x(\xi,\eta)$ is real.

The case where the series $f(x)$ has only real coefficients and λ_1 is real, whereby $\bar{\lambda}_1=\lambda_1, \lambda_2=-\lambda_1=\bar{\lambda}_2$, can be treated analogously. In that case $l_1=1, l_2=2$ and normalization achieves that $\varrho_1=\varrho_2=1$, whereupon $y_1=y_1^*, y_2=y_2^*$. The first two components of $\mathfrak{T}\bar{y}$ then already have the form $\bar{y}_1=\xi+\cdots, \bar{y}_2=\eta+\cdots$, so that this time

$$\mathfrak{C}y(\xi,\eta)=\bar{\mathfrak{C}}\bar{y}(\xi,\eta), \quad \alpha(\xi,\eta)=\bar{\alpha}(\xi,\eta), \quad \beta(\xi,\eta)=\bar{\beta}(\xi,\eta),$$

and, in case of convergence, for ξ,η real also the values $x(\xi,\eta), \alpha(\xi,\eta), \beta(\xi,\eta)$ are real.

Next we again neglect the reality requirement on $f(x)$, and consider the special case of a Hamiltonian system. It will be shown that then $\alpha(\xi,\eta)=-\beta(\xi,\eta)$. Accordingly, we alter the above notation to have the $2n$ variables $z_k=x_k, z_{k+n}=y_k$ $(k=1,\dots,n)$ enter in place of the m variables $y_1,\dots,y_m$, with x_1,y_1 appearing in place of y_1,y_2 and correspondingly $\lambda_{n+1}=-\lambda_1$ replacing $\lambda_2=-\lambda_1$. The Hamiltonian function H is considered with the linear canonical transformation already carried out, so that the quadratic terms have the normal form given in (15; 21). Equations (3), (4) are now replaced by

$$z_\xi \xi \alpha + z_\eta \eta \beta = \mathfrak{J} H_z ,$$

and from this the formally constructed differential $dH = H_\xi d\xi + H_\eta d\eta$ becomes

$$dH = H_z' dz = (\alpha \xi z_\xi' + \beta \eta z_\eta') \mathfrak{J} (z_\xi d\xi + z_\eta d\eta) = (\alpha \xi d\eta - \beta \eta d\xi) z_\xi' \mathfrak{J} z_\eta .$$

Abbreviating $z_\xi' \mathfrak{J} z_\eta = \Delta$, we then have

$$H_\xi = -\beta \eta \Delta, \quad H_\eta = \alpha \xi \Delta$$

(6) $$\alpha \xi H_\xi + \beta \eta H_\eta = 0 ,$$

whereby H is now viewed as a power series in ξ, η.

With the aid of (6) we will show that H is a series in $\omega = \xi \eta$ alone. Assume that it has already been shown that the terms of order $\leqq s$ in H form a polynomial in ω. Since

$$H = \sum_{k=1}^{n} \lambda_k x_k y_k + \cdots = \lambda_1 \omega + \cdots ,$$

this is true for $s = 2$. Let $\gamma \xi^p \eta^q$ $(p + q = s + 1)$ be a term of order $s + 1$. Because $\alpha = \lambda_1 + \cdots$, $\beta = -\lambda_1 + \cdots$, in the identity

$$\lambda_1 (\eta H_\eta - \xi H_\xi) = (\alpha - \lambda_1) \xi H_\xi + (\beta + \lambda_1) \eta H_\eta$$

the factors $\alpha - \lambda_1, \beta + \lambda_1$ are power series in ω with no constant term, and therefore, under the induction hypothesis, the terms of order $\leqq s + 2$ on the right form a polynomial in ω alone, while on the left the coefficient of $\xi^p \eta^q$ is $\lambda_1 (q - p) \gamma$. Therefore $\gamma = 0$ when $p \neq q$, and the assertion is valid for $s + 1$ in place of s.

Now that H depends on ω only, we have

$$H_\xi = \eta H_\omega, \quad H_\eta = \xi H_\omega ,$$

and (6) becomes

$$(\alpha + \beta) \omega H_\omega = 0 .$$

Since the series $\omega H_\omega = \lambda_1 \omega + \cdots$ is not identically 0, if follows that

(7) $$\alpha + \beta = 0 .$$

We return once more to the general case (2) and this time assume that our solution also fulfills (7). It will be shown in the next section that in that case, if $f(x)$ is convergent, also the series $y(\xi, \eta)$, α, β converge for complex ξ, η sufficiently small in absolute value. The differential equations (1) thereby acquire meaning, while (7) now says that

$$\dot{\omega} = \dot{\xi}\eta + \xi\dot{\eta} = (\alpha + \beta)\xi\eta = 0\,,$$

so that ω, α, β are independent of t. Consequently

$$\xi = \xi_0 e^{\alpha t}, \quad \eta = \eta_0 e^{\beta t}\,. \tag{8}$$

If again the series $f(x)$ has only real coefficients and λ_1 is purely imaginary, the initial values ξ_0, η_0 of ξ, η at $t=0$ are chosen so that $\bar{\xi}_0 = \varrho_1 \eta_0$ and $|\xi_0|$ is sufficiently small. By our earlier result the numbers $\alpha = \alpha(\xi, \eta) = \alpha(\xi_0, \eta_0)$ and $\beta = \beta(\xi, \eta) = \beta(\xi_0, \eta_0)$ are then complex conjugates, and therefore, by (7), purely imaginary. Consequently, in view of (8), we have $\bar{\xi} = \varrho_1 \eta$ for all real t, while $|\xi| = |\xi_0|$, so that, by the result mentioned, $x(\xi, \eta)$ is real. Thus (8) gives a family of real periodic solutions of the system (15;1) that depend on the complex parameter ξ_0. Since the right sides of this system of differential equations are independent of t, each solution curve is transformed into itself if t is replaced by $t+c$ for an arbitrary real c, and therefore it is sufficient to choose $\xi_0 = \varrho \geqq 0$. The value of the period is $\tau(\varrho) = \dfrac{2\pi}{|\alpha|}$, with $\alpha = \lambda_1 + \cdots$, so that $\tau(0) = \dfrac{2\pi}{|\lambda_1|}$. In the initial formulation we had $y_1 = \xi + \cdots$, $y_2 = \eta + \cdots$, which shows that, for $\varrho > 0$ sufficiently small, y_1 and y_2 do in fact depend on t, with $\tau(\varrho)$ being the primitive period. In view of (8), the power series obtained can be written as Fourier series in multiples of the angle $|\alpha| t$. Once we carry out the still missing convergence proof in the next section, we will, in particular, have the assertions of the existence theorem.

On the other hand, should $f(x)$ have real coefficients and λ_1 be real, so that also $\lambda_2 = -\lambda_1$ is real, we choose the initial values ξ_0, η_0 real with $|\xi_0|, |\eta_0|$ sufficiently small. In that case the numbers $\alpha = \alpha(\xi, \eta) = \alpha(\xi_0, \eta_0)$ and $\beta = -\alpha$ are real, and by (8) we have ξ, η real for all real t. Thus, for $|t|$ sufficiently small, the series $x(\xi, \eta)$ converges and is real. In the (ξ, η)-plane (8) then defines a family of equilateral hyperbolas that depend on the real parameter $\xi_0 \eta_0$, or, to be more exact, if the signs of ξ_0, η_0 are taken into account, four families of hyperbola branches. This gives four one-parameter families of real solution curves for the system (15; 1). If, say, $\alpha > 0$ and $\beta = -\alpha < 0$, then $e^{\alpha t}$ or $e^{\beta t}$ goes to ∞ as $t \to \infty$ or $t \to -\infty$ respectively. Thus, according to (8), for $\xi_0 \eta_0 \neq 0$ the point ξ, η remains in a given bounded neighborhood of the origin for a finite time interval only, while it moves into the origin along the η-axis if $\xi_0 = 0$, $\eta_0 \neq 0$, $t \to \infty$, or along the ξ-axis if $\xi_0 \neq 0$, $\eta_0 = 0$, $t \to -\infty$. The two possible signs for η_0 or ξ_0 in the latter case then give four solutions $x(t)$ that asymptotically approach the equilibrium solution as $t \to \infty$ or $t \to -\infty$ respectively. In contrast, for $\xi_0 \eta_0 \neq 0$ our solutions $x(t)$ have the property that they remain in a sufficiently small neighborhood of $x=0$ only for a bounded time interval, being, so to speak, first captured and then cast out. This, however, does not mean that there are no periodic solutions among these, but only that their periodicity can no longer be determined by local analysis. With $e^{\alpha t}=q$, the $x_k(t)$ $(k=1,\ldots,m)$ are Laurent series in the variable q and have the purely imaginary period $\frac{2\pi i}{\alpha}$ in t. To examine this result more closely, we introduce for comparison a system constructed analogously to (15; 24); namely,

$$\dot{x}=x+x(xy)^g, \quad \dot{y}=-y+y(xy)^g$$

with g again a given natural number. Then

$$(xy)^{\cdot}=2(xy)^{g+1}, \quad y\dot{x}-x\dot{y}=2xy.$$

For $xy \neq 0$, it follows that

$$xy=(a-2gt)^{-\frac{1}{g}}, \quad x=bye^{2t}$$

with $a, b \neq 0$ two integration constants, and therefore

$$x=b^{\frac{1}{2}}e^{t}(a-2gt)^{-\frac{1}{2g}}, \quad y=b^{-\frac{1}{2}}e^{-t}(a-2gt)^{-\frac{1}{2g}}; \tag{9}$$

while for $xy=0$, there are the special solutions $x=0, y=ce^{-t}$ and $x=ce^{t}, y=0$ with c a constant. Here $\lambda_1=1, \lambda_2=-1$, yet no solution of the two-parameter family (9) has a complex period. Thus, also for λ_1 real we have exhibited an analytic property of solutions to a Hamiltonian system that need not appear for (15; 1) in general.

For a recursive determination of the coefficients of y, α, β by comparison of coefficients, it was essential that none of the $m-2$ quotients $\frac{\lambda_3}{\lambda_1}, \ldots, \frac{\lambda_m}{\lambda_1}$ be an integer, which in the Hamiltonian case is equivalent to none of the $n-1$ quotients $\frac{\lambda_2}{\lambda_1}, \ldots, \frac{\lambda_n}{\lambda_1}$ being an integer. We wish to show by an example that in general this assumption is necessary for the validity of the existence theorem. With the cubic polynomial

$$H=\tfrac{1}{2}(x_1^2+y_1^2)-x_2^2-y_2^2+x_1y_1x_2+\tfrac{1}{2}(x_1^2-y_1^2)\,y_2$$

as our Hamiltonian function, the corresponding system becomes

$$\begin{cases} \dot{x}_1=y_1+x_1x_2-y_1y_2, & \dot{y}_1=-x_1-y_1x_2-x_1y_2, \\ \dot{x}_2=-2y_2+\tfrac{1}{2}(x_1^2-y_1^2), & \dot{y}_2=2x_2-x_1y_1 \end{cases} \tag{10}$$

with the obvious equilibrium solution $x_1=x_2=y_1=y_2=0$ and corresponding eigenvalues $\lambda_1=-\lambda_3=i$, $\lambda_2=-\lambda_4=2i$. If λ_2 is taken as the purely imaginary eigenvalue used in the existence theorem, then the assumption that $\frac{\lambda_1}{\lambda_2} = \frac{1}{2}$ not be an integer is valid, implying the existence of a one-parameter family of periodic solutions with approximate period $\frac{2\pi i}{\lambda_2} = \pi$. These solutions can be readily exhibited. Indeed, the uniqueness theorem for differential equations implies that for initial values $x_1 = y_1 = 0$ the general solution is

$$x_1 = 0, \quad y_1 = 0, \quad x_2 = \alpha \cos 2t - \beta \sin 2t, \quad y_2 = \alpha \sin 2t + \beta \cos 2t$$

with α, β constant, i.e. a circle in the (x_2, y_2)-plane traversed in time π, with no restriction on the radius. These are precisely the solutions given by the existence theorem, and here the period is equal to π exactly, rather than only in the first approximation. On the other hand, if λ_1 is taken as the eigenvalue in the existence theorem, then, as we will show, there is no periodic solution with approximate period $\frac{2\pi i}{\lambda_1} = 2\pi$, apart from the trivial equilibrium solution. Because in this case only one condition from the hypothesis of the existence theorem is violated, namely, that $\frac{\lambda_2}{\lambda_1}$ not be an integer, whereas here we have $\frac{\lambda_2}{\lambda_1} = 2$, we see that the condition is essential. As we have just seen, all solutions with initial values $x_1 = y_1 = 0$ have period π, and we now show that there are no other periodic solutions. By the uniqueness theorem for differential equations, the quantity $p = x_1^2 + y_1^2$ remains positive throughout for the other solutions. With $q = x_2^2 + y_2^2$, a simple calculation, aided by a suitable combination of the terms in complex form, leads from (10) to the differential equation

$$\ddot{p} = 4pq + p^2 .$$

Because $p^2 > 0$, $4pq \geqq 0$, it follows that p is a strictly convex function of t and therefore certainly not periodic. Thus, in fact, apart from the given circular orbits, there are no additional periodic solutions to this system.

§ 17. The Convergence Proof

Using the method of majorants already employed in § 4, we now prove convergence of the formal power series constructed in the previous section. For this it will be assumed that the solution to (16;3) satisfies also the relation $\alpha + \beta = 0$, which according to (16;7) is always the case for a Hamiltonian system.

For $h=h(\xi,\eta)$ a power series in ξ,η, the coefficient of $\xi^p\eta^q$ in $h(\xi,\eta)$ will be denoted by $\{h\}_{pq}$. According to (16;3), with the constructed power series inserted for y and α, we have

$$\{(y_\xi\xi-y_\eta\eta)\,\alpha-\mathfrak{L}y\}_{pq}=\{g(y)\}_{pq},$$

or more exactly, because

$$\alpha=\lambda_1+\sum_{r=1}^{\infty}\{\alpha\}_{rr}(\xi\eta)^r$$

is a power series in $\xi\eta$ alone,

(1) $$((p-q)\,\lambda_1-\lambda_k)\,\{y_k\}_{pq}+\sum_{r=1}^{\infty}(p-q)\,\{\alpha\}_{rr}\{y_k\}_{p-r,q-r}=\{g_k(y)\}_{pq}$$

for $k=1,\ldots,m$, the sum here really breaking off with the term $r=\min(p,q)$. First, suppose that not simultaneously $k=1, p=q+1$ nor $k=2, q=p+1$. Then $(p-q)\,\lambda_1-\lambda_k\neq 0$, and there is a number $c_1>0$ independent of k,p,q such that

$$\left|\frac{1}{(p-q)\,\lambda_1-\lambda_k}\right|<c_1\,,\qquad\left|\frac{p-q}{(p-q)\,\lambda_1-\lambda_k}\right|<c_1\,,$$

whereupon (1) implies that

(2) $$|\{y_k\}_{pq}|\leqq c_1|\{g_k(y)\}_{pq}|+c_1\sum_{r=1}^{\infty}|\{\alpha\}_{rr}\{y_k\}_{p-r,q-r}|\,.$$

In the remaining cases, when $k=1, p=q+1$ or $k=2, q=p+1$, we always have $\{y_k\}_{pq}=0$ if $p+q>1$, while $\{y_1\}_{10}=\{y_2\}_{01}=1$, so that (1) gives

(3) $$\begin{cases}|\{\alpha\}_{qq}|=|\{g_1(y)\}_{pq}| & (p=q+1>1)\,,\\ |\{\alpha\}_{pp}|=|\{g_2(y)\}_{pq}| & (q=p+1>1)\,.\end{cases}$$

For h again a power series in ξ,η, let

$$\boxed{h}=\sum_{p,q}|\{h\}_{pq}|\,\xi^p\eta^q\,.$$

Thus $\boxed{h}$ is obtained from h by replacing all coefficients by their absolute values. Moreover, let $y_1^*=y_1-\xi$, $y_2^*=y_2-\eta$, $y_k^*=y_k$ $(k=3,\ldots,m)$, and $\alpha^*=\alpha-\lambda_1$. Multiplication of (2), (3) by $\xi^p\eta^q$ and summation over k,p,q then leads to the majorant relation

(4) $$(\xi+\eta)\,\boxed{\alpha^*}+\sum_{k=1}^{m}\boxed{y_k^*}\prec c_1\left(\sum_{k=1}^{m}\boxed{g_k(y)}+\boxed{\alpha^*}\sum_{k=1}^{m}\boxed{y_k^*}\right),$$

which no longer contains any derivatives.

We require an estimate for $\overline{|g_k(y)|}$, and will denote by $c_2, \ldots, c_8$ suitable positive constants. The m functions $f_k(x)$ $(k=1, \ldots, m)$ are by assumption regular in a neighborhood of $x=0$, and therefore the functions $g_k(y)$ defined by (16;4) are regular at $y=0$. If $g_k(y)$ is regular for $|y_l| \leqq c_2$ $(l=1, \ldots, m)$ and in absolute value $\leqq c_3$, then Cauchy's estimate gives the relation

$$g_k(y) \prec c_3 \prod_{l=1}^{m} \left(1 - \frac{y_l}{c_2}\right)^{-1},$$

with $y_1, \ldots, y_m$ treated here as independent variables. Moreover, abbreviating $s = y_1 + \cdots + y_m$, we have

$$\prod_{l=1}^{m} \left(1 - \frac{y_l}{c_2}\right)^{-1} \prec \left(1 - \frac{s}{c_2}\right)^{-m} \prec \frac{c_4}{1 - c_5 s},$$

and because $g_k(y)$ begins with quadratic terms, it follows that

$$g_k(y) \prec \frac{c_6 s^2}{1 - c_5 s} \qquad (k=1, \ldots, m). \tag{5}$$

On the other hand, with

$$\sum_{k=1}^{m} \overline{|y_k^*|} = S, \tag{6}$$

certainly

$$\overline{|s|} \prec \xi + \eta + S,$$

and hence (4), (5), (6) imply that

$$(\xi + \eta) \overline{|\alpha^*|} + S \prec c_1 \left(\frac{c_6(\xi + \eta + S)^2}{1 - c_5(\xi + \eta + S)} + \overline{|\alpha^*|}\, S \right). \tag{7}$$

In view of (6) it is sufficient to prove convergence of S and $\overline{|\alpha^*|}$ in a neighborhood of $\xi = 0$, $\eta = 0$, and since here all the coefficients are $\geqq 0$, it is enough to consider the case $\xi = \eta$. Because S also begins with quadratic terms in ξ, η, the expression

$$2\overline{|\alpha^*|} + \xi^{-1} S = U$$

is now a power series in ξ with nonnegative coefficients and having no constant term, while (7) implies that

$$U \prec c_7 \left(\frac{\xi(1+U)^2}{1 - 2c_5 \xi(1+U)} + U^2 \right).$$

For

$$\xi + U + \xi U = V,$$

so that

$$2\xi U + 2\xi U^2 + U^2 \prec V^2 ,$$

this gives

$$V \prec \xi + U + V^2 \prec \xi + V^2 + c_7 \left(\frac{\xi + V^2}{1 - 2c_5 V} + V^2 \right)$$

(8)
$$V \prec c \frac{2\xi + V^2}{4 - cV}, \qquad c = c_8 ,$$

and it is enough to prove that V converges for sufficiently small positive ξ. Instead of (8), let us consider the equation

(9)
$$W = c \frac{2\xi + W^2}{4 - cW}$$

for an unknown power series

$$W = W(\xi) = \sum_{l=1}^{\infty} \gamma_l \xi^l .$$

If the right side of (9) is expanded in powers of W and the series $W(\xi)$ then inserted, comparison of coefficients leads uniquely to a recursive determination of the coefficients γ_l, with the recursion formulas showing that, in view of (8), the series W majorizes V. On the other hand, by (9) we have

$$cW^2 - 2W + c\xi = 0, \qquad (1 - cW)^2 = 1 - c^2 \xi ,$$

so that

$$2cW \prec (1 - cW)^{-2} - 1 = \frac{c^2 \xi}{1 - c^2 \xi} ,$$

which proves convergence of $W(\xi)$ for $|\xi| < c^{-2}$. To make the computations as short as possible, we chose to forgo determining c explicitly. However, by making a few of the estimates more precise one can obtain also a practical result by the above method.

§ 18. An Application to the Solutions of Lagrange

The existence theorem formulated in § 16 will now be applied to the planar three-body problem to prove existence of periodic solutions near the circular orbits of Lagrange. We again use the notation of § 14, whereby the three particles in the given plane have coordinates q_{2k-1}, q_{2k} $(k = 1, 2, 3)$ and, with $E = T - U$ defined by (14;1), the equations of motion in Hamiltonian form read

$$\dot{q}_k = E_{p_k}, \quad \dot{p}_k = -E_{q_k} \quad (k = 1, \ldots, 6) .$$

The canonical transformation (14;3), (14;4) corresponded to a rotation of the coordinate system with constant angular velocity ω in the plane, and the transformed differential equations were

$$(1)\qquad \dot{x}_k = F_{y_k}, \quad \dot{y}_k = -F_{x_k} \quad (k=1,\ldots,6)$$

with

$$(2)\quad F = E + \omega Q = T - U + \omega Q, \quad Q = \sum_{k=1}^{3} (x_{2k}y_{2k-1} - x_{2k-1}y_{2k}),$$

$$(3)\qquad T = \tfrac{1}{2}\sum_{k=1}^{3} m_k^{-1}(y_{2k-1}^2 + y_{2k}^2).$$

Correspondingly, for the equilibrium solutions we had the equilateral and collinear solutions of Lagrange, with ω chosen in accordance with the respective conditions (14;7) and (14;10). There is clearly no loss of generality in assuming from now on that $\omega = 1$.

For the application of the existence theorem in § 16 to the system (1), we must expand the function $F = F(x, y)$ into a power series in a neighborhood of the corresponding equilibrium solution $x_k = x_{k0}$, $y_k = y_{k0}$ $(k=1,\ldots,6)$ and compute the quadratic terms. With $z_k = x_k - x_{k0}$, $z_{k+6} = y_k - y_{k0}$, the Taylor expansion has the form

$$F(x, y) = F(x_0, y_0) + \tfrac{1}{2}\sum_{k,l=1}^{12} s_{kl} z_k z_l + \cdots,$$

wherein the matrix $\mathfrak{S} = (s_{kl})$ is chosen symmetric. According to § 15 the corresponding eigenvalues λ_k are determined by the twelfth degree equation $|\lambda\mathfrak{E} - \mathfrak{J}\mathfrak{S}| = 0$, which can also be expressed as $|\lambda\mathfrak{J} + \mathfrak{S}| = 0$. Subsequently we will actually carry out the computation of this determinant, but for the moment let us anticipate the result. In the equilateral case

$$(4)\qquad |\lambda\mathfrak{J} + \mathfrak{S}| = \lambda^2(\lambda^2+1)^3(\lambda^4+\lambda^2+\gamma),$$

where

$$(5)\qquad \gamma = \frac{27}{4}\,\frac{m_1m_2 + m_2m_3 + m_1m_3}{(m_1+m_2+m_3)^2},$$

while in the collinear case, with P_2 between P_1 and P_3, we have

$$(6)\qquad |\lambda\mathfrak{J} + \mathfrak{S}| = \lambda^2(\lambda^2+1)^3(\lambda^4 + (1-\alpha)\lambda^2 - \alpha(2\alpha+3)),$$

where

$$(7)\qquad \alpha = \frac{m_1(1+\varrho^{-1}+\varrho^{-2}) + m_3(1+\sigma^{-1}+\sigma^{-2})}{m_1 + m_2(\varrho^{-2}+\sigma^{-2}) + m_3},$$

with $\sigma=1-\varrho$, and ϱ again the solution to (14;10). Hence in both cases $\lambda=0$ enters as a double eigenvalue and $\lambda=\pm i$ as a triple eigenvalue, whereas in the existence theorem the λ_k were all assumed to be simple.

The root $\pm i$ could have been expected at the outset without any calculation. Namely, the original coordinates q in the stationary coordinate system for the related equilibrium solution obviously have period 2π with respect to time. Moreover, if the coordinates q_{2k-1}, q_{2k} are replaced by $q_{2k-1}+a$, $q_{2k}+b$ $(k=1,2,3)$ with a and b arbitrary linear functions of t, the equations of motion remain satisfied, so that by virtue of (14;3), (14;4) in this way each equilibrium solution x_{k0}, y_{k0} gives rise to the solution

$$(8)\qquad \begin{cases} x_{2k-1}=x_{2k-1,0}+ac+bs\,, & y_{2k-1}=y_{2k-1,0}+\dot{a}c+\dot{b}s \\ x_{2k}=x_{2k,0}-as+bc\,, & y_{2k}=y_{2k,0}-\dot{a}s+\dot{b}c \\ & (k=1,2,3)\,. \end{cases}$$

If a and b are chosen constant, this solution has period 2π and is trivially seen from (8) to have a power series expansion in $e^{\lambda t}$ and $e^{-\lambda t}$ with $\lambda=i$. On the other hand, the existence theorem in § 16 leads directly to a series expansion of the solutions in powers of $e^{\alpha t}$ and $e^{-\alpha t}$, with $\alpha=\lambda+\cdots$ and λ a purely imaginary eigenvalue. Thus the solution (8) makes the appearance of the eigenvalue $\pm i$ at least plausible. Furthermore, the fact that $\pm i$ is a multiple eigenvalue can be correlated to our being able to take any linear function of t for a and b in (8). This observation, however, cannot be further motivated here, since the theory in § 16 was carried out for simple eigenvalues only. There is an intrinsic reason also for the appearance of the eigenvalue $\lambda=0$, and we will later show that the angular momentum integral is responsible for this.

To be able to apply the existence theorem from § 16, we must try to get rid of the multiple eigenvalues, and this can be achieved by reducing the Hamiltonian system (1) with the aid of the center of mass integrals and the angular momentum integral. First, in analogy to (7;4), (7;5), we introduce relative coordinates of P_1 and P_2 with respect to P_3 by means of the linear substitution

$$(9)\qquad \begin{cases} \xi_{2k-1}=x_{2k-1}-x_5\,, \quad \xi_{2k}=x_{2k}-x_6 & (k=1,2)\,, \\ \xi_5=x_5\,, \quad \xi_6=x_6\,, \quad \eta_k=y_k & (k=1,\ldots,4)\,, \\ \eta_5=y_1+y_3+y_5\,, \quad \eta_6=y_2+y_4+y_6\,, \end{cases}$$

whereupon the new equations of motion become

$$(10)\qquad \dot{\xi}_k=F_{\eta_k}\,, \quad \dot{\eta}_k=-F_{\xi_k} \quad (k=1,\ldots,6)\,.$$

Because U depends only on the differences between the original coordinates, the new variables ξ_5, ξ_6 do not appear in U, while from (2), (3) one finds that $F = T - U + Q$ with

$$(11) \qquad Q = \sum_{k=1}^{3} (\xi_{2k}\eta_{2k-1} - \xi_{2k-1}\eta_{2k}),$$

$$(12) \quad T = \tfrac{1}{2}m_3^{-1}\{(\eta_5 - \eta_3 - \eta_1)^2 + (\eta_6 - \eta_4 - \eta_2)^2\} + \tfrac{1}{2}\sum_{k=1}^{2} m_k^{-1}(\eta_{2k-1}^2 + \eta_{2k}^2).$$

In particular, since $Q_{\xi_6} = \eta_5$, $Q_{\xi_5} = -\eta_6$, it follows from (10) that

$$(13) \qquad \dot{\eta}_5 = \eta_6, \quad \dot{\eta}_6 = -\eta_5,$$

which governs the motion of the center of mass in the rotating coordinate system. If in the original coordinate system the center of mass is now assumed to be at rest, and this is indeed the case for Lagrange's solution, then $\eta_5 = 0$, $\eta_6 = 0$ and solving (10) simplifies to integrating the reduced Hamiltonian system

$$(14) \qquad \dot{\xi}_k = F_{\eta_k}, \quad \dot{\eta}_k = -F_{\xi_k} \quad (k = 1, \ldots, 4),$$

with ξ_5, ξ_6 subsequently determined from

$$(15) \quad \dot{\xi}_5 = F_{\eta_5} = \xi_6 - m_3^{-1}(\eta_3 + \eta_1), \; \dot{\xi}_6 = F_{\eta_6} = -\xi_5 - m_3^{-1}(\eta_4 + \eta_2)$$

by quadrature. It now follows from (13), (15) without further calculation that the numbers $i, -i$ appear as simple eigenvalues for the equilibrium solution of the reduced system (14), and therefore after this reduction the factor $(\lambda^2 + 1)^3$ in the characteristic polynomials (4) and (6) is replaced by the first power of $\lambda^2 + 1$. The troublesome factor λ^2, however, still remains, but it will be removed next with the aid of the angular momentum integral, which now, because $\eta_5 = \eta_6 = 0$, reduces to

$$(16) \qquad Q = \sum_{k=1}^{2} (\xi_{2k}\eta_{2k-1} - \xi_{2k-1}\eta_{2k}).$$

First we wish to show that the double root $\lambda = 0$ has its origin in the existence of the time-independent integral Q. Consider again the general system

$$(17) \qquad \dot{x}_k = f_k(x) \quad (k = 1, \ldots, m)$$

with $x = 0$ an equilibrium solution and the $f_k(x)$ regular at $x = 0$, so that we have a series expansion $f(x) = \mathfrak{A}x + \cdots$. Furthermore, assume that there exists an integral $\psi(x)$ for (17) that is analytic at $x = 0$ and does not depend explicitly on t. If $\psi(x) = \psi(0) + cx + \cdots$ is the series for $\psi(x)$, wherein c denotes a row vector, comparison of coefficients in the linear terms of the partial differential equation

$$\sum_{k=1}^{m} \psi_{x_k} f_k(x) = 0$$

shows that $c\mathfrak{A} = 0$. If $c \neq 0$, it follows that $|\mathfrak{A}| = 0$ and the characteristic equation $|\lambda\mathfrak{E} - \mathfrak{A}| = 0$ has $\lambda = 0$ as a root. Now from the expressions (2), (11), (16) for Q it is easily seen that the first order partial derivatives do not all vanish at the equilibrium solution, so that the condition $c \neq 0$ is fulfilled. This explains the appearance of the factor λ in (4) and (6), and because $|\lambda\mathfrak{J} + \mathfrak{S}|$ is an even function of λ, actually the factor λ^2 must appear. It should be added that in the rotating coordinate system the center of mass integrals are no longer independent of t, and therefore cannot take the place of Q in this reasoning.

To remove the factor λ^2 we will attempt a further reduction of the Hamiltonian system of equations of motion with the aid of the angular momentum integral. For this we will find a canonical transformation that introduces Q as a new variable – an idea carried out already by Jacobi even for the three-body problem in space, where the process is known as elimination of the node. The substitution to be used played an important role already in § 13. To explain the idea, we consider an arbitrary Hamiltonian system $\dot{x}_k = H_{y_k}$, $\dot{y}_k = -H_{x_k}$ with unknown functions x_k, y_k $(k = 1, \ldots, n)$, and assume that it possesses an integral $\psi(x, y)$ independent of t. We now pass from x, y to new variables ξ, η using a canonical transformation generated, as in (3; 4), by $w = w(\xi, y)$ via

$$\eta_k = w_{\xi_k}, \quad x_k = w_{y_k} \quad (k = 1, \ldots, n), \quad |w_{\xi_k y_l}| \neq 0. \tag{18}$$

In the process we wish to have $\eta_n = \psi(x, y)$. This leads to the partial differential equation

$$w_{\xi_n} = \psi(w_y, y) \tag{19}$$

for the generating function w. Suppose now that we have a solution to (19) which satisfies the condition $|w_{\xi_k y_l}| \neq 0$. If the columns with the respective variables x, y and ξ, η are denoted by z and ζ, and the Jacobian matrix ζ_z by $\mathfrak{M}$, then $\mathfrak{M}$ is symplectic, so that $\mathfrak{M}\mathfrak{J}\mathfrak{M}' = \mathfrak{J}$, while $H_z = \mathfrak{M}' H_\zeta$, $\psi_z = \mathfrak{M}' \psi_\zeta$. Hence the expression

$$\sum_{k=1}^{n} (\psi_{x_k} H_{y_k} - \psi_{y_k} H_{x_k}) = \psi_z' \mathfrak{J} H_z \tag{20}$$

remains invariant under the transformation from z to ζ; on the other hand, $\psi(x, y)$ being an integral, this expression is identically 0. By (18), (19), however, $\psi = \eta_n$ and therefore $\psi_{\eta_n} = 1$, while the other partial derivatives of ψ as a function of ζ are all 0. It thus follows from (20) that $H_{\xi_n} = 0$

and consequently H as a function of ξ, η is independent of ξ_n. If in addition the given system has an equilibrium solution in whose neighborhood the Hamiltonian function and the canonical transformation (18) are analytic, the new differential equations

$$(21) \qquad \dot{\xi}_k = H_{\eta_k}, \quad \dot{\eta}_k = -H_{\xi_k} \quad (k=1,\dots,n)$$

show that in the related matrix $\mathfrak{A}$ the row corresponding to the variable η_n and the column corresponding to the variable ξ_n are both 0. This again exhibits a factor λ^2 in $|\lambda\mathfrak{E}-\mathfrak{A}|$, which then disappears upon passage to the reduced system

$$(22) \qquad \dot{\xi}_k = H_{\eta_k}, \quad \dot{\eta}_k = -H_{\xi_k} \quad (k=1,\dots,n-1).$$

Finally, with (22) integrated, the missing function ξ_n is obtained from the differential equation $\dot{\xi}_n = H_{\eta_n}$ by quadrature.

We apply this to the Hamiltonian system (14), whereby $\xi, \eta, u, v, Q, 4$ will appear in place of x, y, ξ, η, ψ, n. Since by (11) the function Q is bilinear in ξ, η, we seek $w(u,\eta)$ in the linear form

$$w = \sum_{k=1}^{4} g_k \eta_k, \quad g_k = g_k(u),$$

with the partial differential equation (19) then becoming

$$\sum_{k=1}^{4} g_{k u_4} \eta_k = \sum_{k=1}^{2} (g_{2k}\eta_{2k-1} - g_{2k-1}\eta_{2k})$$

or

$$(23) \qquad g_{2k-1,u_4} = g_{2k}, \quad g_{2k,u_4} = -g_{2k-1} \quad (k=1,2).$$

This has the particular solution

$$g_1 = u_1 c, \quad g_2 = -u_1 s, \quad g_3 = u_2 c + u_3 s, \quad g_4 = -u_2 s + u_3 c,$$
$$c = \cos u_4, \quad s = \sin u_4,$$

with $|w_{u_k \eta_l}| = |g_{l u_k}| = -u_1$, so that the condition $|w_{u_k \eta_l}| \neq 0$ is satisfied if $u_1 \neq 0$. Under this assumption the desired canonical transformation becomes

$$(24) \qquad \xi_1 = u_1 c, \quad \xi_2 = -u_1 s, \quad \xi_3 = u_2 c + u_3 s, \quad \xi_4 = -u_2 s + u_3 c,$$

$$(25) \qquad \begin{cases} v_1 = \eta_1 c - \eta_2 s, \quad v_2 = \eta_3 c - \eta_4 s, \quad v_3 = \eta_3 s + \eta_4 c, \\ v_4 = \sum_{k=1}^{2} (\xi_{2k}\eta_{2k-1} - \xi_{2k-1}\eta_{2k}) = Q, \end{cases}$$

by (18). The last equation transforms to

$$v_4 = u_3 v_2 - u_2 v_3 - u_1(\eta_1 s + \eta_2 c),$$

and defining

(26) $$\eta_1 s + \eta_2 c = v_0$$

we have

$$v_0 = u_1^{-1}(u_3 v_2 - u_2 v_3 - v_4).$$

The fact that in the new coordinates u, v the Hamiltonian function F no longer depends on u_4 can be easily shown also by direct calculation. Namely, (25), (26) show that

$$v_2^2 + v_3^2 = \eta_3^2 + \eta_4^2, \quad v_1^2 + v_0^2 = \eta_1^2 + \eta_2^2,$$
$$(v_1 + v_2)^2 + (v_3 + v_0)^2 = (\eta_1 + \eta_3)^2 + (\eta_2 + \eta_4)^2,$$

and because $\eta_5 = \eta_6 = 0$, formula (12) becomes

$$T = \tfrac{1}{2}\{m_1^{-1}(v_1^2 + v_0^2) + m_2^{-1}(v_2^2 + v_3^2) + m_3^{-1}((v_1 + v_2)^2 + (v_3 + v_0)^2)\}.$$

Hence T and $Q = v_4$ are free of u_4. To show the same for U, note that according to (24) the transformation from the ξ_k $(k = 1, \dots, 4)$ to the u_k corresponds to a rotation through an angle $-u_4$ with the particle P_3 as pivot, whereby the points (ξ_1, ξ_2), (ξ_3, ξ_4) go over to $(u_1, 0)$, (u_2, u_3). Thus $(u_1, 0)$, (u_2, u_3) are the coordinates of P_1, P_2 in the rectangular Cartesian system with origin at P_3 and abscissa axis passing through P_1; in particular, $u_1 \neq 0$. Because it depends only on the distances between the three particles, U must be a function of u_1, u_2, u_3 alone. Upon introduction of these new coordinates the Hamiltonian system (14) now decomposes into the further reduced system

(27) $$\dot{u}_k = F_{v_k}, \quad \dot{v}_k = -F_{u_k} \quad (k = 1, 2, 3)$$

and the trivial residue

(28) $$\dot{u}_4 = F_{v_4}, \quad \dot{v}_4 = 0.$$

The constant for v_4 is chosen in accordance with the original equilibrium solution, whereupon, after (27) is integrated, u_4 is obtained from (28) by quadrature. For the Hamiltonian system (27) the characteristic polynomials corresponding to the equilateral and collinear equilibrium solutions evidently are

$$(\lambda^2 + 1)(\lambda^4 + \lambda^2 + \gamma) \quad \text{and} \quad (\lambda^2 + 1)(\lambda^4 + (1 - \alpha)\lambda^2 - \alpha(2\alpha + 3)),$$

where the values of γ and α are given by (5) and (7).

First we discuss the equilateral case. Setting

$$\sqrt{\tfrac{1}{4} - \gamma} = \varrho, \quad a_1 = \tfrac{1}{2} + \varrho, \quad a_2 = \tfrac{1}{2} - \varrho,$$

we have

$$(\lambda^2 + 1)(\lambda^4 + \lambda^2 + \gamma) = (\lambda^2 + 1)(\lambda^2 + a_1)(\lambda^2 + a_2).$$

Because $\gamma > 0$, a multiple eigenvalue appears only when $\gamma = \frac{1}{4}$, and this case will be excluded. For $\gamma > \frac{1}{4}$ the numbers a_1, a_2 are distinct complex conjugates, while for $\gamma < \frac{1}{4}$ they satisfy $0 < a_2 < a_1 < 1$. By the existence theorem in § 16, to the eigenvalues $\lambda_3 = i$, $\lambda_6 = -i$ there corresponds a one parameter family of periodic solutions to (27) that lie near the equilibrium solution and have the approximate period 2π. These solutions, however, are already known to us from the generalized solutions of Lagrange at the end of § 14, emanating there from elliptic orbits in the vicinity of Lagrange's circular solutions. Namely, from the known formulas for the solution to the two-body problem it is easily seen that corresponding to the prescribed fixed value of the angular momentum integral v_4 there is still a family of elliptic solutions that can be parametrized by, say, the time of traversal τ. With $c = \cos(t - u_4)$, $s = \sin(t - u_4)$, a simple computation leads from (14; 3), (14; 4), (9), (24) to the formulas

$$q_1 - q_5 = u_1 c, \quad q_2 - q_6 = u_1 s, \quad q_3 - q_5 = u_2 c - u_3 s, \quad q_4 - q_6 = u_2 s + u_3 c,$$

$$p_1 = v_1 c - v_0 s, \quad p_2 = v_1 s + v_0 c, \quad p_3 = v_2 c - v_3 s, \quad p_4 = v_2 s + v_3 c,$$

so that $u_1, c, s, u_2, u_3, v_1, v_2, v_3$ do indeed have period τ. From now on we may therefore restrict ourselves to the two other purely imaginary eigenvalue pairs $\lambda_1, \lambda_4 = -\lambda_1$ and $\lambda_2, \lambda_5 = -\lambda_2$ present when $\gamma < \frac{1}{4}$; that is, when

$$27(m_1 m_2 + m_2 m_3 + m_3 m_1) < (m_1 + m_2 + m_3)^2 .$$

This inequality is manifestly a definite restriction on m_1, m_2, m_3, being violated, for example, when $m_1 = m_2 = m_3$. Even then, however, it does not follow that there are no additional periodic solutions, only that they cannot be obtained from the formulation in § 14 and § 16.

With $\lambda_1^2 = -a_1$, $\lambda_2^2 = -a_2$, we now wish to check whether one can satisfy the additional condition that $\dfrac{\lambda_k}{\lambda_1}$ not be an integer for $k = 2, 3$. It follows from $-\lambda_3^2 = 1 > -\lambda_1^2 = a_1 > \dfrac{1}{2} > a_2 = -\lambda_2^2 > 0$ that $0 < \left(\dfrac{\lambda_2}{\lambda_1}\right)^2 < 1$ and $1 < \left(\dfrac{\lambda_3}{\lambda_1}\right)^2 < 2$, so that the condition is indeed met, and the existence theorem then gives a family of periodic solutions with the approximate period $\dfrac{2\pi i}{\lambda_1}$. Along the same lines, to examine the ratio $\dfrac{\lambda_k}{\lambda_2}$ for $k = 1, 3$, let $\dfrac{\lambda_3}{\lambda_2} = \varkappa_2$, whereupon $\lambda_2^2 = -\varkappa_2^{-2}$ and $\varkappa_2^{-4} - \varkappa_2^{-2} + \gamma = 0$. Consequently the inequality

$$\gamma \neq g^{-2} - g^{-4} \tag{29}$$

is required to hold for all integers $g > 1$. Analogously, if $\frac{\lambda_1}{\lambda_2} = \varkappa$ then $\lambda_1^2 = \varkappa^2 \lambda_2^2$, hence $\varkappa^2 > 1$ and $(\varkappa\lambda_2)^4 + (\varkappa\lambda_2)^2 + \gamma = 0$, which in conjunction with $\lambda_2^4 + \lambda_2^2 + \gamma = 0$ gives $(\varkappa^4 - 1)\lambda_2^4 + (\varkappa^2 - 1)\lambda_2^2 = 0, (\varkappa^2 + 1)\lambda_2^2 + 1 = 0,$ $(\varkappa^2 + 1)^{-2} - (\varkappa^2 + 1)^{-1} + \gamma = 0$. This leads to the added requirement that

$$(30) \qquad \gamma \neq (g + g^{-1})^{-2},$$

again for all integers $g > 1$. Thus, if γ satisfies the countably many conditions (29), (30), the existence theorem yields a second family of periodic solutions with the approximate period $\frac{2\pi i}{\lambda_2}$.

A similar analysis leads to periodic solutions in the vicinity of a collinear equilibrium solution. First of all, we have, as before, the family of Lagrange's elliptic solutions near a circular one, these associated with the eigenvalue pair $i, -i$. The other eigenvalues come from the roots λ_1^2, λ_2^2 of the quadratic equation

$$(31) \qquad x^2 + (1 - \alpha)x - \alpha(2\alpha + 3) = 0$$

with α defined by (7). Because $\alpha > 0$, the roots of this equation are real and of opposite sign, say $\lambda_1^2 < 0, \lambda_2^2 > 0$. Thus, in addition to $\pm\lambda_3 = i$, there is one other purely imaginary eigenvalue pair, namely $\pm\lambda_1$. Since for $x = -1$ the left side of (31) has the negative value $-2\alpha(\alpha + 1)$, the negative root satisfies $\lambda_1^2 < -1 = \lambda_3^2 < 0$ and therefore the quotient $\frac{\lambda_3}{\lambda_1}$ is not an integer. This then leads to a family of periodic solutions in a neighborhood of the collinear solution of Lagrange which have the approximate period $\frac{2\pi i}{\lambda_1}$. Corresponding to the real eigenvalue pair $\pm\lambda_2$ there are, according to § 16, four solutions to the three-body problem which asymptotically approach the equilibrium solution as $t \to \infty$ and $t \to -\infty$ respectively, as well as a family of solutions that remain in any fixed small neighborhood of the equilibrium solution for a bounded time interval only.

The formulation discussed in § 16 allows these periodic solutions, whose existence we have proved here, to be explicitly expanded into Fourier series.

Finally, it remains to compute the determinant $|\lambda\mathfrak{J} + \mathfrak{S}|$. In the equilateral case we use the relative coordinates ξ_k, η_k $(k = 1, \dots, 6)$ and denote their values for Lagrange's solution by ξ_k^*, η_k^*. After a suitable rotation it may be assumed that $\xi_1^* = -\xi_3^* = \frac{r}{2}, \xi_2^* = \xi_4^* = \frac{r}{2}\sqrt{3}$. Next, replacing ξ, η by $\xi + \xi^*, \eta + \eta^*$, we expand U into powers of the ξ_k

($k=1, \ldots, 6$). For $1 \leqq k < l \leqq 3$, with the abbreviations

$$s_{kl} = 2r^{-2}\{(x_k - x_l)(x_k^* - x_l^*) + (x_{k+3} - x_{l+3})(x_{k+3}^* - x_{l+3}^*)\},$$
$$q_{kl} = r^{-2}\{(x_k - x_l)^2 + (x_{k+3} - x_{l+3})^2\},$$

we have

$$r_{kl}^{-1} = r^{-1}(1 + s_{kl} + q_{kl})^{-\frac{1}{2}} = r^{-1}(1 - \tfrac{1}{2}s_{kl} - \tfrac{1}{2}q_{kl} + \tfrac{3}{8}s_{kl}^2 + \cdots), \tag{32}$$

so that the terms of $-2U$ quadratic in the ξ_k are given by the expression

$$V = \frac{m_1 m_3}{4r^3}(\xi_1^2 - 6\sqrt{3}\,\xi_1\xi_2 - 5\xi_2^2) + \frac{m_2 m_3}{4r^3}(\xi_3^2 + 6\sqrt{3}\,\xi_3\xi_4 - 5\xi_4^2) + \frac{m_1 m_2}{r^3}\{(\xi_2 - \xi_4)^2 - 2(\xi_1 - \xi_3)^2\}.$$

By (11), (12) then $\mathfrak{S}$ is the matrix of the quadratic form $V + 2Q + 2T$ in the twelve variables ξ_k, η_k ($k=1, \ldots, 6$). Setting $m_k^{-1} = \mu_k$ ($k=1,2,3$) and introducing the four-by-four matrices

$$\mathfrak{A} = \frac{m_1 m_2 m_3}{4r^3}\begin{pmatrix} \mu_2 - 8\mu_3 & -3\sqrt{3}\mu_2 & 8\mu_3 & 0 \\ -3\sqrt{3}\mu_2 & 4\mu_3 - 5\mu_2 & 0 & -4\mu_3 \\ 8\mu_3 & 0 & \mu_1 - 8\mu_3 & 3\sqrt{3}\mu_1 \\ 0 & -4\mu_3 & 3\sqrt{3}\mu_1 & 4\mu_3 - 5\mu_1 \end{pmatrix},$$

$$\mathfrak{B} = \begin{pmatrix} \lambda & -1 & 0 & 0 \\ 1 & \lambda & 0 & 0 \\ 0 & 0 & \lambda & -1 \\ 0 & 0 & 1 & \lambda \end{pmatrix}, \quad \mathfrak{C} = \begin{pmatrix} -\lambda & 1 & 0 & 0 \\ -1 & -\lambda & 0 & 0 \\ 0 & 0 & -\lambda & 1 \\ 0 & 0 & -1 & -\lambda \end{pmatrix},$$

$$\mathfrak{D} = \begin{pmatrix} \mu_1 + \mu_3 & 0 & \mu_3 & 0 \\ 0 & \mu_1 + \mu_3 & 0 & \mu_3 \\ \mu_3 & 0 & \mu_2 + \mu_3 & 0 \\ 0 & \mu_3 & 0 & \mu_2 + \mu_3 \end{pmatrix},$$

we see that

$$|\lambda\mathfrak{J} + \mathfrak{S}| = (\lambda^2 + 1)^2 \begin{vmatrix} \mathfrak{A} & \mathfrak{B} \\ \mathfrak{C} & \mathfrak{D} \end{vmatrix},$$

and since

$$\begin{pmatrix} \mathfrak{E} & -\mathfrak{B}\mathfrak{D}^{-1} \\ 0 & \mathfrak{E} \end{pmatrix}\begin{pmatrix} \mathfrak{A} & \mathfrak{B} \\ \mathfrak{C} & \mathfrak{D} \end{pmatrix} = \begin{pmatrix} \mathfrak{A} - \mathfrak{B}\mathfrak{D}^{-1}\mathfrak{C} & 0 \\ \mathfrak{C} & \mathfrak{D} \end{pmatrix},$$

we have

$$\begin{vmatrix} \mathfrak{A} & \mathfrak{B} \\ \mathfrak{C} & \mathfrak{D} \end{vmatrix} = |\mathfrak{D}\mathfrak{A} - \mathfrak{D}\mathfrak{B}\mathfrak{D}^{-1}\mathfrak{C}|.$$

A direct computation of the last four-by-four determinant now leads to the result given in (4), (5).

In the collinear case one may take the values found in (14;8) for the coordinates $x_{2k-1} = x^*_{2k-1}$ $(k=1,2,3)$ of the equilibrium solution, while $x^*_{2k}=0$. With

$$(33) \quad \begin{cases} u_k = x_{2k-1} - x^*_{2k-1}, & u_{k+3} = x_{2k} - x^*_{2k}, \\ & \qquad\qquad (k=1,2,3) \\ u_{k+6} = y_{2k-1} - y^*_{2k-1}, & u_{k+9} = y_{2k} - y^*_{2k}, \end{cases}$$

the expansion of F in powers of $u_1, \ldots, u_{12}$ takes the form

$$F(x,y) = F(x^*, y^*) + \tfrac{1}{2} \sum_{k,l=1}^{12} r_{kl} u_k u_l + \cdots,$$

where $(r_{kl}) = \mathfrak{R}$ is a symmetric matrix of twelve rows. Since the linear substitution (33) is canonical, we have $|\lambda\mathfrak{J} + \mathfrak{S}| = |\lambda\mathfrak{J} + \mathfrak{R}|$, while on the other hand, using (32), one finds that

$$\lambda\mathfrak{J} + \mathfrak{R} = \begin{pmatrix} -2\mathfrak{W} & 0 & \lambda\mathfrak{E} & -\mathfrak{E} \\ 0 & \mathfrak{W} & \mathfrak{E} & \lambda\mathfrak{E} \\ -\lambda\mathfrak{E} & \mathfrak{E} & \mathfrak{M}^2 & 0 \\ -\mathfrak{E} & -\lambda\mathfrak{E} & 0 & \mathfrak{M}^2 \end{pmatrix},$$

where

$$\mathfrak{W} = m_1 m_2 m_3 a^{-3} \begin{pmatrix} \mu_2 + \mu_3 \varrho^{-3} & -\mu_3\varrho^{-3} & -\mu_2 \\ -\mu_3\varrho^{-3} & \mu_3\varrho^{-3} + \mu_1\sigma^{-3} & -\mu_1\sigma^{-3} \\ -\mu_2 & -\mu_1\sigma^{-3} & \mu_2 + \mu_1\sigma^{-3} \end{pmatrix},$$

$$\mathfrak{E} = \begin{pmatrix} 1 & 0 & 0 \\ 0 & 1 & 0 \\ 0 & 0 & 1 \end{pmatrix}, \qquad \mathfrak{M} = \begin{pmatrix} \mu_1^{\frac{1}{2}} & 0 & 0 \\ 0 & \mu_2^{\frac{1}{2}} & 0 \\ 0 & 0 & \mu_3^{\frac{1}{2}} \end{pmatrix},$$

and a, ϱ, σ have the same meaning as in § 14. The quadratic form in the three real variables w_1, w_2, w_3 constructed from the matrix $(m_1 m_2 m_3)^{-1} a^3 \mathfrak{W}$ is given by

$$\mu_1\sigma^{-3}(w_2 - w_3)^2 + \mu_2(w_3 - w_1)^2 + \mu_3\varrho^{-3}(w_1 - w_2)^2 \geqq 0,$$

and is therefore nonnegative, albeit degenerate, since it vanishes for $w_1 = w_2 = w_3$. It follows that $|\mathfrak{W}| = 0$. For the twelve-by-twelve diagonal matrix

$$\mathfrak{N} = \begin{pmatrix} \mathfrak{M} & 0 & 0 & 0 \\ 0 & \mathfrak{M} & 0 & 0 \\ 0 & 0 & \mathfrak{M}^{-1} & 0 \\ 0 & 0 & 0 & \mathfrak{M}^{-1} \end{pmatrix}$$

we have $|\mathfrak{N}| = 1$ and

$$|\lambda\mathfrak{J} + \mathfrak{R}| = |\mathfrak{N}(\lambda\mathfrak{J} + \mathfrak{R})\,\mathfrak{N}| = \begin{vmatrix} -2\mathfrak{G} & 0 & \lambda\mathfrak{E} & -\mathfrak{E} \\ 0 & \mathfrak{G} & \mathfrak{E} & \lambda\mathfrak{E} \\ -\lambda\mathfrak{E} & \mathfrak{E} & \mathfrak{E} & 0 \\ -\mathfrak{E} & -\lambda\mathfrak{E} & 0 & \mathfrak{E} \end{vmatrix}, \qquad \mathfrak{G} = \mathfrak{M}\mathfrak{W}\mathfrak{M}.$$

Since the above twelve-by-twelve determinant is made up of three-by-three matrices that mutually commute, it may formally be treated as a four-by-four determinant, whereupon an elementary computation gives

$$(34) \qquad \begin{cases} |\lambda\mathfrak{J} + \mathfrak{S}| = |(\lambda^2+1)^2\mathfrak{E} + (1-\lambda^2)\,\mathfrak{G} - 2\mathfrak{G}^2| \\ \qquad = \prod_{k=1}^{3} \left((\lambda^2+1)^2 + (1-\lambda^2)\,\gamma_k - 2\gamma_k^2\right), \end{cases}$$

with $\gamma_1, \gamma_2, \gamma_3$ the eigenvalues of $\mathfrak{G}$. Because $|\mathfrak{W}| = 0$, also $|\mathfrak{G}| = 0$, so that one of the eigenvalues is 0, say $\gamma_3 = 0$. To determine γ_1 and γ_2 we recall that from the existence of the angular momentum integral one was able to deduce that $|\lambda\mathfrak{J} + \mathfrak{S}|$ vanishes for $\lambda = 0$. Hence, after suitable ordering of γ_1, γ_2, we have

$$0 = 1 + \gamma_2 - 2\gamma_2^2 = (1 + 2\gamma_2)\,(1 - \gamma_2)\,.$$

Now together with $\mathfrak{W}$ also $\mathfrak{G}$ is nonnegative, so that $\gamma_2 \geqq 0$, and therefore $\gamma_2 = 1$. On the other hand, the trace of $\mathfrak{G}$ is given by

$$(35) \quad \gamma_1 + \gamma_2 + \gamma_3 = \gamma = a^{-3}\{m_1(1+\varrho^{-3}) + m_2(\varrho^{-3} + \sigma^{-3}) + m_3(1+\sigma^{-3})\}\,,$$

so that $\gamma_1 = \gamma - 1$. Inserting these values of $\gamma_1, \gamma_2, \gamma_3$ into (34) and setting $\alpha = \gamma - 2$, one then obtains the formula

$$(36) \qquad |\lambda\mathfrak{J} + \mathfrak{S}| = \lambda^2(\lambda^2+1)^3\,\{\lambda^4 + (1-\alpha)\,\lambda^2 - \alpha(2\alpha+3)\}\,.$$

Since $x_3^* - x_1^* = \varrho a$, $x_5^* - x_3^* = \sigma a$, from (14;8) we have

$$(37) \qquad -1 = m_3(\varrho a)^{-1}(\sigma a)^{-2} - m_3(\varrho a)^{-1}a^{-2} - (m_1 + m_2)\,(\varrho a)^{-3}\,,$$

$$(38) \qquad -1 = m_1(\sigma a)^{-1}(\varrho a)^{-2} - m_1(\sigma a)^{-1}a^{-2} - (m_2 + m_3)\,(\sigma a)^{-3}\,,$$

and recalling that $\varrho + \sigma = 1$, by adding (35), (37), (38) we obtain the relation

$$(39) \qquad \alpha = m_1 a^{-3}(1+\varrho^{-1}+\varrho^{-2}) + m_3 a^{-3}(1+\sigma^{-1}+\sigma^{-2}).$$

Finally, addition of the two equations in (14;9) gives

$$(40) \qquad a^3 = m_1 + m_2(\varrho^{-2}+\sigma^{-2}) + m_3\,,$$

and (36), (39), (40) combine to give the result in (6), (7).

§ 19. Hill's Problem

We seek periodic solutions to the three-body problem other than those found in the previous sections. Restricting ourselves again to planar orbits, we first neglect the particle P_2 and consider only the motion of two particles P_1, P_3. These travel along conic sections and, in particular, we will assume that they describe circles about their common center of mass P_0. We now replace the two particles P_1, P_3 by P_0 and reintroduce the third particle P_2. Of the possible orbits for P_0 and P_2 we again consider the particular case of circles. If P_2 is far enough from the other two particles, starting with this approximate solution we will arrive at an actual solution to the three-body problem.

A limiting case of the above problem which is easier to handle arises from considering not the general three-body problem, but the so called restricted three-body problem. This is a special case of the planar three-body problem in which the mass of P_3 is taken to be 0 and P_1, P_2 describe circular orbits. To obtain the differential equations of motion for P_3, we introduce in the given plane a rotating coordinate system with origin at the center of mass of P_1 and P_2, so that in the new system P_1 and P_2 are at rest. Without loss of generality the rotational velocity may be taken as $\omega = 1$, so that by (14;5) the rectangular coordinates x_{2k-1}, x_{2k} of P_k $(k=1,2,3)$ in the rotating coordinate system satisfy the differential equations

$$\dot{x}_{2k-1} = m_k^{-1} y_{2k-1} + x_{2k}\,, \qquad \dot{y}_{2k-1} = U_{x_{2k-1}} + y_{2k}\,,$$
$$\dot{x}_{2k} = m_k^{-1} y_{2k} - x_{2k-1}\,, \qquad \dot{y}_{2k} = U_{x_{2k}} - y_{2k-1}\,.$$

Upon elimination of y_{2k-1}, y_{2k} these lead to the second order differential equations

$$(1) \qquad \begin{cases} \ddot{x}_{2k-1} = 2\dot{x}_{2k} + x_{2k-1} + m_k^{-1} U_{x_{2k-1}}\,, \\ \ddot{x}_{2k} = -2\dot{x}_{2k-1} + x_{2k} + m_k^{-1} U_{x_{2k}}\,, \end{cases} \qquad (k=1,2,3)$$

wherein m_3 is not yet assumed to be 0 nor need P_1, P_2 be at rest. Now

$$(2)\qquad \begin{cases} m_k^{-1} U_{x_{2k-1}} = \sum\limits_{l \neq k} m_l (x_{2l-1} - x_{2k-1}) r_{kl}^{-3}, \\ m_k^{-1} U_{x_{2k}} = \sum\limits_{l \neq k} m_l (x_{2l} - x_{2k}) r_{kl}^{-3}, \end{cases}$$

and here the right sides make sense also when $m_3 = 0$. In that case formulas (1) with $k = 1, 2$ reduce to precisely the differential equations of the two-body problem for the particles P_1, P_2. Normalizing $m_1 + m_2 = 1$, we set $m_1 = \mu$, $m_2 = 1 - \mu$ with $0 < \mu < 1$ and obtain $x_1 = 1 - \mu$, $x_2 = 0$, $x_3 = -\mu$, $x_4 = 0$ as a particular solution that corresponds to circular orbits for P_1, P_2 in the stationary coordinate system. The equations of motion for the third point P_3 with coordinates $x_5 = x$, $x_6 = y$ then become

$$(3)\qquad \ddot{x} = 2\dot{y} + x + F_x, \quad \ddot{y} = -2\dot{x} + y + F_y$$

with

$$F = \frac{1-\mu}{r_{23}} + \frac{\mu}{r_{13}} = (1-\mu)\{(x+\mu)^2 + y^2\}^{-\frac{1}{2}} + \mu\{(x+\mu-1)^2 + y^2\}^{-\frac{1}{2}}.$$

These are the differential equations for the restricted three-body problem. Although this is only a fourth order system, we are still far from knowing a complete solution. It is more convenient to express equations (3) in terms of the conjugate complex variables

$$(4)\qquad p = (x + \mu - 1) + iy, \quad q = \bar{p} = (x + \mu - 1) - iy,$$

where p represents the vector from P_1 to P_3 in the complex plane. Then

$$F = \frac{\mu}{\sqrt{pq}} + \frac{1-\mu}{\sqrt{(1+p)(1+q)}}, \quad F_x = F_p + F_q, \quad F_y = i(F_p - F_q),$$

hence

$$\ddot{p} = -2i\dot{p} + p - \mu + 1 + 2F_q, \quad \ddot{q} = 2i\dot{q} + q - \mu + 1 + 2F_p,$$

and with

$$G = pq + (1-\mu)(p+q) + 2F = pq + (1-\mu)(p+q) + \frac{2\mu}{\sqrt{pq}} + \frac{2-2\mu}{\sqrt{(1+p)(1+q)}}$$

the differential equations assume the abbreviated form

$$(5)\qquad \ddot{p} = -2i\dot{p} + G_q, \quad \ddot{q} = 2i\dot{q} + G_p.$$

To obtain a periodic solution of (5) we introduce yet another simplification which is due to Hill and has its basis in astronomy.

Namely, if P_2 is taken to be the sun, P_1 the earth, and P_3 the moon, then the mass μ of the earth is small relative to the mass $1-\mu$ of the sun; moreover, the sun and earth move approximately in circles about their common center of mass, while the moon remains nearly in a plane containing these circles. In addition, the mass of the moon is small relative to that of the earth and is taken to be $m_3=0$. Thus we seek a periodic solution to (5) for small values of μ. In particular, since $|p|$ corresponds to the distance between the moon and the earth and is therefore small relative to the distance 1 between the sun and the earth, we seek a periodic solution for which $|p|$ is small. If as a first attempt we neglect the terms $-2i\dot{p}$, $2i\dot{q}$ in (5) and in addition retain only the leading term $2\mu(pq)^{-\frac{1}{2}}$ of G, we obtain the system

$$\ddot{p}=-\mu p(pq)^{-\frac{3}{2}}, \quad \ddot{q}=-\mu q(pq)^{-\frac{3}{2}}. \tag{6}$$

This is again the complex form of the differential equations of the two-body problem for P_1, P_3, encountered already in (14;12), having the circular orbits $p=\mu^{\frac{1}{3}}e^{it}$, $q=\bar{p}=\mu^{\frac{1}{3}}e^{-it}$, $|p|=|q|=\mu^{\frac{1}{3}}$ as a particular solution. This suggests the transformation of variables

$$p=\mu^{\frac{1}{3}}u, \quad q=\mu^{\frac{1}{3}}v, \tag{7}$$

whereby (5) becomes

$$\ddot{u}=-2i\dot{u}+H_v, \quad \ddot{v}=2i\dot{v}+H_u \tag{8}$$

with

$$H=\mu^{-\frac{2}{3}}G=uv+\mu^{-\frac{1}{3}}(1-\mu)(u+v)+\frac{2}{\sqrt{uv}}+\frac{2\mu^{-\frac{2}{3}}(1-\mu)}{\sqrt{(1+\mu^{\frac{1}{3}}u)(1+\mu^{\frac{1}{3}}v)}}.$$

The expansion of H in ascending powers of $\mu^{\frac{1}{3}}$ takes the form

$$\begin{aligned} H &= uv+\mu^{-\frac{1}{3}}(u+v) \\ &\quad +2\mu^{-\frac{2}{3}}(1-\tfrac{1}{2}\mu^{\frac{1}{3}}u+\tfrac{3}{8}\mu^{\frac{2}{3}}u^2)(1-\tfrac{1}{2}\mu^{\frac{1}{3}}v+\tfrac{3}{8}\mu^{\frac{2}{3}}v^2)+2(uv)^{-\frac{1}{2}}+\cdots \\ &= 2\mu^{-\frac{2}{3}}+\tfrac{3}{4}(u+v)^2+2(uv)^{-\frac{1}{2}}+\cdots, \end{aligned}$$

where the remaining terms contain only positive powers of $\mu^{\frac{1}{3}}$. Because μ is assumed to be small, we neglect these additional terms, and accordingly, instead of (8), consider the system

$$\ddot{u}=-2i\dot{u}+\tfrac{3}{2}(u+v)-u(uv)^{-\frac{3}{2}}, \quad \ddot{v}=2i\dot{v}+\tfrac{3}{2}(u+v)-v(uv)^{-\frac{3}{2}}. \tag{9}$$

These are known as Hill's differential equations, and even here the general solution is not known. We will, however, determine periodic solutions to (9) through a power series expansion similar to that in § 16.

To discover a form for this expansion, we consider again the simplified system

$$\text{(10)}\qquad \ddot{u} = -u(uv)^{-\frac{3}{2}}, \quad \ddot{v} = -v(uv)^{-\frac{3}{2}},$$

which in analogy to (6) is obtained from (9) by neglecting the rest of the terms on the right. We seek periodic solutions for which $\bar{v} = u$, since then by (4), (7) the coordinates x, y are real. One such solution is given by the circular orbits $u = u_0 e^{\lambda t}$, $v = v_0 e^{-\lambda t}$ with $\lambda^2 = -(uv)^{-\frac{3}{2}} = -(u_0 v_0)^{-\frac{3}{2}}$, $v_0 = \bar{u}_0$. Setting $u = \xi^4$, $v = \eta^4$ so as to eliminate roots, we then have $\xi = \xi_0 e^{\alpha t}$, $\eta = \eta_0 e^{-\alpha t}$, $\dot{\xi} = \alpha\xi$, $\dot{\eta} = -\alpha\eta$ with $\alpha = \frac{\lambda}{4} = \pm\frac{i}{4}(\xi_0\eta_0)^{-3}$, $\eta_0 = \bar{\xi}_0$. With this in mind, we now seek an exact solution to (9) in the form of power series

$$\text{(11)}\qquad u = \xi^4\left(1 + \sum_{k,l} a_{kl}\xi^{3k+4l}\eta^{3k-4l}\right), \quad v = \eta^4\left(1 + \sum_{k,l} a_{kl}\xi^{3k-4l}\eta^{3k+4l}\right)$$

in two new variables ξ, η, with k, l running over all pairs of integers satisfying $3k \geqq 4|l|$, $k > 0$, and the coefficients a_{kl} as yet undetermined. The special form of this expression will be justified later. In accordance with the previous heuristic reasoning, the new unknowns ξ, η are to satisfy the differential equations

$$\text{(12)}\qquad \dot{\xi} = \alpha\xi, \quad \dot{\eta} = -\alpha\eta, \quad \alpha = \pm\frac{i}{4}(\xi\eta)^{-3},$$

and, since then $(\xi\eta)^{\cdot} = \alpha\xi\eta + \xi(-\alpha\eta) = 0$, it follows that

$$\text{(13)}\qquad \xi = \xi_0 e^{\alpha t}, \quad \eta = \eta_0 e^{-\alpha t}, \quad \alpha = \pm\frac{i}{4}(\xi_0\eta_0)^{-3},$$

with ξ_0, η_0 nonzero constants.

The derivatives of the series for u and v are obtained by formally differentiating each term with respect to t and then using (12) to express $\dot{\xi}, \dot{\eta}$ again in terms of ξ, η. With the abbreviation

$$\text{(14)}\qquad \zeta_{kl} = \xi^{3k+4l}\eta^{3k-4l}$$

this gives

$$\dot{u} = 4\alpha\xi^4\{1 + \Sigma\,(2l+1)a_{kl}\zeta_{kl}\}, \qquad \dot{v} = -4\alpha\eta^4\{1 + \Sigma\,(2l+1)a_{kl}\zeta_{k,-l}\}$$

$$\ddot{u} = (4\alpha)^2\xi^4\{1 + \Sigma\,(2l+1)^2 a_{kl}\zeta_{kl}\}, \quad \ddot{v} = (4\alpha)^2\eta^4\{1 + \Sigma\,(2l+1)^2 a_{kl}\zeta_{k,-l}\}.$$

Here, as well as in the rest of this section, the sum is always taken over the previously specified pairs of integers k, l, so that the exponents $3k+4l$, $3k-4l$ in (14) are nonnegative and their sum $6k$ is positive. We will refer to k as the order of ζ_{kl}. In addition, let

$$A = -\Sigma\, a_{kl}\zeta_{kl}, \quad B = -\Sigma\, a_{kl}\zeta_{k,-l}.$$

To get rid of the negative exponents we multiply the first equation in (9) by $\xi^2\eta^6$ and, observing that $4\alpha = \pm i(\xi\eta)^{-3}$, obtain for the resulting individual terms the series

$$
(15)\quad \begin{cases} -\ddot{u}\xi^2\eta^6 = 1 + \Sigma\,(2l+1)^2 a_{kl}\zeta_{kl}, \\ -2i\dot{u}\xi^2\eta^6 = \pm 2\{\zeta_{10} + \Sigma\,(2l+1)a_{kl}\zeta_{k+1,l}\}, \\ \tfrac{3}{2}(u+v)\xi^2\eta^6 = \tfrac{3}{2}(\zeta_{20} + \Sigma\, a_{kl}\zeta_{k+2,l}) + \tfrac{3}{2}(\zeta_{2,-1} + \Sigma\, a_{kl}\zeta_{k+2,-l-1}), \\ -u^{-\frac{1}{2}}v^{-\frac{3}{2}}\xi^2\eta^6 = -(1-A)^{-\frac{1}{2}}(1-B)^{-\frac{3}{2}}. \end{cases}
$$

Here the expression

$$
(16)\quad \begin{cases} (1-A)^{-\frac{1}{2}}(1-B)^{-\frac{3}{2}} \\ \quad = (1+\tfrac{1}{2}A+\cdots)(1+\tfrac{3}{2}B+\cdots) = 1+\tfrac{1}{2}A+\tfrac{3}{2}B+\cdots \end{cases}
$$

is to be expanded in powers of A, B and the latter then replaced by their respective series in the ζ_{kl}. Since $\zeta_{kl}\zeta_{gh} = \zeta_{k+g,l+h}$, the right side of the last row in (15) also becomes a series of the form $\Sigma\, c_{kl}\zeta_{kl}$, as is already the case for the other rows. Thus, to satisfy the first equation in (9) formally, we have to determine the constants a_{kl} so that the sum of the expressions in (15) vanish identically in the ζ_{kl}.

Multiplying the second equation in (9) by $\xi^6\eta^2$, one obtains for the individual terms again series analogous to (15). These, however, can be obtained directly from (15) by interchanging ξ and η, or equivalently, by writing $\zeta_{k,-l}$ in place of ζ_{kl}. It should be noticed that on the left side of the second row the term $-2i\dot{u}\xi^2\eta^6$ is to be replaced by $2i\dot{v}\xi^6\eta^2$, so that the sign $\pm$ on the right also remains unchanged. Thus, comparison of coefficients for ζ_{kl} in the first equation of (9) leads to exactly the same conditions on the a_{kl} as the corresponding comparison for $\zeta_{k,-l}$ in the second equation of (9). It is therefore sufficient to determine the unknown coefficients a_{kl} so that the sum of the four terms on the right side in (15) vanishes identically in the ζ_{kl}. We will show that this can be done in exactly one way, and that the a_{kl} are all rational numbers.

To prove this assertion we proceed by induction on k. It is clear from our expression that the coefficients of the constant term do indeed match. Now let $r \geqq 1$, and suppose that it has already been shown that for $0 < k \leqq r-1$ the a_{kl} can be uniquely determined so that the coefficients for each of the terms of order $1, 2, \ldots, r-1$ match and, furthermore, that these a_{kl} are rational numbers. For $r=1$ the assumption is vacuous. To show that the inductional hypothesis is valid also for the a_{rl} $(4|l| \leqq 3r)$, we observe that the series expansion of

$$
(17)\quad D = (1-A)^{-\frac{1}{2}}(1-B)^{-\frac{3}{2}} - 1 - \tfrac{1}{2}A - \tfrac{3}{2}B
$$

in terms of A, B begins with quadratic terms. If A and B in this expression are replaced by their respective series in the ζ_{kl}, the coefficient of ζ_{rl} in D will then be a polynomial in the a_{ks} with $k<r$, which have already been determined and are rational. Moreover, the coefficients of this polynomial are well defined rational numbers. Consequently the coefficient of ζ_{rl} on the right side of the last row in (15) is the sum of $\frac{1}{2}a_{rl}+\frac{3}{2}a_{r,-l}$ and an already known rational number. Comparison of coefficients of the corresponding term in the first three rows of (15) now leads to the condition

$$\{(2l+1)^2+\tfrac{1}{2}\}a_{rl}+\tfrac{3}{2}a_{r,-l}=\varrho_{rl}, \tag{18}$$

where ϱ_{rl} is an already uniquely determined rational number. For $l=0$ this gives

$$3a_{r0}=\varrho_{r0}, \tag{19}$$

which fixes a_{r0} as a rational number. On the other hand, for $l \neq 0$ we obtain a second equation

$$\tfrac{3}{2}a_{rl}+\{(-2l+1)^2+\tfrac{1}{2}\}a_{r,-l}=\varrho_{r,-l} \tag{20}$$

from (18) by changing the sign of l. Since the two linear equations (18), (20) for $a_{rl}, a_{r,-l}$ have the positive determinant

$$\{(2l+1)^2+\tfrac{1}{2}\}\,\{(-2l+1)^2+\tfrac{1}{2}\}-(\tfrac{3}{2})^2=4l^2(4l^2-1) \tag{21}$$

and their coefficients are rational, they uniquely determine $a_{rl}, a_{r,-l}$, and indeed as rational numbers. This completes the induction.

We will subsequently prove that these series for u, v converge absolutely if $|\xi|, |\eta|$ are sufficiently small. If in addition $\eta=\bar{\xi}$, then, since the a_{kl} are real, the two quantities u, v are complex conjugates, and therefore the original coordinates x, y are real. This corresponds to choosing $\eta_0=\bar{\xi}_0$ in (13).

As in § 17, convergence is proved by using the method of majorants. For

$$\phi=\Sigma\, c_{kl}\zeta_{kl}=\Sigma\, c_{kl}\xi^{3k+4l}\eta^{3k-4l}$$

a formal power series with constant coefficients c_{kl}, we abbreviate $c_{kl}=[\phi]_{kl}$. Writing down once more the formulas used for comparing coefficients, from (15), (17) we obtain the relation

$$\left\{\begin{aligned} \varrho_{kl} = [D \mp 2(\zeta_{10}+\Sigma\,(2l+1)a_{kl}\zeta_{k+1,l})+{} \\ -\tfrac{3}{2}\zeta_{20}(1+A)-\tfrac{3}{2}\zeta_{2,-1}(1+B)]_{kl}, \end{aligned}\right. \tag{22}$$

while, on the other hand,

$$(23)\qquad a_{k0}=\frac{1}{3}\varrho_{k0},\quad a_{kl}=\frac{\{(1-2l)^2+\frac{1}{2}\}\varrho_{kl}-\frac{3}{2}\varrho_{k,-l}}{4l^2(4l^2-1)}\qquad (l\neq 0).$$

To prove absolute convergence of the series for u and v in a complex neighborhood of $\xi=0, \eta=0$, it is enough to consider the particular case of $\xi=\eta$. Then $\zeta_{kl}=\xi^{3k+4l}\eta^{3k-4l}=\xi^{6k}=\zeta^k$, where $\zeta=\xi^6$, and by (11) it is sufficient to prove convergence of

$$Z=\Sigma\,|a_{kl}|\,\zeta^k$$

for some $\zeta>0$. With this in mind we first majorize D. Because $\xi=\eta$, we have $A=B\prec Z$, where ζ is now treated as an indeterminate. Since, moreover, the expansion of $(1-A)^{-\frac{1}{2}}(1-B)^{-\frac{3}{2}}$ in powers of A, B has only positive coefficients, it follows that

$$(24)\quad D\prec(1-Z)^{-2}-1-2Z=\frac{3Z^2}{(1-Z)^2}-\frac{2Z^3}{(1-Z)^2}\prec\frac{3Z^2}{(1-Z)^2}\prec\frac{3Z^2}{1-2Z},$$

while in addition

$$(25)\qquad -\tfrac{3}{2}\zeta_{20}(1+A)-\tfrac{3}{2}\zeta_{2,-1}(1+B)\prec 3\zeta^2(1+Z).$$

Noting that $|2l+1|\geqq 1$, we then obtain from (22), (24), (25) the estimate

$$(26)\quad \sum_{k,l}\left|\frac{\varrho_{kl}}{2l+1}\right|\zeta^k\prec\frac{3Z^2}{1-2Z}+3\zeta^2(1+Z)+2\zeta(1+Z).$$

One observes next that the two quotients

$$\frac{(2l+1)\{(1-2l)^2+\frac{1}{2}\}}{4l^2(4l^2-1)},\qquad \frac{\frac{3}{2}(2l+1)}{4l^2(4l^2-1)}$$

remain bounded over all integers $l\neq 0$, so that (23) and (26) lead to the relation

$$Z\prec c_1\left\{\frac{Z^2}{1-2Z}+\zeta^2(1+Z)+\zeta(1+Z)\right\}.$$

Here c_1, and later also $c_2,\ldots,c_5$, denotes a positive universal constant. With the aid of $1+Z\prec(1-2Z)^{-1}$ it follows that

$$Z\prec c_1\frac{Z^2+\zeta^2+\zeta}{1-2Z}\prec c_1\frac{\zeta+(\zeta+Z)^2}{1-2(\zeta+Z)},$$

and for $V=\zeta+Z$ one obtains

$$V\prec c_2\frac{2\zeta+V^2}{4-c_2V}.$$

As we saw in the concluding argument of § 17, this implies convergence of V for $\zeta < c_3$. Thus the power series for u and v are absolutely convergent in $|\xi| < c_4, |\eta| < c_4$.

Let us now summarize this result. In the solutions

$$\xi = \xi_0 e^{\alpha t}, \quad \eta = \eta_0 e^{-\alpha t} \tag{27}$$

of the differential equations (12), choose $\eta_0 = \bar{\xi}_0$ and $0 < |\xi_0| = \varrho < c_4$. Then the quantity

$$\alpha = \pm \frac{i}{4}(\xi_0 \eta_0)^{-3} = \pm \frac{i}{4} \varrho^{-6}$$

is purely imaginary, and therefore $\eta = \bar{\xi}, |\xi| = \varrho$ remains valid for all real t. Consequently the power series for u and v converge, while $v = \bar{u}$, so that in the original coordinates x, y the solution is real. Furthermore ξ, η as well as $\xi^4 = \xi_0^4 e^{4\alpha t}, \eta^4 = \eta_0^4 e^{-4\alpha t}$, and $\zeta_{kl} = \xi_0^{3k+4l} \eta_0^{3k-4l} e^{8l\alpha t}$ are periodic functions of t. Inserting these functions into the series (11), one obtains u, v and also x, y as periodic functions of t with period $\left|\frac{2\pi i}{4\alpha}\right| = 2\pi\varrho^6$. In view of (27), by a suitable translation of the time variable t, one evidently can achieve that $\xi_0 = \eta_0 = \varrho$. Setting $\varrho^2 = \sigma$, we can choose this quantity as a free parameter in the interval $0 < \sigma < c_4^2 = c_5$. The series for x, y are Fourier series in $e^{4\alpha t}$, with the coefficients being power series in σ that likewise converge absolutely for $|\sigma| < c_5$. If t is replaced by $-t$, then ξ and η are interchanged, as are u and v, so that the point x, y goes over to its mirror image $x, -y$ relative to the x-axis. In the Fourier series this is expressed by x being a cosine series and y a sine series. In particular, the orbits lie symmetrically about the x-axis. One thus has a family of such symmetric solutions, which depend on the real parameter σ and have period $2\pi\sigma^3$. Admitting the two possible signs for α in (12), which manifest themselves also in the $\pm$ sign in the second equation (15), we obtain two different families of periodic solutions corresponding to the two directions in the moon's P_3 motion about the earth P_1. Namely, for the positive sign the direction is the same as of the earth rotating about the sun, while for the negative sign it is opposite. The orbits in the two families of solutions are actually distinct from one another.

The above solutions u, v of the differential equations (9) were discovered in 1878 by Hill [1]. However, he found them in a somewhat different way in that he introduced the period of the solutions as a parameter and directly set up a Fourier series with undetermined coefficients. By comparison of coefficients he then obtained infinitely many equations with infinitely many unknowns in each, whereupon through a power series expansion with respect to the parameter he was led to recursion formulas equivalent to (18), (20). However, Hill did not prove that the series he constructed actually converge. A convergence proof was given in 1925 by Wintner [2]. Another existence proof for these periodic solutions is due to E. Hopf [3]. He made use of Poincaré's continuation method, which will be discussed in § 21, and obtained the solutions as convergent series in fractional powers of the period.

§ 20. A Generalization of Hill's Problem

Whereas in the previous section we found only approximate solutions to the three-body problem, we will now seek exact solutions for which Hill's solution is a limiting case. Here we will restrict ourselves to the three-body problem in the plane and start from the heuristic considerations found in the beginning of § 19. Replacing the particles P_1, P_3 by their common center of mass P_0 with mass $m_1+m_3=\mu$, we may consider, in particular, circular orbits of P_0, P_2 traversed with angular velocity $\omega=1$. We also normalize the unit of mass so that $m_1+m_2+m_3=1$, hence $m_2=1-\mu$ and $0<\mu<1$. As the earlier reasoning shows, the distance between P_0 and P_2 must then be chosen equal to 1. Furthermore, we take the distance between P_1 and P_3 to be small relative to 1, and assume that P_1, P_3 also describe circular orbits about their center of mass P_0. It will now be shown that in a neighborhood of these circular orbits there are actual solutions to the three-body problem.

Since we are dealing with a problem in the plane, we introduce complex coordinates z_k $(k=0, 1, 2, 3)$ for the positions of the particles, as we found occasion to do already in § 14, so that the real and imaginary parts of z_k denote the abscissa and ordinate of P_k. The coordinate axes are assumed to rotate about the center of mass of the three particles P_1, P_2, P_3 with angular velocity 1. Thus

$$m_1 z_1 + m_2 z_2 + m_3 z_3 = 0, \quad \mu z_0 = m_1 z_1 + m_3 z_3 = -m_2 z_2,$$

whereby

$$z_2 = \mu(z_2 - z_0), \quad \mu(z_0 - z_3) = m_1(z_1 - z_3),$$

while from (19; 1), (19; 2) the equations of motion become

$$\ddot{z}_k + 2i\dot{z}_k - z_k = \sum_{l \neq k} m_l(z_l - z_k)\,|z_l - z_k|^{-3} \quad (k=1, 2, 3). \tag{1}$$

We will assume that the distance r_{13} is small relative to 1, the distances r_{12}, r_{23} are approximately equal to 1, and that the points P_0, P_2 in the rotating coordinate system are nearly at rest. This suggests introducing new complex variables x, y by

$$z_1 - z_3 = x, \quad z_0 - z_2 = 1 + y\,.$$

The variables z_0, z_1, z_2, z_3 are then linear functions of x, y, namely,

$$z_0 = (1-\mu)(1+y), \quad z_1 = (1-\mu)(1+y) + \frac{m_3}{\mu}\,x\,,$$

$$z_2 = -\mu(1+y), \qquad z_3 = (1-\mu)(1+y) - \frac{m_1}{\mu}\,x\,,$$

and for $x = y = 0$ we have $z_0 = z_1 = z_3 = 1-\mu$, $z_2 = -\mu$. As a generalization of Hill's problem, one seeks periodic solutions to (1) for which the absolute values of x, y remain sufficiently small. If one introduces the abbreviations

$$m_1 = \delta_1\mu, \quad m_3 = \delta_3\mu, \quad \delta_3 = \delta, \quad \delta_1 = 1-\delta\,,$$

then $0 < \delta < 1$ and

$$z_1 - z_2 = 1 + y + \delta_3 x, \quad z_3 - z_2 = 1 + y - \delta_1 x\,.$$

From (1) the differential equations for x, y are given by

$$\begin{aligned} \ddot{x} + 2i\dot{x} - x &= m_2(z_2 - z_1)r_{12}^{-3} + m_2(z_3 - z_2)r_{23}^{-3} + \mu(z_3 - z_1)r_{13}^{-3}\,, \\ \ddot{y} + 2i\dot{y} - y &= 1 + \delta_1(z_2 - z_1)r_{12}^{-3} + \delta_3(z_2 - z_3)r_{23}^{-3}\,, \end{aligned}$$

wherein the right sides are to be expressed in terms of x, y. Since

$$r_{13}^2 = x\bar{x}, \quad r_{23}^2 = (1 + y - \delta_1 x)(1 + \bar{y} - \delta_1\bar{x})\,,$$
$$r_{12}^2 = (1 + y + \delta_3 x)(1 + \bar{y} + \delta_3\bar{x})\,,$$

we obtain the series expansions

$$\begin{aligned} (z_3 - z_2)\,r_{23}^{-3} &= (1 + y - \delta_1 x)^{-\frac{1}{2}}(1 + \bar{y} - \delta_1\bar{x})^{-\frac{3}{2}} \\ &= 1 - \tfrac{1}{2}(y - \delta_1 x) - \tfrac{3}{2}(\bar{y} - \delta_1\bar{x}) + \cdots\,, \\ (z_1 - z_2)\,r_{12}^{-3} &= (1 + y + \delta_3 x)^{-\frac{1}{2}}(1 + \bar{y} + \delta_3\bar{x})^{-\frac{3}{2}} \\ &= 1 - \tfrac{1}{2}(y + \delta_3 x) - \tfrac{3}{2}(\bar{y} + \delta_3\bar{x}) + \cdots\,. \end{aligned}$$

Because $\delta_3 = \delta$, $\delta_1 = 1 - \delta$, these series converge for $|x| + |y| < 1, 0 \leqq \delta \leqq 1$, and indeed uniformly in x, y, δ for $|x| + |y| \leqq \vartheta$, for each fixed positive $\vartheta < 1$. After insertion of these power series into (1), an easy computation gives

$$(2) \qquad \begin{cases} \ddot{x} + 2i\dot{x} + \tfrac{1}{2}(\mu - 3)\,x + \tfrac{3}{2}(\mu - 1)\,\bar{x} + \mu x^{-\frac{1}{2}}\bar{x}^{-\frac{3}{2}} = P\,, \\ \ddot{y} + 2i\dot{y} - \tfrac{3}{2}(y + \bar{y}) = Q\,, \end{cases}$$

where P and Q are power series in $x, y, \bar{x}, \bar{y}$ that begin with quadratic terms and converge for $|x|+|y|<1$. The coefficients of these series are given as polynomials in μ, δ having rational numbers as coefficients.

We will construct periodic solutions to the system (2), and to that end express $x, \bar{x}, y, \bar{y}$ again as power series in two new variables ξ, η. As in the previous section, ξ, η shall satisfy the differential equations

$$\dot{\xi}=\alpha\xi, \quad \dot{\eta}=-\alpha\eta, \quad \alpha=\pm\frac{i}{4}(\xi\eta)^{-3}. \tag{3}$$

This time we introduce the notation

$$\zeta_{kl}=\xi^{k+2l}\eta^{k-2l}, \quad \zeta_k=\zeta_{k0}=(\xi\eta)^k$$

and assume an expansion in the form

$$x=\mu^{\frac{1}{3}}(1\mp 2\zeta_3)^{\frac{1}{3}}\xi^4\Big(1+\sum_{k>4} a_{kl}\zeta_{kl}\Big), \quad \bar{x}=\mu^{\frac{1}{3}}(1\mp 2\zeta_3)^{\frac{1}{3}}\eta^4\Big(1+\sum_{k>4} a_{kl}\zeta_{k,-l}\Big),$$

$$y=\mu^{\frac{2}{3}}\sum_{k>3} b_{kl}\zeta_{kl}, \qquad \bar{y}=\mu^{\frac{2}{3}}\sum_{k>3} b_{kl}\zeta_{k,-l},$$

where the summation is always to be taken over k and all l with $2|l|\leqq k$. The choice of sign in the factor

$$(1\mp 2\zeta_3)^{\frac{1}{3}}=\gamma$$

appearing in x and $\bar{x}$ is governed by the choice of sign for α in (3). The special form of this expression will justify itself in the matching of coefficients that follows. We now construct the derivatives of x, y with respect to t, expressing $\dot{\xi}, \dot{\eta}$ again in terms of ξ, η, in accordance with (3), and noticing in the process that the factor γ is independent of t. This gives

$$\dot{x}=\mu^{\frac{1}{3}}\gamma\xi^4(\pm i\zeta_3^{-1})\Big(1+\sum_{k>4}(l+1)\,a_{kl}\zeta_{kl}\Big),$$

$$\ddot{x}=\mu^{\frac{1}{3}}\gamma\xi^4(-\zeta_6^{-1})\Big(1+\sum_{k>4}(l+1)^2 a_{kl}\zeta_{kl}\Big),$$

$$\dot{y}=\mu^{\frac{2}{3}}(\pm i\zeta_3^{-1})\sum_{k>3} l b_{kl}\zeta_{kl}, \quad \ddot{y}=\mu^{\frac{2}{3}}(-\zeta_6^{-1})\sum_{k>3} l^2 b_{kl}\zeta_{kl}.$$

We insert the series for $x, y, \bar{x}, \bar{y}$ and the derivatives of x, y into the differential equations (2) and multiply the first equation by $-\mu^{-\frac{1}{3}}\gamma^2\xi^{-4}\zeta_6$ and the second by $-\mu^{-\frac{2}{3}}\zeta_6$. After some slight manipulation we obtain

$$(1\mp 2\zeta_3)\Big(\sum_{k>4}(l+1)^2 a_{kl}\zeta_{kl}\pm 2\sum_{k>4}(l+1)a_{kl}\zeta_{k+3,l}\Big)+\tfrac{1}{2}A+\tfrac{3}{2}B=f, \tag{4}$$

$$\sum_{k>3} l^2 b_{kl}\zeta_{kl}\pm 2\sum_{k>3} l b_{kl}\zeta_{k+3,l}+\tfrac{3}{2}\sum_{k>3}(b_{kl}+b_{k,-l})\,\zeta_{k+6,l}=g, \tag{5}$$

where

$$A=\sum_{k>4} a_{kl}\zeta_{kl}\,,\qquad B=\sum_{k>4} a_{kl}\zeta_{k,-l}\,,$$

$$(6)\quad \begin{cases} f=\{(1+A)^{-\frac12}(1+B)^{-\frac32}-1+\tfrac12 A+\tfrac32 B\} \\ \quad +4\zeta_6+\tfrac12(1\mp 2\zeta_3)\,\{(\mu-3)\,\zeta_6(1+A)+3(\mu-1)\,\zeta_{6,-2}(1+B)\}+ \\ \quad -\mu^{-\frac13}(1\mp 2\zeta_3)^{\frac23}\zeta_{4,-1}P\,, \end{cases}$$

$$(7)\quad g=-\mu^{-\frac23}\zeta_6 Q\,.$$

If in addition one expands $(1+A)^{-\frac12}(1+B)^{-\frac32}$ in powers of A, B and introduces the series expressions for $x, y, \overline{x}, \overline{y}$ into P, Q, then f and g likewise become power series in ξ, η of the form

$$f=\sum_{k\geqq 0} f_{kl}\zeta_{kl}\,,\qquad g=\sum_{k\geqq 0} g_{kl}\zeta_{kl}\,.$$

Since the expansions of P, Q in terms of $x, y, \overline{x}, \overline{y}$ begin with quadratic terms and in our setup the expressions $x, \overline{x}$ contain $\mu^{\frac13}$ as a factor while $y, \overline{y}$ contain $\mu^{\frac23}$, the coefficients f_{kl}, g_{kl} are polynomials in the $a_{\varkappa\lambda}, b_{\varkappa\lambda}, \mu^{\frac13}, \delta$ with rational numbers as coefficients. We wish to look at these polynomials more closely.

The expression $\zeta_{kl}=\xi^{k+2l}\eta^{k-2l}$ has degree $2k$, and we will refer to k as the weight of ζ_{kl}. Since P as a power series in $x, y, \overline{x}, \overline{y}$ begins with terms of at least second degree, while on the other hand the expansions of $x, y, \overline{x}, \overline{y}$ in powers of ξ, η begin with terms of weight at least 2, the expansion of P in powers of ξ, η does not contain any terms of weight <4. The same is true for Q, and therefore by (7) we have $g_{kl}=0$ for $k<10$. Correspondingly, the expansion of $\zeta_{4,-1}\,P$ begins with terms of weight $\geqq 8$. Moreover, since by definition A and B have no terms of weight <5, the expression in the first curved bracket in (6) begins with terms of weight $\geqq 10$. It then follows from (6) that we have $f_{kl}=0$ for $k<8$, with the exception of f_{60} and $f_{6,-2}$. We still wish to determine on which of the $a_{\varkappa\lambda}, b_{\varkappa\lambda}$ the coefficients f_{kl}, g_{kl} depend. To that end we recall that the P, Q as power series in $x, y, \overline{x}, \overline{y}$ begin with quadratic terms, while on the other hand $x, y, \overline{x}, \overline{y}$ as series in ξ, η begin with terms of weight $\geqq 2$. Thus, if the series for $x, y, \overline{x}, \overline{y}$ are inserted into $\zeta_6 Q$, a $b_{\varkappa\lambda}$ that appears in y or $\overline{y}$ can enter $\zeta_6 Q$ only through a term whose weight is at least $\varkappa+6+2=\varkappa+8$. For the $b_{\varkappa\lambda}$ entering into g_{kl} we therefore have the inequality $\varkappa\leqq k-8$. Noting also that the series for x contains the factor $\xi^4=\zeta_{21}$ and for $\overline{x}$ the factor $\eta^4=\zeta_{2,-1}$, one sees on the other hand that $a_{\varkappa\lambda}$ can enter into g_{kl} only if $\varkappa\leqq k-10$. Thus, if we denote by $\mathfrak{P}(r,s)$ any polynomial in $a_{\varkappa\lambda}, b_{kl}, \mu^{\frac13}, \delta$ for $\varkappa\leqq r, k\leqq s$ with rational coefficients, we have

$$(8)\quad g_{kl}=\mathfrak{P}(k-10,k-8)\quad (k\geqq 10)\,.$$

In exactly the same way one concludes that the coefficient of ζ_{kl} in $\zeta_{4,-1}P$ has the form $\mathfrak{P}(k-8, k-6)$. Since A, B contain only the a_{kl}, and these only for $k \geqq 5$, the coefficients of ζ_{kl} in the first curved bracket in (6) have the form $\mathfrak{P}(k-5, 0)$. It therefore follows from (6) that

(9) $$f_{kl} = \mathfrak{P}(k-5, k-6) \quad (k \geqq 6).$$

We now compare coefficients in (4) and (5). Abbreviating

(10) $$F_{kl} = \{(l+1)^2 + \tfrac{1}{2}\} a_{kl} + \tfrac{3}{2} a_{k,-l} \mp 2l(l+1) a_{k-3,l} - 4(l+1) a_{k-6,l} - f_{kl},$$

(11) $$G_{kl} = l^2 b_{kl} \pm 2l b_{k-3,l} + \tfrac{3}{2}(b_{k-6,l} + b_{k+6,-l}) - g_{kl},$$

we have to satisfy the conditions

(12) $$F_{kl} = 0, \quad G_{kl} = 0$$

for all integers k, l with $2|l| \leqq k$. Accordingly, one sets $a_{\varkappa\lambda} = 0$, $b_{\varkappa\lambda} = 0$ for $2|\lambda| > \varkappa$ and, in view of our original expression, also $a_{\varkappa\lambda} = 0$ for $\varkappa < 5$, $b_{\varkappa\lambda} = 0$ for $\varkappa < 4$. Since $g_{kl} = 0$ for $k < 10$, the conditions $G_{k0} = 0$ are automatically satisfied when $k < 10$, as are the conditions $G_{kl} = 0$ when $k < 4$. Analogously, $F_{kl} = 0$ already holds for $k < 5$. Equations (10), (11) say in particular that

(13) $$F_{k0} = 3a_{k0} - 4a_{k-6,0} - f_{k0}, \quad G_{k0} = 3b_{k-6,0} - g_{k0}$$

and

(14) $$\begin{cases} F_{k1} = \tfrac{9}{2} a_{k1} + \tfrac{3}{2} a_{k,-1} \mp 4a_{k-3,1} - 8a_{k-6,1} - f_{k1}, \\ F_{k,-1} = \tfrac{3}{2} a_{k1} + \tfrac{1}{2} a_{k,-1} - f_{k,-1} \end{cases}$$

so that also

(15) $$F_{k+3,1} - 3F_{k+3,-1} = \mp 4a_{k1} - 8a_{k-3,1} - f_{k+3,1} + 3f_{k+3,-1}.$$

For $l \neq 0, \pm 1$, in addition to (10) we consider

(16) $$\begin{cases} F_{k,-l} = \tfrac{3}{2} a_{kl} + \{(1-l)^2 + \tfrac{1}{2}\} a_{k,-l} + \\ \qquad \pm 2l(1-l) a_{k-3,-l} - 4(1-l) a_{k-6,-l} - f_{k,-l} \end{cases}$$

and treat (10), (16) as two linear equations for $a_{kl}, a_{k,-l}$. Analogous to (19; 21), their determinant is

$$\{(l+1)^2 + \tfrac{1}{2}\}\{(1-l)^2 + \tfrac{1}{2}\} - (\tfrac{3}{2})^2 = l^2(l^2-1) > 0 \quad (l^2 > 1).$$

The induction argument now proceeds as follows. Let r be a natural number, and consider the equations

(17) $$G_{kl} = 0 \quad (l \neq 0), \quad G_{k+6,0} = 0,$$

(18) $$F_{kl} = 0 \quad (l \neq 1), \quad F_{k+3,1} - 3F_{k+3,-1} = 0$$

for $k<r$. According to (8), (9), (10), (11), (13), (14), (15) their left sides are polynomials in the $a_{\varkappa\lambda}$, $b_{\varkappa\lambda}$ with $\varkappa<r$, and we assume that these equations are already known to have a unique solution. For $r<5$ this assumption is trivially satisfied by our specification that $a_{\varkappa\lambda}=0\,(\varkappa<5)$, $b_{\varkappa\lambda}=0\,(\varkappa<4)$, since $g_{kl}=0\,(k<10)$, $f_{kl}=0\,(k<6)$, $f_{6l}=0\,(l\neq 0,-2)$. For $r=5$ the equations $G_{4l}=0\,(l\neq 0)$ and $G_{10,0}=0$ uniquely determine $b_{4l}\,(l\neq 0)$ and b_{40}, while because $f_{7l}=0$, equations (18) with $k=4$ are again satisfied trivially. Now let $r>5$. In view of (8), (11), (13), equations (17) with $k=r$ again uniquely determine $b_{rl}\,(l\neq 0)$ and b_{r0}, while by (9), (10), (13), (16) the $a_{rl}\,(l\neq \pm 1)$ are uniquely determined from $F_{rl}=0\,(l\neq \pm 1)$, as by (9), (14), (15) are also $a_{r1}, a_{r,-1}$ from $F_{r+3,1}-3F_{r+3,-1}=0$, $F_{r,-1}=0$. This verifies the inductive assumption for $r+1$ in place of r, and it is therefore valid for all r. From (17), (18) it now follows that $G_{kl}=0$ for $k\geqq 0$, $l\neq 0$ and for $k\geqq 6$, $l=0$, as well as $F_{kl}=0$ for $k\geqq 0$, $l\neq 1$ and for $k\geqq 3$, $l=1$, while for the remaining cases $k<6$, $l=0$ and $k<3$, $l=1$ conditions (12) are trivially satisfied. Since the f_{kl}, g_{kl} were polynomials in the $a_{\varkappa\lambda}$, $b_{\varkappa\lambda}$, $\mu^{\frac{1}{3}}$, δ with rational numbers as coefficients, and since the recursive calculation of the a_{kl}, b_{kl} from (17), (18) involved only the solving of linear equations with rational coefficients in their homogeneous parts, it follows that the a_{kl}, b_{kl} are all uniquely determined as polynomials in $\mu^{\frac{1}{3}}$, δ with rational numbers as coefficients, so that, in particular, they are all real.

Convergence of the above series for $x, y, \bar{x}, \bar{y}$ can be proved by the method of majorants just as in the case of Hill's solution, and since this would not involve anything new the proof will not be carried out here. The series are found to converge absolutely and uniformly in the region $0\leqq\mu\leqq 1$, $0\leqq\delta\leqq 1$, $|\xi|<c$, $|\eta|<c$, where c is a positive universal constant [1].

We now insert the solution

$$\xi=\varrho e^{\alpha t}, \quad \eta=\varrho e^{-\alpha t}, \quad \alpha=\pm\frac{i}{4}\varrho^{-6} \quad (0<\varrho<c)$$

of (3) into the power series for $x, y, \bar{x}, \bar{y}$, whereby we still have an arbitrary translation of the time variable at our disposal. Then ξ, η are complex conjugates, so that $\zeta_{k,-l}=\bar{\zeta}_{kl}$, and, because the coefficients a_{kl}, b_{kl} are real, the values of the series $\bar{x}, \bar{y}$ are likewise complex conjugate to those of x, y. Thus, for each sign of α we have found a family of solutions to the planar three-body problem that depend on the parameter ϱ $(0<\varrho<c)$ and have period $\left|\frac{\pi i}{2\alpha}\right|=2\pi\varrho^6$ in the rotating coordinate system. It should be emphasized that these solutions exist for any choice of μ, δ in the intervals $0\leqq\mu\leqq 1$, $0\leqq\delta\leqq 1$, so that there is no restriction on the ratios of the three masses. For example, the case $\mu=\frac{2}{3}$, $\delta=\frac{1}{2}$, which corresponds to all the masses being equal, is admissible.

Here the limiting case $\mu=0$, $\delta=0$ gives the solution of Hill derived in the previous section. There the discussion of the recursion formulas was simpler because $2l$ appeared in place of the l here, and so the special case $l=\pm 1$ was absent. The case $\delta=0$, $0<\mu<1$ corresponds to the restricted three-body problem, in which the mass of the moon is taken to be 0. For this case the periodic solution was constructed by Brown [2] using the method of Hill. The generalized solution given here was also found by Moulton [3] and Perron [4] using a different method, namely, the continuation method due to Poincaré.

The result of this section can be generalized to give periodic solutions of the four-body problem. Starting with Lagrange's equilateral solution of the three-body problem, one may replace one of these particles by a pair which are very close to each other and perform approximately circular motion about the removed point. Crandall [5] has shown that near this configuration there are actual solutions of the four-body problem which are periodic in a rotating coordinate system. He also found another type of periodic solution of the four-body problem by starting with a circular solution of the two-body problem and replacing one of the particles by three that lie on a small equilateral triangle rotating about the removed point. Similar periodic solutions, obtained by splitting one particle into two, had been discussed by Perron [6, 7], who applied Poincaré's continuation method which will be discussed in the next section.

§ 21. The Continuation Method

In the first several sections of this chapter we dealt with a method for determining periodic solutions to a Hamiltonian system by assuming an expansion in power series, whereas in the preceding two sections we used a similar method to obtain periodic solutions to the three-body problem. In the ensuing section we will discuss still another method for determining periodic solutions to a system of differential equations. Although for the most part the results that follow can be obtained under weaker assumptions, for the sake of simplicity we will retain the assumption of regularity.

We consider a system of differential equations

$$\dot{x}_k = f_k(x, \alpha) \qquad (k=1, \ldots, m) \tag{1}$$

depending on a parameter α. The right sides f_k are assumed to be regular functions of the $m+1$ complex variables x_l, α in a region

$$|x_l - \xi_l^*| < r \qquad (l=1, \ldots, m), \qquad \alpha \in G \tag{2}$$

where G is a domain in the complex α-plane. Moreover, in this region the functions are assumed to satisfy

$$(3) \qquad |f_k(x,\alpha)| \leqq M \quad (k=1,\ldots,m).$$

Before turning to the continuation method, we will first investigate the dependence of the solutions to (1) on the parameter α and the initial values $\xi_1, \ldots, \xi_m$.

Let $\xi_1, \ldots, \xi_m$ be arbitrary complex numbers subject to the restrictions

$$(4) \qquad |\xi_l - \xi_l^*| < \frac{r}{2} \quad (l=1,\ldots,m).$$

The $f_k(x,\alpha)$ are then regular functions of the variables x_l, α in the region

$$|x_l - \xi_l| < \frac{r}{2}, \qquad \alpha \in G$$

where they satisfy the estimate (3). By the existence theorem of Cauchy in § 4, the system (1) has a solution $x(t,\xi,\alpha)$ for which $x(0,\xi,\alpha)=\xi$ and the $x_k(t,\xi,\alpha)$ $(k=1,\ldots,m)$ are regular analytic functions of the complex variable t in the disk

$$|t| < \frac{r}{2(m+1)M} = \varrho.$$

This is true for each choice of ξ in the region (4) and each α in G. We will now show that the $x_k(t,\xi,\alpha)$ are regular functions of all $m+2$ independent complex variables t, ξ_l, α in the region

$$(5) \qquad |t| < \varrho, \quad |\xi_l - \xi_l^*| < \frac{r}{2} \; (l=1,\ldots,m), \qquad \alpha \in G.$$

This follows from the proof of Cauchy's theorem given in § 4. Namely, in the comparison of coefficients there, the coefficients α_{kn} in the series expansions of $x_k(t,\xi,\alpha)$ in powers of t were found to be polynomials in the coefficients of the Taylor expansions for the $f_l(x,\alpha)$ $(l=1,\ldots,m)$ in powers of $x_1-\xi_1, \ldots, x_m-\xi_m$, while by Taylor's formula the latter coefficients are analytic functions of $\xi_1, \ldots, \xi_m, \alpha$ in the region $|\xi_h - \xi_h^*| < \frac{r}{2}$ $(h=1,\ldots,m)$, $\alpha \in G$. On the other hand, the expansions of the $x_k(t,\xi,\alpha)$ in powers of t were majorized by a series whose coefficients depended only on M and r, so that the former converge in each closed subdomain of $|t|<\varrho$ uniformly with respect to the ξ_h, α. Consequently, by a theorem of Weierstrass, the $x_k(t,\xi,\alpha)$ are regular functions of all $m+2$ variables in the region (5).

If for t in the interval $0<t<\varrho$, say for example $t=\varrho/2$, one chooses $x_k(t,\xi,\alpha)$ again as an initial value and proceeds in the above manner, possibly one can analytically continue the solution beyond the point $t=\varrho$. Suppose that for a fixed $\xi=\xi^*, \alpha=\alpha^*$ the solution $x(t,\xi,\alpha)$ as a function of t can be continued in this way over the interval $0 \leqq t \leqq t_1$.

If in the process the curve $x(t, \xi^*, \alpha^*)$ for $0 \leqq t \leqq t_1$ remains completely in the region of regularity for $f_1(x, \alpha), \ldots, f_m(x, \alpha)$, a repeated application of the previous reasoning together with a compactness argument will show that there exists a sufficiently small neighborhood U of $\xi = \xi^*$, $\alpha = \alpha^*$ wherein the solution $x(t, \xi, \alpha)$ can be continued over the interval $0 \leqq t \leqq t_1$ to give a function regular in all the variables t, ξ, α. Of course, the larger one takes t_1, the smaller in general are the neighborhoods U in which this regularity is guaranteed. It should be observed that the above analysis can also be carried out for systems of differential equations

$$\dot{x}_k = f_k(x, t, \alpha) \quad (k = 1, \ldots, m) \tag{6}$$

whose right sides contain the independent variable t explicitly, for by introducing an additional unknown x_0 and replacing (6) by

$$\dot{x}_0 = 1, \quad \dot{x}_k = f_k(x, x_0, \alpha) \quad (k = 1, \ldots, m)$$

one is led to a system of $m + 1$ equations in which the variable t no longer appears on the right.

Since for $0 \leqq t \leqq t_1$ the $x_k(t, \xi, \alpha)$ are regular functions of the ξ_l in a neighborhood of $\xi = \xi^*$, they possess in particular the partial derivatives $x_{k\xi_l}(t, \xi, \alpha)$. Because in the upcoming discussion the dependence on α plays no role, α being considered fixed, for the time being we will leave the symbol α out of the functional notation. If now for a fixed system ξ_l^* $(l = 1, \ldots, m)$ of initial values the solution $x(t, \xi^*)$ is known, the partial derivatives $x_{k\xi_l} = x_{k\xi_l}(t, \xi^*)$ can be determined as follows from the so-called equations of variation. Since the variables ξ_l, t in $x_k(t, \xi)$ are to be treated as independent, differentiation of the equations (1) with respect to ξ_l shows that

$$\dot{x}_{k\xi_l} = \sum_{r=1}^{m} f_{kx_r} x_{r\xi_l}, \quad f_{kx_r} = f_{kx_r}(x(t, \xi^*)) \quad (k, l = 1, \ldots, m).$$

Introducing the $m \times m$ matrices $\mathfrak{X} = (x_{k\xi_l})$, $\mathfrak{F} = (f_{kx_l})$, one then obtains for $\mathfrak{X}$ the equation of variation

$$\dot{\mathfrak{X}} = \mathfrak{F}\mathfrak{X}, \tag{7}$$

where $\mathfrak{F}$ is known. Because $x(0, \xi) = \xi$, we have $\mathfrak{X} = \mathfrak{E}$ at $t = 0$. The matrix $\mathfrak{X} = \mathfrak{X}(t)$ is thus obtained by integrating the linear differential equation (7) with the initial condition $\mathfrak{X}(0) = \mathfrak{E}$. This integration can be carried out by, say, the method of successive approximations via the integral equation

$$\mathfrak{X} = \mathfrak{E} + \int_0^t \mathfrak{F}\mathfrak{X}\,dt,$$

whereby one constructs the sequence of matrices

$$\mathfrak{X}_0 = \mathfrak{E}, \quad \mathfrak{X}_n = \mathfrak{E} + \int_0^t \mathfrak{F}\mathfrak{X}_{n-1}\,dt \quad (n = 1, 2, \ldots),$$

or also by comparison of coefficients as in the existence theorem of Cauchy.

For the determinant $\Delta = |\mathfrak{X}|$ we have

$$\dot{\Delta} = \sum_{k,l=1}^{m} \dot{x}_{k\xi_l} X_{lk}, \tag{8}$$

where X_{lk} denotes the cofactor of the element $x_{k\xi_l}$ in the matrix $\mathfrak{X}$, that is, the determinant of the minor corresponding to $x_{k\xi_l}$ adjusted by the factor $(-1)^{k+l}$. Upon introduction of the matrix $\mathfrak{Y} = (X_{kl})$ with the cofactors of the elements of $\mathfrak{X}$ as entries, (8) can be expressed in the form

$$\dot{\Delta} = \sigma(\dot{\mathfrak{X}}\mathfrak{Y}),$$

where the symbol σ denotes the trace function. It then follows from (7) that

$$\dot{\Delta} = \sigma(\mathfrak{F}\mathfrak{X}\mathfrak{Y}),$$

and because $\mathfrak{X}\mathfrak{Y} = \Delta\mathfrak{E}$, this gives

$$\dot{\Delta} = \Delta\sigma(\mathfrak{F}) = \Delta\sigma \tag{9}$$

with

$$\sigma = \sigma(\mathfrak{F}) = \sum_{k=1}^{m} f_{kx_k}(x)$$

and $x = x(t, \xi^*)$. Upon integrating (9) and recalling that $\mathfrak{X}(0) = \mathfrak{E}$, so that the initial value $\Delta(0)$ of $\Delta = \Delta(t, \xi^*) = \Delta(t)$ is equal to 1, we finally have

$$\log\Delta = \int_0^t \sigma\,dt.$$

The system (1) can be interpreted as the differential equations of a flow, with the x_k $(k = 1, \dots, m)$ being the coordinates of an element of fluid. Since the right sides do not contain the variable t explicitly, the flow is stationary. For $t = 0$ the position of an element of fluid is described by the coordinates ξ_k, and in time t it moves from ξ to $x(t, \xi)$. This results in a mapping from ξ to x whose Jacobian matrix is precisely $(x_{k\xi_l}) = \mathfrak{X}(t, \xi)$ with determinant Δ. The mapping will be volume-preserving, i.e. the flow incompressible, if Δ is identically 1, or, in view of (9), if

$$\sigma = \sum_{k=1}^{m} f_{kx_k} = 0. \tag{10}$$

For a Hamiltonian system

$$\dot{x}_k = E_{y_k}, \quad \dot{y}_k = -E_{x_k} \quad (k = 1, \dots, n)$$

we have

$$\sigma = \sum_{k=1}^{n} \{(E_{y_k})_{x_k} + (-E_{x_k})_{y_k}\} = 0,$$

so that (10) is indeed satisfied.

The continuation method, due to Poincaré [1], is concerned with the following problem. Let the solutions $x(t, \xi, \alpha)$ of the system (1) be again considered as dependent on ξ, α, and assume that a periodic solution is known for $\alpha = \alpha^*$. Let its initial value be $\xi = \xi^*$, so that $x = x(t, \xi^*, \alpha^*)$ is this periodic solution which we assume not to be an equilibrium solution. If $\tau^* > 0$ is a period of $x(t, \xi^*, \alpha^*)$ with respect to t, not necessarily the smallest positive period, let us also assume that the curve $x(t, \xi^*, \alpha^*)$ for $0 \leqq t \leqq \tau^*$ lies entirely in the region of regularity of $f_1, \ldots, f_m$ as functions of x and α, which in view of

$$x(t + \tau^*, \xi^*, \alpha^*) = x(t, \xi^*, \alpha^*) \tag{11}$$

will then be true for all real t. Moreover, by the uniqueness theorem for differential equations, (11) will hold for all t as soon as it holds for one value, say $t = 0$. We now ask whether the system (1) also has periodic solutions corresponding to certain initial values ξ, α near ξ^*, α^*.

First we seek periodic solutions with the same period τ^*. By the uniqueness theorem, for $x(t, \xi, \alpha)$ to have period τ^* it is necessary and sufficient that $x(\tau^*, \xi, \alpha) = x(0, \xi, \alpha) = \xi$. Thus, setting

$$\phi_k(\xi, \alpha) = x_k(\tau^*, \xi, \alpha) - \xi_k, \tag{12}$$

we have to satisfy the m analytic equations

$$\phi_k(\xi, \alpha) = 0 \quad (k = 1, \ldots, m). \tag{13}$$

This is an implicit system, which in view of the periodicity of our original solution is satisfied when $\xi = \xi^*, \alpha = \alpha^*$. If the $m \times m$ Jacobian determinant $|\phi_{k\xi_l}|$ did not vanish at $\xi = \xi^*, \alpha = \alpha^*$, then by the known existence theorem for implicit functions the system (13) would indeed have solutions in a neighborhood of $\alpha = \alpha^*$, with the differences $\xi_k - \xi_k^*$ expressable as power series in $\alpha - \alpha^*$ without constant term. This contention, however, is vacuous here, since the determinant $|\phi_{k\xi_l}|$ is necessarily always 0 at the point in question. Nevertheless, by a slight modification of our reasoning this difficulty can be sidestepped, and to that end we investigate why the determinant must always vanish. With $\alpha = \alpha^*$, setting $\mathfrak{X} = (x_{k\xi_l}(t, \xi))$ and introducing the matrix $\mathfrak{C}(t, \xi) = \mathfrak{X} - \mathfrak{E}$, from (12) we have

$$(\phi_{k\xi_l}) = \mathfrak{C}(\tau^*, \xi^*) = \mathfrak{C}. \tag{14}$$

Now any point ξ on the solution curve $x(t, \xi^*)$ can be expressed as

(15) $$\xi = x(t', \xi^*)$$

for a suitable choice of t', and as t' varies this defines ξ as a function of t'. Since the right sides of the differential equations (1) do not depend explicitly on t, it follows that

(16) $$x(t+t', \xi^*) = x(t, \xi).$$

Differentiating (16) with respect to t' and using (1), (15), one is led to the equations

$$f_k(x(t,\xi)) = \sum_{l=1}^{m} x_{k\xi_l}(t,\xi) f_l(\xi) \quad (k=1, \ldots, m),$$

so that

$$f(x) = \mathfrak{X} f(\xi), \quad f(x) - f(\xi) = \mathfrak{C}(t, \xi) f(\xi),$$

where $f(x)$ is the vector with components $f_1(x), \ldots, f_m(x)$ and $x = x(t, \xi)$. In particular, for $\xi = \xi^*, t = \tau^*$ we have $f(x) = f(\xi)$ and therefore

(17) $$\mathfrak{C} f = 0, \quad f = f(\xi^*).$$

Since the original periodic solution $x(t, \xi^*)$ is not an equilibrium solution, the vector $f(\xi^*)$ is not null, and it follows that $|\mathfrak{C}| = 0$. The reason behind the vanishing of the determinant of $\mathfrak{C}$ lies in the fact that translation of the initial value ξ along the solution curve $x(t, \xi^*)$ again leads to a periodic solution, namely the same orbit with the variable t shifted by a constant t'. To avoid this, we restrict the initial values ξ to vary over an $(m-1)$-dimensional plane not tangent to the solution curve at the initial point ξ^*. Having seen that $f(\xi^*)$ is not the null vector, we may assume that $f_m(\xi^*) \neq 0$, in which case, by (1), the plane $x_m = \xi_m^*$ will have the desired property. Thus we set $\xi_m = \xi_m^*$ and allow only the $m-1$ initial values $\xi_1, \ldots, \xi_{m-1}$ to vary. However, since we have to satisfy m equations in (13), we will also consider the period τ of the solution we seek as variable.

Setting

(18) $$\phi_k(\tau, \xi, \alpha) = x_k(\tau, \xi, \alpha) - \xi_k,$$

we then have to satisfy the m equations

(19) $$\phi_k(\tau, \xi, \alpha) = 0 \quad (k = 1, \ldots, m)$$

subject to the additional condition that $\xi_m = \xi_m^*$, whereby $\tau = \tau^*, \xi = \xi^*$, $\alpha = \alpha^*$ is a known solution. Treating $\xi_1, \ldots, \xi_{m-1}$ and τ in (19) as unknowns and α as the independent variable, we now have to look at the corresponding $m \times m$ Jacobian matrix $\mathfrak{B}$ which is obtained from $\mathfrak{C}$ by replacing the last column ϕ_{ξ_m} by the column $\phi_\tau = \phi_\tau(\tau^*, \xi^*, \alpha^*)$. In view of (1), (18) this latter column is given by

$$\phi_\tau = \dot{x}(\tau, \xi, \alpha) = f(x(\tau, \xi, \alpha), \alpha)$$

evaluated at $\tau = \tau^*, \xi = \xi^*, \alpha = \alpha^*$, whereby

$$\phi_\tau = f(\xi^*, \alpha^*) = f$$

and

$$\mathfrak{B} = (\phi_{\xi_1} \cdots \phi_{\xi_{m-1}} f), \tag{20}$$

with the first $m-1$ columns being $\phi_{\xi_k} = \phi_{\xi_k}(\tau^*, \xi^*, \alpha^*)$ $(k = 1, \ldots, m-1)$. If the determinant $|\mathfrak{B}| \neq 0$, then the system of equations (19) with $\xi_m = \xi_m^*$ can be solved for $\tau, \xi_1, \ldots, \xi_{m-1}$ in a neighborhood of $\alpha = \alpha^*$ and the differences $\tau - \tau^*, \xi_1 - \xi_1^*, \ldots, \xi_{m-1} - \xi_{m-1}^*$ expressed as power series in $\alpha - \alpha^*$ without constant term. Thus, for all parameter values α sufficiently near α^* one then can determine initial values $\xi_1, \ldots, \xi_{m-1}, \xi_m = \xi_m^*$, as well as a period τ, such that the corresponding solutions with these initial values are periodic with period τ. According to (14), to compute the matrix $\mathfrak{B}$ one need only integrate the linear system (7) of the equations of variation, wherein $x = x(t, \xi^*, \alpha^*)$ is to be taken as the periodic solution with initial values $\xi = \xi^*, \alpha = \alpha^*$ which was assumed to be known. Simple examples will show that, as opposed to $|\mathfrak{C}|$, the determinant $|\mathfrak{B}|$ is not always 0.

Poincaré applied his method also to the more general case where the right sides of the given differential equations depend explicitly on t, although they must then be periodic functions of t. It is then assumed that there exists a periodic solution with the same period, and examples show that now the analogously constructed determinant $|\mathfrak{C}|$ need no longer vanish. This is indeed plausible, since the earlier grounds for the vanishing of this determinant were strongly based on the flow being stationary. We will not, however, go any further into the important and interesting questions arising in the theory of differential equations with periodic coefficients. The known methods and results can for the most part be interpreted in terms of our case of stationary flow, while for the as yet unsolved problems one generally begins with the simplest non-trivial case anyway.

We will show next that the knowledge of a time-independent integral $\psi(x, \alpha)$ that is not stationary at $x = \xi^*, \alpha = \alpha^*$, i.e. whose gradient does not vanish there, likewise allows us to obtain periodic solutions when α is near α^*, this time with the same period $\tau = \tau^*$ as our initial solution. Let the integral $\psi(x, \alpha)$ be analytic in x, α in a neighborhood of the periodic solution $x(t, \xi^*, \alpha^*)$ and the value $\alpha = \alpha^*$. Being an integral for the system (1), the function $\psi=\psi(x,\alpha)$ satisfies the partial differential equation

$$\sum_{k=1}^{m} \psi_{x_k} f_k(x,\alpha)=0$$

identically in x, α. With ψ_x denoting the row vector with components $\psi_{x_k}(\xi^*,\alpha^*)$, we then have

$$\psi_x f=0, \tag{21}$$

and since ψ was assumed not to be stationary at $x=\xi^*, \alpha=\alpha^*$, the vector ψ_x is not null. On the other hand, we had $f_m(\xi^*,\alpha^*)\neq 0$ so that by (21) also the values $\psi_{x_k}(\xi^*,\alpha^*)$ for $k=1,\dots,m-1$ cannot all be 0, and we can choose the notation so that, say, $\psi_{x_{m-1}}(\xi^*,\alpha^*)\neq 0$. To be an integral, the function $\psi(x,\alpha)$ must be constant along each solution curve, and therefore

$$\psi(x(t,\xi,\alpha),\alpha)=\psi(\xi,\alpha) \tag{22}$$

is an identity in t, ξ, α. Differentiation of this with respect to ξ_l gives the equations

$$\sum_{k=1}^{m} \psi_{x_k}(x,\alpha) x_{k\xi_l}=\psi_{x_l}(\xi,\alpha) \quad (l=1,\dots,m)$$

along each solution curve $x=x(t,\xi,\alpha)$, which for $t=\tau^*, \xi=\xi^*, \alpha=\alpha^*$, and therefore $x=\xi^*$, can be expressed in vector form as

$$\psi_x \mathfrak{X}(\tau^*,\xi^*)-\psi_x=\psi_x \mathfrak{C}=0. \tag{23}$$

It follows from (14), (20), (21), (23) that $\psi_x \mathfrak{B}=0$, so that in case of the existence of a nonstationary integral the determinant $|\mathfrak{B}|$ vanishes and the previous method is not directly applicable. The conditions for $x(t,\xi,\alpha)$ to be a periodic solution with period τ^* are given by the m equations (13). We again stipulate that $\xi_m=\xi_m^*$, which leaves us with $m-1$ unknowns $\xi_1,\dots,\xi_{m-1}$ to fulfill the m equations. We then solve the $m-1$ equations

$$\phi_k(\xi,\alpha)=0 \quad (k\neq m-1) \tag{24}$$

for $\xi_l-\xi_l^*$ $(l=1,\dots,m-1)$ as power series in $\alpha-\alpha^*$ without constant terms, assuming that the corresponding Jacobian determinant does not vanish at the point $\xi=\xi^*, \alpha=\alpha^*$. The $(m-1)\times(m-1)$ Jacobian matrix $\mathfrak{A}$ in question is obtained from $\mathfrak{C}$ by deleting the last column and the next to last row, and under the assumption $|\mathfrak{A}|\neq 0$ the equations (24) are then satisfied in a neighborhood of $\alpha=\alpha^*$. It remains to show that, because of the existence of the integral $\psi(x,\alpha)$, also the remaining equation

$\phi_{m-1}(\xi, \alpha)=0$ is satisfied. If $x=x(t, \xi, \alpha)$ is the solution corresponding to the above $\xi_1, \ldots, \xi_{m-1}$ and $\xi_m=\xi_m^*$, then equation (22) is satisfied along the corresponding solution curve. In particular, setting $t=\tau^*$ and applying the mean value theorem to the function $\psi(x, \alpha)$ with respect to the variable x_{m-1}, in view of (24) we have

$$0=\psi(x(\tau^*, \xi, \alpha), \alpha)-\psi(\xi, \alpha)=\psi_{x_{m-1}}(\tilde{x}, \alpha)\phi_{m-1}(\xi, \alpha),$$

where $\tilde{x}_k=\xi_k$ $(k \neq m-1)$ and $\tilde{x}_{m-1}$ lies between ξ_{m-1} and $x_{m-1}(\tau^*, \xi, \alpha)$. From the assumption $\psi_{x_{m-1}}(\xi^*, \alpha^*) \neq 0$ it follows that also $\psi_{x_{m-1}}(\tilde{x}, \alpha) \neq 0$ if α is sufficiently near α^*, in which case we must have $\phi_{m-1}(\xi, \alpha)=0$, as was to be shown. Thus, under the assumption that $|\mathfrak{A}| \neq 0$, there exist periodic solutions in a neighborhood of α^* with the fixed period τ^*.

On the other hand, one may treat τ instead of α as a parameter and fix $\alpha=\alpha^*$, whereby if the $m-1$ periodicity conditions

$$\phi_k(\tau, \xi, \alpha^*)=x_k(\tau, \xi, \alpha^*)-\xi_k=0 \quad (k \neq m-1) \tag{25}$$

are satisfied, the remaining one will follow as before. The relevant Jacobian determinant, at $\tau=\tau^*, \xi=\xi^*$, is again $|\mathfrak{A}|$, and consequently, under the assumption $|\mathfrak{A}| \neq 0$, there exist also for the fixed parameter value α^* periodic solutions corresponding to each prescribed period τ sufficiently near τ^*, with the initial values of these solutions having power series expansions in $\tau-\tau^*$. For some purposes it is useful, instead of τ, to introduce as a new variable the value $\psi(x, \alpha)=\gamma$ of the integral along the indicated closed orbits. Let $\gamma=\gamma^*$ for the original orbit $x=x(t, \xi^*, \alpha^*)$. To the $m-1$ equations (25) we must then add

$$\psi(\xi, \alpha^*)-\gamma=0, \tag{26}$$

giving us m equations for the m unknowns $\xi_1, \ldots, \xi_{m-1}, \tau$. The Jacobian matrix of this system at the point $\xi=\xi^*, \tau=\tau^*$ is obtained by deleting the m-th column and the $(m-1)$-th row from the matrix

$$\mathfrak{D}=\begin{pmatrix} \mathfrak{C} & f \\ \psi_x & 0 \end{pmatrix} \tag{27}$$

of $m+1$ rows. In view of the relations (17), (21), (23) the $(m-1)$-th row of $\mathfrak{D}$ is dependent on the other rows, as is the m-th column upon the other columns, and therefore the nonvanishing of our Jacobian determinant corresponds to the matrix $\mathfrak{D}$ having rank m. Under this assumption the above system of equations can be solved in a neighborhood of $\gamma=\gamma^*$ for $\xi_k-\xi_k^*$ $(k=1, \ldots, m-1)$ and $\tau-\tau^*$ as power series in $\gamma-\gamma^*$, and validity of the equation $\phi_{m-1}(\tau, \xi, \alpha^*)=0$, and with it periodicity, then follows from the existence of the integral.

Finally, with α treated as a variable and $\gamma=\gamma^*$ held fixed, under the same assumptions one obtains corresponding series expansions in powers of $\alpha-\alpha^*$. This gives for each α near α^* a periodic solution near the original solution having the same value $\gamma=\gamma^*$ for the integral.

The computation of the various power series expansions appearing in this discussion requires, however, the knowledge of the full solution $x(t,\xi)$ in a neighborhood of $\xi=\xi^*$, $t=\tau^*$. On the other hand, to investigate the nonvanishing of the relevant Jacobian determinants one only has to integrate the linear system (7) along the original periodic solution.

The knowledge of more time-independent integrals allows the above method to be modified accordingly, but we do not wish to pursue this question any further.

We will now apply the continuation method to the restricted three-body problem. Let the points P_1, P_2, P_3 have masses $m_1=\mu$, $m_2=1-\mu$, $m_3=0$, with $0<\mu<1$, and let P_1, P_2 rotate about their center of mass with angular velocity 1. As in § 19, we introduce a rotating coordinate system in which the coordinates of P_1, P_2, P_3 are $(1-\mu,0)$, $(-\mu,0)$, (x,y). If in addition we set $x_1=x$, $x_2=y$, $x_3=\dot{x}$, $x_4=\dot{y}$, by (19;3) the equations of motion for P_3 become

$$\text{(28)}\quad \dot{x}_1=x_3,\ \dot{x}_2=x_4,\ \dot{x}_3=2x_4+x_1+F_{x_1},\ \dot{x}_4=-2x_3+x_2+F_{x_2}$$

with

$$\text{(29)}\qquad F=(1-\mu)\{(x_1+\mu)^2+x_2^2\}^{-\frac{1}{2}}+\mu\{(x_1+\mu-1)^2+x_2^2\}^{-\frac{1}{2}},$$

which is a system of the form (1) with the parameter $\alpha=\mu$ and $m=4$. For $\mu=0$ also the mass of P_1 vanishes, whereby

$$\text{(30)}\quad x_1=rc,\ x_2=rs,\ x_3=-r\omega s,\ x_4=r\omega c,\ c=\cos(\omega t),\ s=\sin(\omega t)$$

gives a periodic solution with period $\tau^*=2\pi|\omega|^{-1}$, where $\omega\neq 0$ is a real constant and $r^3(\omega+1)^2=1$. We choose this as our initial solution corresponding to $\alpha^*=0$. It is also necessary to stipulate that $r\neq 1$, for otherwise P_3 would pass through the location $(1,0)$ of P_1, while, on the other hand, the point $x_1=1-\mu$, $x_2=0$ for $\mu\neq 0$ is a singularity of the system (28) which in the limit as $\mu\to 0$ coincides with P_1. We therefore assume that $\omega\neq -2, -1, 0$ and seek periodic solutions to (28) for sufficiently small positive values of μ.

If the right sides of the system (28) are denoted by $f_k(x,\mu)$ $(k=1,\dots,4)$ and the initial values for the original solution at $t=0$ by $\xi_1^*=r$, $\xi_2^*=\xi_3^*=0$, $\xi_4^*=r\omega$, then $f_3(\xi^*,0)=-r\omega^2\neq 0$ and in place of $f_m\neq 0$ we now have $f_3\neq 0$. To apply the continuation method we first solve the equation of variation (7). Here it can be integrated in terms of elementary functions, with the substitution

$$y_{2k-1}=x_{2k-1}c+x_{2k}s,\qquad y_{2k}=-x_{2k-1}s+x_{2k}c\qquad (k=1,2)$$

proving useful in the process. Then we construct the matrix $\mathfrak{C} = \mathfrak{X} - \mathfrak{E}$ and from it obtain, by (20), the matrix $\mathfrak{B}$, where here the third rather than the last column of $\mathfrak{C}$ has to be replaced by f. The computation results in $|\mathfrak{B}| = 0$, so that the first method discussed does not apply. This is due to the existence of the so-called Jacobian integral

$$\psi(x, \mu) = \tfrac{1}{2}(x_3^2 + x_4^2 - x_1^2 - x_2^2) - F \tag{31}$$

for the restricted three-body problem, where in our case $\psi_{x_4}(\xi^*, 0) = r\omega \neq 0$, giving us $\psi_{x_4} \neq 0$ in place of $\psi_{x_{m-1}} \neq 0$. For $\mathfrak{A}$ the three-by-three matrix obtained by deleting the third column and fourth row from $\mathfrak{C}$, a direct computation gives

$$|\mathfrak{A}| = 24\pi \sin^2 \frac{\pi}{\omega},$$

and to have this determinant nonzero we must stipulate that in addition to $\omega \neq -2, -1, 0$ also

$$\omega \neq g^{-1} \qquad (g = \pm 1, \pm 2, \ldots). \tag{32}$$

For the radius r of our original solution this excludes the corresponding values $(g^{-1} + 1)^{-\frac{2}{3}}$ and their limit 1. Under these assumptions, there exist for sufficiently small values of μ periodic solutions to the system (28) with the period $\tau = \tau^* = 2\pi|\omega|^{-1}$.

Now suppose that $\mu = \mu^*$ is a sufficiently small positive number for which the existence of a periodic solution to (28) with period $\tau = \tau^* = 2\pi|\omega|^{-1}$ has been established, and let γ^* be the corresponding value of the Jacobian integral (31). Using Poincaré's method we will show that for each value γ of the integral sufficiently near γ^* there exists a periodic solution whose period is near τ^*. For this we look at the rank of the matrix $\mathfrak{D}$ in (27), which in this case has five rows. If the third column and fourth row of $\mathfrak{D}$ are deleted, the corresponding sub-determinant has the value $4r^2\omega^3 \sin^2 \dfrac{\pi}{\omega}$ at $\xi = \xi^*$, $\tau = \tau^*$, $\mu = 0$ and therefore, under assumption (32), is likewise not 0. Consequently, for such a fixed ω, if the positive number μ^* is chosen sufficiently small, the rank of the matrix $\mathfrak{D}$ corresponding to $\mu = \mu^*$ and the periodic solution in question will still be equal to 4. Thus for each such value of μ^* there exists a family of periodic solutions depending on γ whose period τ can be expanded in a neighborhood of γ^* as a power series in $\gamma - \gamma^*$ and has the value τ^* for $\gamma = \gamma^*$. In this way, starting for $\mu = 0$ from a circular orbit with period $\tau^* = 2\pi|\omega|^{-1}$ $(\omega \neq 0, -2, g^{-1})$, for small positive values of μ we first obtained periodic solution to (28) having the same period, and then for μ fixed, a family of periodic solutions depending on the parameter γ and having a period τ in general $\neq \tau^*$.

Initially, the continuation method gives periodic solutions only for the parameter values γ restricted to sufficiently small neighborhoods, and it is of interest to investigate the behavior of these solutions under analytic continuation with respect to γ. For this we consider the case of a Hamiltonian system

$$(33) \qquad \dot{x}_k = E_{y_k}, \quad \dot{y}_k = -E_{x_k} \quad (k=1,\ldots,n),$$

whereby $m=2n$ and $\psi = E(x,y)$ is an integral. In particular, for $n=2$ and $y_1 = x_3 - x_2, y_2 = x_4 + x_1$ this corresponds to the system (28) with ψ given by (31). Let G now be a domain in the real (x,y)-space with the function E regular and having no stationary points there, and assume that we begin with a periodic orbit C that lies completely in G and consequently is not an equilibrium solution. Then also the integral $\psi = E$ is nowhere stationary along this solution. In addition, we assume that the rank of the corresponding matrix $\mathfrak{D}$ in (27) is equal to m. For γ^* the value of the parameter $E=\gamma$ along the given solution, the continuation method now leads to a family of periodic orbits C_γ that depend on the parameter γ and whose initial values $x=\xi, y=\eta$ and period τ can be expanded in a neighborhood of γ^* as a power series in $\gamma - \gamma^*$. The original curve C then corresponds to C_{γ^*}, while if $\gamma - \gamma^*$ is sufficiently small in absolute value also C_γ will remain in G. Let us now analytically continue these solutions along the real γ-axis by a repeated application of the continuation method. Assuming that this has been done for $\gamma^* \leqq \gamma < \gamma_0$ and that in the process the solutions C_γ remain in G, we wish to investigate their behavior as $\gamma \to \gamma_0$. We will say that the C_γ leave G as $\gamma \to \gamma_0$ if for each compact subset H of G there exists an $\varepsilon > 0$ such that no C_γ for $\gamma_0 - \varepsilon < \gamma < \gamma_0$ lies completely in H. Let us assume that this is not the case. Then by compactness of H there exists a sequence $\gamma \to \gamma_0$ such that the corresponding C_γ remain completely in H and their initial values ξ, η converge to a point ξ_0, η_0 in H. It is conceivable that for each such sequence the corresponding periods $\tau = \tau_\gamma$ of C_γ tend to ∞, but let us exclude also this possibility, in which case we can find a subsequence for which the τ_γ converge to a finite limit τ_{γ_0}. This limit is not 0, for otherwise the continuous dependence of solutions on initial values would imply that the point ξ_0, η_0 is an equilibrium solution to (33), whereas the stationary points of E were excluded from G. By the same theorems on continuous dependence the C_γ corresponding to the above sequence must then converge to a solution C_{γ_0} through ξ_0, η_0, with period τ_{γ_0}, which also lies in H and therefore certainly in G.

If for the solution C_{γ_0} the rank of $\mathfrak{D}$ is again equal to m, then the solutions C_γ can evidently be continued beyond γ_0. It remains to consider the case when the rank is less than m. Thus, keeping the earlier notation, we have to consider the m analytic equations (25), (26) in the vicinity of a point $\gamma=\gamma_0, \tau, \xi_k$ $(k \neq m-1)$ at which now the Jacobian determinant with respect to τ, ξ_k vanishes. Moreover, we already have a one-parameter family of real solutions for $\gamma \to \gamma_0$. With the aid of the Weierstrass preparation theorem it can be shown that there exists then a solution which is a power series in $(\gamma_0-\gamma)^{1/p}$ with real coefficients, where p is a certain natural number chosen as small as possible. Thus we have a branch-point of order $p-1$ and can continue beyond $\gamma=\gamma_0$. If p is odd, one again obtains in this way real values for the power series of τ, ξ_k also for $\gamma>\gamma_0$. On the other hand, if p is even then for $\gamma<\gamma_0$ there are two different real values of the root $(\gamma_0-\gamma)^{1/p}$, and by reversing direction of γ at γ_0 we can continue the solution at γ_0 by going into the other real branch of $(\gamma_0-\gamma)^{1/p}$. This leads to a second family of periodic solutions, different from the first, and the process can be continued also in this case.

The analogous analysis can be carried out for γ decreasing from γ^*. Furthermore, one may proceed again starting with γ_0 in place of γ^* and obtain in this way additional branch-points $\gamma_1, \gamma_2, \ldots$ through real valued analytic continuation along the intervals between γ_{k-1} and γ_k $(k=1,2,\ldots)$. The process will cease only when one of the excluded cases is encountered: namely, when either the C_γ leave the domain G or the τ_γ never remain bounded.

In the restricted three-body problem G can be taken as the space of all real x_1, x_2, y_1, y_2 from which the singularities $x_1=-\mu, x_2=0$ and $x_1=1-\mu, x_2=0$, as well as the five stationary points of E, have been removed. If the solution curves C_γ leave the domain G as $\gamma \to \gamma_1$, it means that the removed points are accumulation points of points on C_γ. For the stationary points of E this leads to equilibrium solutions in the limit. For the singular points, on the other hand, a suitable regularizing transformation shows that in the limit one obtains collision orbits and that the analytic continuation can even be carried out beyond these. The process of continuing periodic solutions of the restricted three-body problem has been carried out numerically by E. Strömgren and his collaborators, while the theoretical questions arising in the process have been investigated more closely by Wintner [2].

§ 22. The Fixed-Point Method

The following method for determining periodic solutions also goes back to Poincaré. We consider again a system of differential equations

$$(1) \qquad \dot{x}_k = f_k(x) \quad (k=1,\ldots,m),$$

which this time need not depend on an additional parameter. The functions $f_k(x)$ are assumed to be regular in a domain G of the real x-space, and $x_k(t, \xi)$ $(k=1, \ldots, m)$ will again denote the solution with initial values $x_k(0, \xi)=\xi_k$. For $\xi=\xi^*$ let $x(t, \xi^*)$ be a periodic solution that lies completely in G, is not an equilibrium solution, and whose period is $\tau^*>0$. Since not all the $f_k(\xi^*)$ are equal to 0, we may assume that, say, $f_m(\xi^*) \neq 0$. The periodic solution $x(t, \xi^*)$ then intersects the plane $x_m=\xi_m^*$ at the times $t=0$ and $t=\tau^*$ in the point $x=\xi^*$. If the initial values ξ_k are now slightly altered within the plane $\xi_m=\xi_m^*$, the corresponding solution $x(t, \xi)$ will intersect the plane $x_m=\xi_m^*$ at time $t=0$ and again at a time $t=\tau$ approximately equal to τ^*. By the theorems on continuous dependence for solutions to differential equations, this defines a mapping of a neighborhood of the point $x=\xi^*$ in the plane $x_m=\xi_m^*$ into another such neighborhood, with a periodic solution corresponding to a fixed-point of this mapping.

We wish to generalize the above somewhat. Namely, let us not assume that the original orbit $x(t, \xi^*)$ is necessarily closed, but only that it intersects the plane $x_m=\xi_m^*$ again at a time $t=\tau^*>0$. By this we mean that $x_m(\tau^*, \xi^*)=\xi_m^*$, while also $f_m(x(\tau^*, \xi^*)) \neq 0$. Moreover, let the solution $x(t, \xi^*)$ for $0 \leqq t \leqq \tau^*$ lie completely in G. For ξ sufficiently near ξ^* and $\xi_m=\xi_m^*$ the solutions $x(t, \xi)$ will then intersect the plane $x_m=\xi_m^*$ after a time $t=\tau$ that differs only slightly from τ^*, and in this way one obtains an analytic mapping from a neighborhood of the point ξ^* in the plane $x_m=\xi_m^*$ to a neighborhood of the point $x(\tau^*, \xi^*)$ in the same plane.

This situation can be generalized still further by passing through the endpoints $\xi^*, x(\tau^*, \xi^*)$ of a segment of a solution curve $x(t, \xi^*)$ $(0 \leqq t \leqq \tau^*)$, which lies completely in G, two arbitrary smooth $(m-1)$-dimensional surfaces not tangent to the curve. Reasoning heuristically, let us assume that we have a smooth surface F in G, small enough so that for each point ξ in F the solution $x(t, \xi)$ lies completely in G and for $t>0$ intersects F at least once more, always non-tangentially. If $t=\tau>0$ is the first point in time for which $x(t, \xi)$ again meets F, then the correspondence between ξ and $x(\tau, \xi)=S\xi$ defines a topological mapping S of F into itself. Should the solution $x(t, \xi)$ be periodic, there must exist a natural number n such that $S^n\xi=\xi$, so that ξ is a fixed-point of the mapping S^n for a suitable n. The search for periodic solutions is thereby reduced to determining fixed-points for the iterates of S. However, as simple examples of analytic mappings S of certain surfaces will show, it may well happen that all the S^n are fixed-point free. On the other hand, Poincaré recognized that simple additional assumptions will guarantee the existence of a fixed-point for S. He imposed the following conditions. Let F be an annulus in the plane, with the two rims C_1 and C_2 included, and let the correspondence $\xi \to S\xi$ define an area-preserving topological mapping of F onto itself which leaves each of the two rims invariant. To each point ξ of the annulus we assign, as a continuous function in ξ, the angle $\phi(\xi)$ formed by the lines from the center of the annulus to ξ and $S\xi$ respectively. This defines ϕ uniquely up to a multiple of 2π independent of ξ, and upon fixing a suitable branch we assume that $\phi(\xi) \geqq 0$ on C_1 and $\phi(\xi) \leqq 0$ on C_2. Evidently, this says that the mapping rotates the two rims in opposite directions relative to one another. Poincaré [1] asserted that under these conditions the mapping S must have at least two fixed-points, an assertion that was first proved by Birkhoff [2] after Poincaré's death. This fixed-point theorem is of interest for the restricted three-body problem where, as was claimed already by Poincaré and later shown by Birkhoff, for sufficiently small values of the mass μ, introduced as a parameter, and a fixed Jacobian constant γ, one can find a surface F having the required properties. Moreover, Poincaré believed that his theorem would also imply the existence of at least two periodic solutions to the restricted three-body problem for arbitrary μ in the interval $0<\mu<1$, although up to now it has not been possible to prove the existence of the necessary surface F in this generality. We will not go into Poincaré's fixed-point theorem any further, but instead will give a complete treatment of a related theorem by Birkhoff that appears to be more useful for applications.

As a preparation for this, let us investigate more closely the property of a mapping being volume-preserving. It will be assumed that for the solution $x(t, \xi)$ of (1) the mapping $\xi \to x(t, \xi)$ for each fixed t is volume-preserving, for which, as we saw in (10) of the previous section, it is necessary and sufficient that

$$\sum_{k=1}^{m} f_{k x_k} = 0. \tag{2}$$

We assume again that $f_m(\xi^*) \neq 0$ and the solution $x(t, \xi^*)$ lies entirely in G and cuts the plane $x_m = \xi_m^*$ once more for $t = \tau^* > 0$. Choosing a sufficiently small neighborhood U of ξ^* in the plane $x_m = \xi_m^*$ we trace the solution curves emanating at $t=0$ from U, and for each of these, after a time approximately equal to τ^*, come to another point of intersection with the plane in question. The new points form a neighborhood U_1 of $x(\tau^*, \xi^*)$ in $x_m = \xi_m^*$ which is the image of U under the mapping described above. For $t_0 > 0$ sufficiently small, we further denote by B and B_1 the respective regions in G defined by the conditions $x = x(t, \xi)$, $0 \leqq t \leqq t_0$, $\xi \in U$ and $\xi \in U_1$, whereby B and B_1 can be viewed as cylinders over the bases U and U_1. Consider now the tube R formed by the solution curves connecting U and U_1. After time t_0 the region R goes over to the region $R + B_1 - B$, and from the preservation of volume it follows that R

and $R+B_1-B$, hence also B and B_1, have the same m-dimensional volume. Denoting the volume element $dx_1 \dots dx_m$ by dx, we thus have

(3) $$\int_B dx = \int_{B_1} dx\,.$$

Let $\xi_1, \dots, \xi_{m-1}$ and t be introduced via the substitution $x_k = x_k(t, \xi)$ $(k=1, \dots, m;\ \xi_m = \xi_m^*)$ as new variables of integration in place of $x_1, \dots, x_m$. The corresponding Jacobian matrix has as its rows $x_{k\xi_l}$ $(l=1, \dots, m-1)$, f_k for $k=1, \dots, m$, and since the $m \times m$ matrix $(x_{k\xi_l})$ at $t=0$ is just the identity $\mathfrak{E}$, the Jacobian determinant there has the value $f_m(\xi) \neq 0$. Dividing (3) by t_0 and letting $t_0 \to 0$, one thus obtains in the limit

$$\int_U f_m(\xi)d\xi = \int_{U_1} f_m(\xi)d\xi \qquad (d\xi = d\xi_1 \dots d\xi_{m-1})\,.$$

Let us further assume that $\psi(x)$ is a time-independent integral for the system (1) with the derivative $\psi_{x_{m-1}} \neq 0$ in U and U_1. Then $\psi(x) = \gamma$ is constant along each solution curve, and if γ is introduced via the substitution $\psi(\xi) = \gamma$ as a new variable in place of ξ_{m-1}, we have

$$\psi_{x_{m-1}}(\xi)d\xi_{m-1} = d\gamma\,.$$

In particular, choose U to be the product of an $(m-2)$-dimensional neighborhood F of $\xi_k = \xi_k^*$ $(k=1, \dots, m-2)$ with an interval containing the point $\psi(\xi^*) = \gamma^*$. By the invariance of $\psi(x)$, in the mapping from U to U_1 this interval remains pointwise fixed, while corresponding to $\psi(\xi) = \gamma$ the image of F is a certain set $F_1 = F_1(\gamma)$. Introducing in addition

$$g = g(\xi_1, \dots, \xi_{m-2}, \gamma) = \frac{f_m(\xi)}{\psi_{x_{m-1}}(\xi)} \qquad (\xi_m = \xi_m^*)\,,$$

we then have

(4) $$\int_F g\,dv = \int_{F_1} g\,dv \qquad (dv = d\xi_1 \dots d\xi_{m-2})\,.$$

For a Hamiltonian system

$$\dot{x}_k = E_{y_k}, \quad \dot{y}_k = -E_{x_k} \qquad (k=1, \dots, n)$$

one has $m=2n$ with condition (2) satisfied, while the integral can be chosen as $\psi(x,y) = E(x,y)$. After a suitable ordering of the coordinates then $f_m = -E_{x_n}$, $\psi_{x_n} = E_{x_n}$, whereupon $g=-1$ and (4) implies that the mapping from F to F_1 is volume-preserving. In particular, if the orbit corresponding to the initial values $x=\xi^*$, $y=\eta^*$ is closed and has period τ^*, assuming that $E_{x_n}(\xi^*, \eta^*) \neq 0$, one can eliminate t, ξ_n, η_n from the equations

$$y_n(t, \xi, \eta) = \eta_n^*, \quad \eta_n = \eta_n^*, \quad E(\xi, \eta) = E(\xi^*, \eta^*)$$

and obtain a volume-preserving mapping

$$(5) \qquad \xi_k, \eta_k \to x_k(t, \xi, \eta), \; y_k(t, \xi, \eta) \qquad (k=1, \ldots, n-1)$$

in a neighborhood of the fixed-point ξ_k^*, η_k^*. Instead of the preservation of volume expressed by (3), one can use an analogous property of certain differential forms introduced by Poincaré [3] to show that the mapping (5) is canonical. For $n=2$ this is equivalent to the preservation of area just proved. From now on we will restrict ourselves to the case $n=2$, which already contains the essential difficulties of the general case. The discussion of area-preserving analytic mappings in a neighborhood of a fixed-point in the plane will be undertaken in the next section.

§ 23. Area-Preserving Analytic Transformations

We consider here a mapping in the (x, y)-plane analytic in a neighborhood of a fixed-point. Without loss of generality this point can be taken as the origin, so that the mapping is of the form

$$(1) \qquad x_1 = f(x, y), \quad y_1 = g(x, y)$$

with

$$(2) \qquad f(x, y) = ax + by + \cdots, \quad g(x, y) = cx + dy + \cdots$$

real power series without constant terms. We will at first, however, work as in § 16 with formal power series, disregarding convergence, and initially the coefficients are to be viewed as arbitrary complex numbers and x, y as indeterminates. The set of all transformations (1) with $ad - bc \neq 0$ forms a group Γ, while the set of transformations that satisfy

$$f_x g_y - f_y g_x = 1$$

as an identity between power series, to which we refer as area-preserving, form a subgroup Δ. The groups Γ_0, Δ_0, consisting of only those series in Γ, Δ respectively that converge in some neighborhood of $x=0, y=0$, are again subgroups of the respective groups Γ, Δ.

Introducing the column vectors $z = \begin{pmatrix} x \\ y \end{pmatrix}$, $z_1 = \begin{pmatrix} x_1 \\ y_1 \end{pmatrix}$, we express the formal transformation (1) symbolically as

$$(3) \qquad z_1 = Sz$$

and make a simultaneous change of variables

$$x=\phi(\xi,\eta)=\alpha\xi+\beta\eta+\cdots,\quad y=\psi(\xi,\eta)=\gamma\xi+\delta\eta+\cdots,$$
$$x_1=\phi(\xi_1,\eta_1),\qquad\qquad y_1=\psi(\xi_1,\eta_1),$$

also expressed symbolically as

$$z=C\zeta,\quad z_1=C\zeta_1,\quad \zeta=\begin{pmatrix}\xi\\ \eta\end{pmatrix},\quad \zeta_1=\begin{pmatrix}\xi_1\\ \eta_1\end{pmatrix},$$

with $\alpha\delta-\beta\gamma\neq 0$ and ϕ,ψ again formal power series without constant terms. If S is area-preserving, the substitution C will in addition be assumed to satisfy

(4)
$$\phi_\xi\psi_\eta-\phi_\eta\psi_\xi=\alpha\delta-\beta\gamma,$$

being then evidently a composition of a linear and an area-preserving substitution. Because $\alpha\delta-\beta\gamma$ does not vanish, C has an inverse C^{-1}, and (3) takes the form

$$\zeta_1=C^{-1}z_1=C^{-1}SC\zeta=T\zeta,\quad T=C^{-1}SC$$

with the transformation T again being in Γ,Δ respectively whenever S is. It is easy to see, however, that $C^{-1}SC$ need not be area-preserving for every area-preserving S unless C satisfies condition (4). Our aim in this section is: Given S, to determine C suitably so that T assumes a certain normal form [1].

First a linear substitution is used to bring the linear terms of (1) into a normal form. Let $\mathfrak{S},\mathfrak{C},\mathfrak{T}$ denote the coefficient matrices of the linear terms in $Sz, C\zeta, T\zeta$, so that

$$\mathfrak{S}=\begin{pmatrix}a & b\\ c & d\end{pmatrix},\quad \mathfrak{C}=\begin{pmatrix}\alpha & \beta\\ \gamma & \delta\end{pmatrix},\quad \mathfrak{T}=\mathfrak{C}^{-1}\mathfrak{S}\mathfrak{C},$$

$$|\mathfrak{S}-\lambda\mathfrak{E}|=\lambda^2-(a+d)\lambda+ad-bc,$$

and the eigenvalues λ,μ of $\mathfrak{S}$ satisfy $\lambda+\mu=a+d$, $\lambda\mu=ad-bc$. In particular, if S is area-preserving then $ad-bc=1$ and therefore $\lambda\mu=1$. From now on we assume that a,b,c,d are real and distinguish between three possible cases: hyperbolic, with λ,μ real and distinct; parabolic, with $\lambda=\mu$; and elliptic, with $\bar{\lambda}=\mu\neq\lambda$. For the sake of simplicity we will exclude the parabolic case, assuming therefore that $\lambda\neq\mu$. Then $\mathfrak{C}$ can be chosen so that $\mathfrak{T}$ has the normal form

$$\mathfrak{T}=\mathfrak{C}^{-1}\mathfrak{S}\mathfrak{C}=\begin{pmatrix}\lambda & 0\\ 0 & \mu\end{pmatrix}.$$

In the hyperbolic case $\mathfrak{C}$ can be taken to be real, while in the elliptic case the two columns of $\mathfrak{C}$ can be chosen to be complex conjugates of one another.

The preliminary linear substitution $z=\mathfrak{C}\zeta$ brings the transformation $z_1=Sz$ into the form

$$(5)\qquad \zeta_1=T\zeta\,,\quad \xi_1=p(\xi,\eta)=\lambda\xi+\cdots,\quad \eta_1=q(\xi,\eta)=\mu\eta+\cdots.$$

If S is real, i.e. if the coefficients of f and g are all real, then in the hyperbolic case T is real, while in the elliptic case we have the relation

$$(6)\qquad \bar{p}(\xi,\eta)=q(\eta,\xi)$$

with $\bar{p}$ the power series obtained from p by complex conjugation of the coefficients. Consequently, by means of a linear substitution, the mapping (1) can be brought into the form (5), and T belongs to Γ,Δ respectively whenever S does. Also convergence of f and g is preserved in the process, and therefore if S belongs to Γ_0,Δ_0 respectively so does T. We again write z, z_1 in place of ζ,ζ_1, so that

$$(7)\quad z_1=Tz\,,\quad x_1=p(x,y)=\lambda x+\sum_{k=2}^{\infty}p_k\,,\quad y_1=q(x,y)=\mu y+\sum_{k=2}^{\infty}q_k$$

with p_k and q_k homogeneous polynomials in x, y of degree k. Let T now be subjected to an arbitrary nonlinear substitution of the form

$$(8)\quad x=\phi(\xi,\eta)=\xi+\sum_{k=2}^{\infty}\phi_k,\ y=\psi(\xi,\eta)=\eta+\sum_{k=2}^{\infty}\psi_k,\ z=C\zeta,\ z_1=C\zeta_1\,,$$

where again ϕ_k and ψ_k are homogeneous polynomials in ξ,η of degree k. The linear part of (7) being already in normal form, the linear terms in this substitution (8) are chosen as in the identity. First, suppose that for all integers p, q with $p\geqq 0, q\geqq 0, p+q>1$ we have

$$(9)\qquad \lambda^p\mu^q\neq\lambda\,,\qquad \lambda^p\mu^q\neq\mu\,.$$

It will be shown that then there is exactly one substitution of the form (8) for which the transformation $U=C^{-1}TC$ assumes the normal form

$$(10)\qquad \xi_1=\lambda\xi\,,\ \eta_1=\mu\eta\,.$$

This is proved by comparing coefficients in the relation $CU=TC$ or

$$(11)\quad \phi(\lambda\xi,\mu\eta)=p(\phi(\xi,\eta),\psi(\xi,\eta))\,,\quad \psi(\lambda\xi,\mu\eta)=q(\phi(\xi,\eta),\psi(\xi,\eta))\,.$$

If the power series from (7), (8) are inserted into (11), the linear terms on both sides are seen to agree. Suppose now that for some $k>1$ the condition that the coefficients of all terms of degree less than k in (11) agree uniquely determines the polynomials ϕ_l,ψ_l $(l=2,\ldots,k-1)$. For $k=2$

this is true, and we will prove the assertion for $k+1$ in place of k. Comparing the terms of degree k in (11) we are led to the conditions

$$(12)\qquad \phi_k(\lambda\xi,\mu\eta)=\lambda\phi_k(\xi,\eta)+\cdots,\qquad \psi_k(\lambda\xi,\mu\eta)=\mu\psi_k(\xi,\eta)+\cdots,$$

where the terms not written down explicitly are homogeneous polynomials of degree k whose coefficients have already been determined. Now if

$$(13)\qquad \phi_k(\xi,\eta)=\sum_{l=0}^{k} a_l\xi^{k-l}\eta^l,\qquad \psi_k(\xi,\eta)=\sum_{l=0}^{k} b_l\xi^{k-l}\eta^l,$$

then

$$(14)\qquad \begin{cases} \phi_k(\lambda\xi,\mu\eta)-\lambda\phi_k(\xi,\eta)=\sum\limits_{l=0}^{k} a_l(\lambda^{k-l}\mu^l-\lambda)\,\xi^{k-l}\eta^l,\\ \psi_k(\lambda\xi,\mu\eta)-\mu\psi_k(\xi,\eta)=\sum\limits_{l=0}^{k} b_l(\lambda^{k-l}\mu^l-\mu)\,\xi^{k-l}\eta^l, \end{cases}$$

and because by (9) the factors $\lambda^{k-l}\mu^l-\lambda$, $\lambda^{k-l}\mu^l-\mu$ are all different from 0, the coefficients a_l, b_l can indeed be uniquely determined so that the conditions (12) are fulfilled.

From now on let us restrict ourselves to area-preserving transformations T. Then $\lambda\mu=1$, so that (9) is certainly not satisfied, and for $U=C^{-1}TC$ we seek a somewhat different normal form. Namely, in place of (10) we set up the more general expression

$$(15)\qquad \xi_1=u\xi,\qquad \eta_1=v\eta,\qquad u=\sum_{k=0}^{\infty}\alpha_{2k}(\xi\eta)^k,\qquad v=\sum_{k=0}^{\infty}\beta_{2k}(\xi\eta)^k$$

with undetermined coefficients α_{2k}, β_{2k}, whereby u and v are power series in the product $\xi\eta=\omega$. For C we again assume the series expansions (8), but instead of (11) we must now satisfy the functional equations

$$(16)\qquad \phi(u\xi,v\eta)=p(\phi(\xi,\eta),\psi(\xi,\eta)),\quad \psi(u\xi,v\eta)=q(\phi(\xi,\eta),\psi(\xi,\eta)).$$

Comparing the linear terms one obtains

$$(17)\qquad \alpha_0=\lambda,\qquad \beta_0=\mu.$$

Let us define $\alpha_l=\beta_l=0$ for $l>0$ odd, and assume that for some $k>1$ the $\phi_l, \psi_l, \alpha_{l-1}, \beta_{l-1}$ $(l<k)$ have already been determined through comparison of coefficients in the terms of degree less than k. For $k=2$ this is true. Comparison of the k-th degree terms then leads to the conditions

$$(18)\qquad \begin{cases} \phi_k(\lambda\xi,\mu\eta)+\alpha_{k-1}(\xi\eta)^{\frac{k-1}{2}}\xi=\lambda\,\phi_k(\xi,\eta)+\cdots,\\ \psi_k(\lambda\xi,\mu\eta)+\beta_{k-1}(\xi\eta)^{\frac{k-1}{2}}\eta=\mu\,\psi_k(\xi,\eta)+\cdots, \end{cases}$$

where again the terms not written down explicitly are homogeneous polynomials of degree k whose coefficients are already known. Because $\lambda\mu = 1$, we also have

$$\lambda^{k-l}\mu^l - \lambda = \lambda(\lambda^{k-2l-1} - 1), \quad \lambda^{k-l}\mu^l - \mu = \lambda^{-1}(\lambda^{k-2l+1} - 1).$$

Furthermore, let it be assumed that λ is not a root of unity, so that $\lambda^{k-2l\mp 1} = 1$ only when $k = 2l \pm 1$. In that case $\alpha_{k-1}, \beta_{k-1}$ and $a_l\left(l \neq \frac{k-1}{2}\right)$, $b_l\left(l \neq \frac{k+1}{2}\right)$ are uniquely determined by (14), (18) while for $k = 2h+1$ odd a_h, b_{h+1} still remain arbitrary. To make the coefficients unique we add the requirement that the power series

$$\phi_\xi - \psi_\eta = \sigma(\xi, \eta), \quad \phi_\xi\psi_\eta - \phi_\eta\psi_\xi - 1 = \tau(\xi, \eta) - 1$$

contain no powers of $\xi\eta = \omega$ alone. Suppose that this is already true for terms of degree below $k-1$. For $k = 2$ it is so, while it is also trivially so for $k+1$ in place of k if k is even. For $k = 2h+1$ odd, the coefficient of ω^h in σ is equal to $(h+1)(a_h - b_{h+1})$, making it necessary that

$$a_h = b_{h+1}. \tag{19}$$

On the other hand, the terms of degree $k-1$ in τ consist of $\phi_{k\xi} + \psi_{k\eta}$ plus a polynomial whose coefficients are already known, and the condition that the coefficient of ω^h in τ vanish therefore determines

$$(h+1)(a_h + b_{h+1}).$$

This in conjunction with (19) now shows that also the remaining coefficients a_h, b_{h+1} are uniquely determined.

In this way, under the above conditions, we have found a unique substitution C that takes T into the normal form $U = C^{-1}TC$ described in (15). It will be shown next that C is area-preserving. For this we use (15) to determine the partial derivatives

$$\xi_{1\xi} = u + u_\xi\xi = u + u_\omega\omega, \quad \xi_{1\eta} = u_\eta\xi = u_\omega\xi^2,$$
$$\eta_{1\xi} = v_\xi\eta = v_\omega\eta^2, \quad \eta_{1\eta} = v + v_\eta\eta = v + v_\omega\omega.$$

Computing the Jacobian determinant and using the relation $CU = TC$ together with the assumption that T is area-preserving, we then obtain the identity

$$\tau(u\xi, v\eta)\{(u + u_\omega\omega)(v + v_\omega\omega) - u_\omega v_\omega\omega^2\} = \tau(\xi, \eta). \tag{20}$$

Now by (15), (17) the expression

$$(u + u_\omega\omega)(v + v_\omega\omega) - u_\omega v_\omega\omega^2 = (uv\omega)_\omega = 1 + \cdots \tag{21}$$

is a power series in ω beginning with 1, while by (8) also the constant term of $\tau(\xi,\eta)$ is equal to 1. Using (20) we will show that actually $\tau(\xi,\eta)=1$. If the series

$$\tau(\xi,\eta)-1=\tau_k(\xi,\eta)+\cdots$$

begins with terms of order $k>0$ and the coefficient of $\omega^{k/2}$ on the right side of (21) is denoted by c, comparison of coefficients in the k-th order terms of (20) leads to the formula

$$\tau_k(\lambda\xi,\mu\eta)+c\omega^{k/2}=\tau_k(\xi,\eta).$$

Since, however, the series $\tau-1$ contains no powers of ω alone, $c=0$ and therefore

$$\tau_k(\lambda\xi,\mu\eta)=\tau_k(\xi,\eta).$$

If then

$$\tau_k(\xi,\eta)=\sum_{l=0}^{k}\gamma_l\xi^{k-l}\eta^l,$$

it follows that

$$\gamma_l(\lambda^{k-2l}-1)=0,$$

and because λ is not a root of unity, we must have $\gamma_l=0$ $(2l\neq k)$. On the other hand, because $\tau-1$ contains no powers of ω alone, $\gamma_l=0$ also for $2l=k$. Consequently $\tau=1$ and C is area-preserving. By (20), (21) this in turn implies that $(uv\omega)_\omega=1$, so that $uv\omega=\omega$ and

$$uv=1. \tag{22}$$

We have thus shown that an area-preserving transformation T of the form (7) can be brought by means of on area-preserving substitution of the form (8) into the normal form $U=C^{-1}TC$ described in (15), provided that the eigenvalue λ is not a root of unity. We still wish to examine to what extent C and U are determined by T. Let V stand for an arbitrary area-preserving substitution of the form (15), that is, $\xi_1=u_0\xi$, $\eta_1=v_0\eta$ with u_0, v_0 power series in $\omega=\xi\eta$ only and

$$(u_0\xi)_\xi(v_0\eta)_\eta-(u_0\xi)_\eta(v_0\eta)_\xi=(u_0v_0\omega)_\omega=1,$$

whereupon in analogy to (22) it follows that $u_0v_0=1$. Consequently (22) becomes a necessary and sufficient condition for (15) to be area-preserving, and therefore one may choose for u_0 an arbitrary power series in ω with nonvanishing constant term and then must take $v_0=u_0^{-1}$. Obviously then $\xi_1\eta_1=\xi\eta$, so that the product $\xi\eta$ is invariant under V. Moreover, if

$$\zeta_1=V_1\zeta,\quad \xi_1=u_1\xi,\quad \eta_1=v_1\eta,\quad u_1v_1=1$$

is another substitution of the same form, then by this invariance also $V_1 V$ has the form

$$\xi_1 = u_1(\omega)\, u(\omega)\, \xi\,, \qquad \eta_1 = v_1(\omega)\, v(\omega)\, \eta\,.$$

It follows that the substitutions V form an abelian group Λ. Suppose now that C_0 is an area-preserving substitution such that $C_0^{-1} T C_0 = U_0$ also has the form (15). Then U_0 is in Λ and, if $C_1 = C_0 V$ for V an arbitrary element of Λ, we again have $C_1^{-1} T C_1 = U_0$. Since the eigenvalues λ, μ are determined by $\mathfrak{S}$ up to their ordering, by possibly interchanging ξ, η we can have the linear terms of the two transformations $\zeta_1 = U\zeta$, $\zeta_1 = U_0 \zeta$ agree, being $\lambda\xi$ and $\mu\eta$ for both. If $\mathfrak{C}_0$ is the coefficient matrix of the linear terms in $C_0 \zeta$, then $\mathfrak{C}_0$ commutes with the diagonal matrix $\mathfrak{T}$ and, because $\lambda \neq \mu = \lambda^{-1}$, the matrix $\mathfrak{C}_0$ must itself be diagonal. We will show that a V of the form (15) with u_0, v_0 in place of u, v can be chosen so that the substitution $C_1 = C_0 V$ satisfies all the conditions that uniquely determined C. First, the requirement that C_1 be of the form (8) fixes the constant term $\varrho = \varrho_0 \neq 0$ of

$$u_0 = \sum_{l=0}^{\infty} \varrho_l \omega^l$$

while leaving the other coefficients ϱ_l $(l > 0)$ free. We claim that these can be uniquely determined by recursive use of the condition that for the power series $\phi(\xi, \eta), \psi(\xi, \eta)$ belonging to C_1 the expression $\phi_\xi - \psi_\eta$ contain no term in ω alone. With

$$v_0 = \sum_{l=0}^{\infty} \sigma_l \omega^l,$$

it follows from $u_0 v_0 = 1$ that $\varrho_0 \sigma_0 = 1$ and

$$\varrho\sigma_k + \varrho^{-1}\varrho_k + \sum_{l=1}^{k-1} \varrho_l \sigma_{k-l} = 0 \quad (k = 1, 2, \ldots)\,. \tag{23}$$

On the other hand, if the series belonging to C_0 are given by

$$\phi^*(\xi, \eta) = \varrho^{-1}\xi + \sum_{k=2}^{\infty} \phi_k^*\,, \quad \psi^*(\xi, \eta) = \varrho\eta + \sum_{k=2}^{\infty} \psi_k^*\,,$$

we have

$$\phi(\xi, \eta) = \phi^*(u_0 \xi, v_0 \eta)\,, \qquad \psi(\xi, \eta) = \psi^*(u_0 \xi, v_0 \eta)$$

and therefore

$$\phi_\xi - \psi_\eta = \phi_\xi^*(u_0\xi, v_0\eta)(u_0\omega)_\omega + \phi_\eta^* v_{0\omega}\eta^2 - \psi_\xi^* u_{0\omega}\xi^2 - \psi_\eta^*(v_0\omega)_\omega\,. \tag{24}$$

Suppose that for some $k > 0$ the coefficients $\varrho_1, \ldots, \varrho_{k-1}$ have already been uniquely determined from the condition that on the right side of

(24) the terms in $\omega, \omega^2, \dots, \omega^{k-1}$ all vanish. Then from (23) also $\sigma_1, \dots, \sigma_{k-1}$ are uniquely determined. The requirement that the coefficient of ω^k in (24) vanish now fixes $(k+1)(\varrho^{-1}\varrho_k - \varrho\sigma_k)$, and this in conjunction with (23) uniquely determines ϱ_k. Thus, our claim is established. Moreover, since C_1 is area-preserving, being the product of C_0 and V, it follows that $\tau = \phi_\xi \psi_\eta - \phi_\eta \psi_\xi = 1$ and, in particular, τ contains no positive powers of ω. Consequently C_1 satisfies all the conditions prescribed for C, and therefore $C_0 V = C_1 = C$, $U_0 = U$. This proves that the normal form U for T, and therefore also for S, is uniquely determined. Moreover, we have found all area-preserving transformations that take S into U. Finally, it follows from the uniqueness of U that two area-preserving transformations for which the eigenvalues λ, μ are not roots of unity can be transformed into one another by means of an area-preserving substitution if and only if they have identical normal forms.

From now on let the original series f, g in (2) be real, so that S is real, and we will discuss how this relates to U and C. In the hyperbolic case T is then also real and, since $\lambda\mu = 1$, $\lambda \neq \mu$ with λ, μ real, it follows that λ is not a root of unity. Moreover, the earlier comparison of coefficients shows that U and C are likewise real and, since $u = \lambda + \cdots$ with $\lambda \neq 0$, there is a unique real power series

$$w = \sum_{k=0}^{\infty} \gamma_k \omega^k$$

with $\gamma_0 \neq 0$ such that

$$u = \pm e^{w}, \quad v = \pm e^{-w}, \quad \lambda = \pm e^{\gamma_0}.$$

The normal form then becomes

$$\xi_1 = \pm e^{w}\xi, \quad \eta_1 = \pm e^{-w}\eta. \tag{25}$$

In the elliptic case $\bar{\lambda} = \mu \neq \lambda$ and $\lambda\mu = 1$, so that $|\lambda| = 1$. The eigenvalue λ is again assumed not to be a root of unity, which incidentally once more excludes the parabolic case $\lambda = \mu = \pm 1$. Now (6) holds, and by going over to complex conjugate coefficients in (16) one obtains the relations

$$\bar{\phi}(\bar{u}\xi, \bar{v}\eta) = q(\bar{\psi}(\xi,\eta), \bar{\phi}(\xi,\eta)), \quad \bar{\psi}(\bar{u}\xi, \bar{v}\eta) = p(\bar{\psi}(\xi,\eta), \bar{\phi}(\xi,\eta)).$$

An interchange of ξ, η leaves $\xi\eta = \omega$, and therefore also $\bar{u}, \bar{v}$, invariant. Consequently one gets another solution to the functional equations (16) by taking $\bar{v}, \bar{u}, \bar{\psi}(\eta,\xi), \bar{\phi}(\eta,\xi)$ in place of the previous solution $u, v, \phi(\xi,\eta)$, $\psi(\xi,\eta)$. Moreover, the series

$$\bar{\psi}(\eta,\xi)_\xi - \bar{\phi}(\eta,\xi)_\eta = -\bar{\sigma}(\eta,\xi), \quad \bar{\psi}(\eta,\xi)_\xi \bar{\phi}(\eta,\xi)_\eta - \bar{\psi}(\eta,\xi)_\eta \bar{\phi}(\eta,\xi)_\xi = \bar{\tau}(\eta,\xi)$$

contain no positive powers of ω, while $\bar{u}, \bar{v}$ are again series in ω alone. From uniqueness of the solution it follows that

$$\bar{\phi}(\eta, \xi) = \psi(\xi, \eta), \quad \bar{u} = v, \tag{26}$$

and since $uv = 1$, also

$$u\bar{u} = 1. \tag{27}$$

Setting

$$\lambda = e^{i\gamma_0}, \quad -\pi < \gamma_0 < \pi, \tag{28}$$

we determine a unique power series

$$w = \sum_{k=0}^{\infty} \gamma_k \omega^k$$

this time by requiring that

$$e^{iw} = u, \; e^{-iw} = v.$$

In view of (27) then

$$e^{i(w-\bar{w})} = 1, \quad w - \bar{w} = \sum_{k=1}^{\infty} (\gamma_k - \bar{\gamma}_k)\omega^k,$$

which implies that $w = \bar{w}$ and the coefficients γ_k are all real. Thus in the elliptic case the normal form of S becomes

$$\xi_1 = e^{iw}\xi, \quad \eta_1 = e^{-iw}\eta \tag{29}$$

with $w = w(\omega)$ a real power series. To express this normal form entirely in terms of real variables, we make the simultaneous linear substitution

$$\xi = r + is, \quad \eta = r - is, \quad \xi_1 = r_1 + is_1, \quad \eta_1 = r_1 - is_1 \tag{30}$$

which takes (29) into

$$r_1 = r\cos w - s\sin w, \quad s_1 = r\sin w + s\cos w, \quad w = \sum_{k=0}^{\infty} \gamma_k (r^2 + s^2)^k, \tag{31}$$

where for $\cos w, \sin w$ one inserts their respective power series. The original unknowns x, y in (1) are related to the new ones through the substitution

$$z = \begin{pmatrix} x \\ y \end{pmatrix} = \mathfrak{C}\begin{pmatrix} \phi(\xi,\eta) \\ \psi(\xi,\eta) \end{pmatrix} = \bar{\mathfrak{C}}\begin{pmatrix} \psi(\xi,\eta) \\ \phi(\xi,\eta) \end{pmatrix}.$$

Since by (26) we have $\psi(r + is, r - is) = \bar{\phi}(r - is, r + is)$, the passage from r, s to x, y is evidently effected by a real substitution with constant Jacobian determinant $\varepsilon = -2i|\mathfrak{C}| \neq 0$. One may still normalize $|\mathfrak{C}| = i/2$, making ε equal to 1, so that S can be brought into the normal form (31) by means of a real area-preserving substitution. Thus in the hyperbolic and elliptic cases, under the assumption that λ is not a root of unity, we have found for the given real area-preserving transformation $z_1 = Sz$ a real normal form relative to the group Δ.

Up to now the substitution C was required to be area-preserving. If this requirement is dropped the normal form can be simplified even further. We will consider two mappings equivalent in the more generalized sense if they can be transformed into one another by a substitution in Γ. The earlier considerations can be easily extended to give again the normal form (15), while now, instead of the previous normalization used to make C unique, we require that $\phi_\xi - 1, \psi_\eta - 1$ not contain powers of the product $\xi\eta = \omega$. This leads to a unique substitution C in Γ taking T into a normal form (15).

If T is area-preserving then we again have the identity $uv = 1$. This is rather surprising since, as we shall see, the power series u may be different for transformations equivalent in this wider sense. To prove the identity, we first show that any substitution

$$\xi' = \phi(\xi, \eta), \quad \eta' = \psi(\xi, \eta)$$

which takes a transformation of the form (15) into another one of this normal form can be expressed as

$$\xi' = \xi\Phi(\xi\eta), \quad \eta' = \eta\Psi(\xi\eta), \tag{32}$$

where Φ, Ψ are power series in the product $\xi\eta$ with nonvanishing constant terms. This is again proved by comparing coefficients in the equations (16), where p, q have the form

$$p(\xi, \eta) = \xi P(\xi\eta), \quad q(\xi, \eta) = \eta Q(\xi\eta)$$

with the constant terms of P, Q being λ, μ respectively. We have to show that $\phi(\xi, \eta), \psi(\xi, \eta)$ contain only terms of the form $\xi(\xi\eta)^r, \eta(\xi\eta)^r$ respectively, or, equivalently, that $\phi(\lambda\xi, \mu\eta) - \lambda\phi(\xi, \eta) = 0, \psi(\lambda\xi, \mu\eta) - \mu\psi(\xi, \eta) = 0$. Assuming these relations to hold for ϕ_l, ψ_l when $l < k$ $(k = 1, 2, \ldots)$, which is certainly the case for $k = 1$, we compare the coefficients of degree k and are led to the equations (18), where one readily verifies that the terms not written down explicitly are a multiple of $\xi(\xi\eta)^{\frac{k-1}{2}}$ in the first equation and a multiple of $\eta(\xi\eta)^{\frac{k-1}{2}}$ in the second. Thus, also ϕ_k, ψ_k are of the desired form.

We now turn to the proof of the identity $uv = 1$, which we know holds if C is area-preserving. On the other hand, if C' is any other substitution such that $C'^{-1}TC'$ assumes the normal form

$$\xi_1' = \xi' u', \quad \eta_1' = \eta' v',$$

then the substitution $C'^{-1}C$ takes this normal form into (15) with $uv=1$, and therefore, as we have just seen, $C'^{-1}C$ must be of the form (32). Consequently

$$\xi_1'\eta_1' = \xi_1\eta_1\Phi_1\Psi_1 = \xi\eta\Phi\Psi u'v',$$

with $\Phi_1 = \Phi(\xi_1\eta_1)$, $\Psi_1 = \Psi(\xi_1\eta_1)$, and, since $uv=1$, we have $\xi_1\eta_1 = \xi\eta$ and hence $\Phi_1\Psi_1 = \Phi\Psi$. It follows that $u'v'=1$ as was to be shown.

We mentioned that within this wider class of substitutions the normal form can be reduced even further. Indeed, consider the hyperbolic case (25) with $u = \pm e^w$ and

$$w = \gamma_0 + \gamma_l\omega^l + \cdots, \quad \gamma_l \neq 0$$

a real power series in $\omega = \xi\eta$. Interchanging ξ, η if necessary, we may assume that $\gamma_l > 0$. Let this transformation now be subjected to a substitution of the form

$$\xi' = \xi a(\xi\eta), \quad \eta' = \eta a(\xi\eta)$$

where a is a real power series in $\xi\eta$ with nonvanishing constant term. The resulting transformation takes the form

$$\xi_1' = \pm\xi' e^p, \quad \eta_1' = \pm\eta' e^{-p},$$

where $p = p(\xi'\eta')$ and $a = a(\xi\eta)$ satisfy the relation

(33) $$w(\omega) = p(\omega a^2(\omega)).$$

Setting

$$p(\omega) = \gamma_0 + \omega^l$$

we choose the real power series $a(\omega)$ in accordance with (33) as one of the real roots of

$$a^{2l} = \omega^{-l}(w(\omega) - \gamma_0) = \gamma_l + \cdots.$$

Thus, with ξ, η again written in place of ξ', η', in the hyperbolic case the normal form reduces to (25) with

(34) $$w = \gamma_0 + (\xi\eta)^l \quad \text{or} \quad w = \gamma_0.$$

Similarly, in the elliptic case the normal form can be reduced to (31), where now

(35) $$w = \gamma_0 + (r^2 + s^2)^l \quad \text{or} \quad w = \gamma_0.$$

Consequently, for a fixed eigenvalue λ, not a root of unity, the equivalence classes in this wider sense are described by a positive integer l, except for the linear case $w = \gamma_0$ which we may associate with $l = \infty$. This is in contrast to the equivalence classes within the group Δ

of area-preserving transformations which are characterized by the infinitely many coefficients of w.

So far we have considered all power series in a formal sense, ignoring the question of convergence. Now we require S to belong to Δ_0, so that it is represented by power series which converge in some neighborhood of the origin. Since the linear transformation $z = \mathfrak{C}\zeta$ preserves convergence, it remains to investigate whether the substitution C obtained from (16) by comparison of coefficients is also in Δ_0. In other words, we wish to know if the formally constructed series $\phi(\xi, \eta), \psi(\xi, \eta)$ converge in some neighborhood of the origin.

In the elliptic case this question is quite delicate, and it can be shown that in general one has divergence [2]. This phenomenon is related to the presence of the "small divisors" $\lambda^{k-l}\mu^l - \lambda, \lambda^{k-l}\mu^l - \mu$ which are introduced into the coefficients of ϕ, ψ when solving (18). On the other hand, it is obvious that convergence can occur; indeed, one merely takes $T = CUC^{-1}$ for arbitrary C, U in Δ_0. However, no general method is known for determining if corresponding to a given S the series ϕ, ψ converge or diverge. Another unsolved problem is whether two convergent transformations that have the same normal form with respect to Δ can be transformed into one another by a transformation in Δ_0. In particular, this contains the problem of whether, together with T and the normal form U, also C lies in Δ_0. On the other hand, convergence of C can be established for the special case when the normal form is linear, provided that λ lies outside a certain set of measure 0 on the unit circle [3]. This observation may lead one to hope for convergence if the set of admissible eigenvalues is properly restricted. That this is not so will be seen from an example in § 34. In other words, convergence of ϕ, ψ does not depend on λ alone, but on the nonlinear terms as well.

In the hyperbolic case, however, one always has convergence. That is, the substitution C found in Δ actually lies in Δ_0 if S does. This is due partially to the fact that the quantities $\lambda^{k-l}\mu^l - \lambda$ entering into the construction of ϕ are bounded away from 0 when $k \neq 2l+1$. However, it also depends crucially on the fact that $uv = 1$, which is a consequence of the area-preserving character of S.

We present the convergence proof for the hyperbolic case, without, however, requiring the substitution C to be area-preserving. More precisely, we will show that the unique substitution C in Γ that takes S in Δ_0 into its normal form and is normalized so that $\phi_\xi - 1, \psi_\eta - 1$ contain no powers of $\xi\eta$ alone, actually belongs to Γ_0. The slightly more difficult convergence proof for area-preserving C appears in [4, 5].

For

$$F(\xi, \eta) = \sum_{k,l=0}^{\infty} a_{kl}\xi^k \eta^l$$

an arbitrary formal power series, let

$$F_n(\xi,\eta)=\sum_{k-l=n} a_{kl}\xi^k\eta^l\,,$$

whereby F is decomposed as the sum of the F_n $(n=0,\pm 1,\pm 2,\ldots)$. Observe that for any power series $u=u(\xi\eta)$ in the product $\xi\eta$ with non-zero constant term one has

$$F_n(u\xi,u^{-1}\eta)=u^n F_n(\xi,\eta)\,,$$

so that the above corresponds to the eigenfunction decomposition for the operator taking $F(\xi,\eta)$ into $F(u\xi,u^{-1}\eta)$.

We recall that ϕ,ψ are formal power series that satisfy the functional equation (16). Separating out the linear parts of p and q in (7), we have

$$p(x,y)-\lambda x=P(x,y)\prec\frac{c(x+y)^2}{1-c(x+y)}=G(x,y)$$

$$q(x,y)-\mu y=Q(x,y)\prec\frac{c(x+y)^2}{1-c(x+y)}=G(x,y)$$

for some positive constant c, where the series P,Q,G are defined by the above equations in which they appear. Equation (16) can now be written in the form

$$\text{(36)}\qquad \begin{cases}(u^n-\lambda)\phi_n=(P(\phi,\psi))_n\\ (u^n-\mu)\psi_n=(Q(\phi,\psi))_n\end{cases}\quad (n=0,\pm 1,\pm 2,\ldots)\,,$$

where the relation $uv=1$ reflecting the area-preserving character of S has been used. In addition, the required normalization becomes

$$\phi_1=\xi,\qquad \psi_{-1}=\eta\,,$$

so that by (36) for $n=\pm 1$ we have

$$(u-\lambda)\xi=(P(\phi,\psi))_1,\quad (v-\mu)\eta=(Q(\phi,\psi))_{-1}\,.$$

The latter equations together with (36) are basic for the convergence proof. These equations are already known to possess a unique formal solution ϕ,ψ,u,v with $uv=1$, and it is necessary only to estimate the coefficients. Since $uv=1$ it suffices to prove convergence of ϕ,ψ, and v, and we therefore disregard the first of the latter pair of equations.

The proof is by the method of majorants and closely resembles that in § 17, from where we also borrow some of the notation. Being in the hyperbolic case, we may assume $|\lambda|>1>|\mu|$. First we derive the majorant relations

$$
(37)\qquad \begin{cases} (u^n - \lambda)^{-1} \prec \dfrac{c_1}{1 - c_1 s} & (n \neq 1), \\[2ex] (u^n - \mu)^{-1} \prec \dfrac{c_1}{1 - c_1 s} & (n \neq -1), \end{cases}
$$

where c_1 is a suitable constant depending on λ, μ but not on n, and $s = \lceil v - \mu \rceil$ is the smallest majorant for $v - \mu$. To verify the first relation, for $n \leqq 0$ we have

$$
(u^n - \lambda)^{-1} = -\lambda^{-1}(1 - \lambda^{-1} v^{|n|})^{-1} \prec |\mu| \sum_{k=0}^{\infty} (\lceil v \rceil^{|n|} |\mu|)^k
$$

$$
\prec \frac{1}{1 - \lceil v \rceil} = \frac{1}{1 - |\mu| - s} \prec \frac{c_1}{1 - c_1 s}
$$

if $c_1 \geqq \dfrac{1}{1 - |\mu|}$. For $n > 1$, on the other hand,

$$
(u^n - \lambda)^{-1} = u^{-n}(1 - \lambda u^{-n})^{-1} = v^n \sum_{k=0}^{\infty} (\lambda v^n)^k
$$

$$
\prec \lceil v \rceil^n \sum_{k=0}^{\infty} (|\lambda|^{\frac{1}{n}} \lceil v \rceil)^{nk}
$$

$$
\prec |\lambda|\, \lceil v \rceil^n \sum_{k=0}^{\infty} (|\lambda|^{\frac{1}{n}} \lceil v \rceil)^{nk} \prec \frac{1}{1 - |\lambda|^{\frac{1}{n}} \lceil v \rceil},
$$

and since $|\lambda|^{\frac{1}{n}} \leqq |\lambda|^{\frac{1}{2}}$, we have

$$
(u^n - \lambda)^{-1} \prec \frac{1}{1 - |\lambda|^{\frac{1}{2}} \lceil v \rceil} \prec \frac{1}{1 - |\lambda|^{\frac{1}{2}} |\mu| - |\lambda|^{\frac{1}{2}} s} \prec \frac{c_1}{1 - c_1 s}
$$

if $c_1 \geqq \dfrac{|\lambda|^{\frac{1}{2}}}{1 - |\lambda|^{\frac{1}{2}} |\mu|} = \dfrac{|\lambda|^{\frac{1}{2}}}{1 - |\lambda|^{-\frac{1}{2}}}$. This proves the first relation in (37), and the second can be established similarly. From (36), (37) we next obtain for $\phi - \xi = \sum_{n \neq 1} \phi_n$, $\psi - \eta = \sum_{n \neq -1} \psi_n$, and $\eta(v - \mu)$ the majorant relations

$$
\lceil \phi \rceil - \xi, \ \lceil \psi \rceil - \eta \prec \frac{c_1}{1 - c_1 s} G(\lceil \phi \rceil, \lceil \psi \rceil),
$$

$$
\eta s = \eta \lceil v - \mu \rceil \prec G(\lceil \phi \rceil, \lceil \psi \rceil).
$$

Setting $\xi = \eta$ and introducing the power series

$$
W(\xi) = \frac{1}{\xi} (\lceil \phi \rceil - \xi + \lceil \psi \rceil - \eta) + s
$$

without constant term, we finally obtain from the last three relations the estimates

$$\xi W \prec \frac{c_2}{1-c_1 W} G(\xi(W+1), \xi(W+1))$$

or

$$W \prec \frac{c_2}{1-c_1 W} \frac{4c\xi(1+W)^2}{1-2c\xi(1+W)} \prec \frac{c_3\xi(1+W)^2}{1-c_1 W-2c\xi(1+W)}$$

for suitable positive constants c_2, c_3. As in §17, this majorant relation implies convergence of W, and therefore of ϕ, ψ, v. It should be emphasized that the identity $uv=1$ was essential for the proof. If T is a general transformation in Γ_0 for which $\lambda\mu=1$, one can again construct a similar normal form for which uv need not equal 1, but a convergence proof is not available for that case.

We now wish to discuss the normal forms still further, this time under the assumption of convergence. First consider the hyperbolic case. By (25) we have $\xi_1\eta_1=\xi\eta$, so that if ξ, η and ξ_1, η_1 are interpreted as the rectangular coordinates of two points P_0 and P_1, then for P_0 in the region of convergence of the series w, and different from $(0,0)$, both P_0 and its image $UP_0=P_1$ lie on an equilateral hyperbola. Because $|\lambda| \neq 1$, one also has $e^w \neq 1$ in a sufficiently small neighborhood G of the origin, so that P_0 and P_1 do not coincide there. If, in addition, the points $P_k=UP_{k-1}=U^kP_0$ for $k=1,\dots,n$ all lie in G, they must all be distinct from P_0, and it follows that in this neighborhood there is no point other than the origin that is a fixed-point of a power U^n and whose images under $U,\dots,U^{n-1}$ lie in G. Indeed, as long as the sequence P_k remains in G it lies on an equilateral hyperbola along which w is constant, so that $P_k=((\pm 1)^k e^{kw}\xi, (\pm 1)^k e^{-kw}\eta)$. If then $\xi\eta \neq 0$, since either $e^w>1$ or $e^{-w}>1$, it follows that not all the P_k $(k=1,2,\dots)$ can lie in G. If on the other hand $\xi\eta=0$ and, say, $e^w=|\lambda|>1$, then for $\xi \neq 0$ the sequence P_k again cannot remain in G, while for $\xi=0$ it converges to the origin. In terms of the original mapping this says that the points P for which all the $T^kP=P_k$ $(k=1,2,\dots)$ remain in a sufficiently small neighborhood of the fixed-point are confined to a real analytic curve through this point, in this case the image $x=\phi(0,\eta)$, $y=\psi(0,\eta)$ of the η-axis under the transformation C. Moreover, for P on this curve the P_k approach the fixed-point as $k\to\infty$. The curve is transformed into itself by T and is called the stable invariant curve of T. Similarly one defines the unstable invariant curve of T as the stable invariant curve of the inverse T^{-1}, which, being the image under C of the other axis, is also real analytic. Finally, returning to the normal form, we observe that if one considers also the inverse mapping U^{-1} and its powers, then for no $P_0 \neq (0,0)$ do all the images $P_k = U^k P_0$ $(k=0, \pm 1, \pm 2, \ldots)$ lie in G.

The last assertion can also be proved directly, without use of the normal form. Starting from (1), (2), suppose that $f(x,y)$, $g(x,y)$ converge. Then there exists a sufficiently small circle of radius $R>0$ about the origin and functions $\vartheta_1, \vartheta_2, \vartheta_3, \vartheta_4$ uniformly bounded in this circle such that

$$(38) \qquad x_1 = f(x,y) = ax + by + \vartheta_1 r^2\,, \qquad y_1 = g(x,y) = cx + dy + \vartheta_2 r^2$$

whenever $x^2+y^2=r^2 \leqq R^2$, while for the inverse mapping

$$(39) \qquad x = dx_1 - by_1 + \vartheta_3 r_1^2\,, \qquad y = -cx_1 + ay_1 + \vartheta_4 r_1^2$$

whenever $x_1^2 + y_1^2 = r_1^2 \leqq R^2$. Suppose now that for each ϱ with $0<\varrho \leqq R$ there is a point $P_\varrho \neq (0,0)$ whose images $S^k P_\varrho$ $(k=0, \pm 1, \pm 2, \ldots)$ all lie in the circle $x^2+y^2 \leqq \varrho^2$. The same then holds for the accumulation points of this sequence, and therefore for the closure H_ϱ of the images. Evidently $SH_\varrho = H_\varrho$, so that H_ϱ is invariant under S. Let $(x,y) = Q_\varrho$ be a point in H_ϱ for which $x^2+y^2=r^2$ is as large as possible. For $SQ_\varrho = (x_1, y_1)$, $S^{-1}Q_\varrho = (x_{-1}, y_{-1})$ we have by (38), (39) the relations

$$x_1 + x_{-1} = (a+d)\,x + (\vartheta_1 + \vartheta_3)\,r^2\,, \qquad y_1 + y_{-1} = (a+d)\,y + (\vartheta_2 + \vartheta_4)\,r^2\,,$$

so that

$$(x_1 + x_{-1})^2 + (y_1 + y_{-1})^2 = (a+d)^2 r^2 + o(r^2) \qquad (0 < r \leqq \varrho \to 0)\,,$$

while for $r_{-1}^2 = x_{-1}^2 + y_{-1}^2$, on the other hand, the triangle inequality gives

$$(x_1 + x_{-1})^2 + (y_1 + y_{-1})^2 \leqq (r_1 + r_{-1})^2 \leqq 4r^2\,.$$

In the limit as $\varrho \to 0$ this says that

$$(a+d)^2 \leqq 4\,,$$

and since $a+d = \lambda + \lambda^{-1}$ it follows that

$$(\lambda - \lambda^{-1})^2 \leqq 0\,.$$

For the hyperbolic case, however, the above is false, and consequently there does exist a circle $x^2+y^2 \leqq \varrho^2$ $(\varrho > 0)$ in the region of convergence of S and S^{-1} such that for no point $P \neq (0,0)$ all the images $S^k P$ $(k=0, \pm 1, \pm 2, \ldots)$ remain in this circle. In particular, it cannot happen that the images $S^k P$ for $k=0, \ldots, n-1$ lie in this circle while $S^n P = P$. In conclusion, we note that for this argument we did not need f and g to be analytic, but only to satisfy (38), (39).

Turning to the elliptic case, assume that the series w in (31) converges for $r^2+s^2=\varrho^2 \leqq R^2$. Then the transformation (31) takes each circle of radius $\leqq R$ centered at the origin into itself while rotating it through an angle w that depends on ϱ. Should this circle contain fixed-points of the n-times iterated map U^n, then the corresponding angle of rotation nw must be a multiple $2m\pi$ of 2π, in which case indeed the whole circle remains pointwise fixed under U^n. If the coefficients $\gamma_1, \gamma_2, \ldots$ in the power series w do not all vanish, i.e. if w is not constant, then by continuity of w as a function of the radius there are infinitely many values $\varrho \leqq R$ for which $\dfrac{w}{2\pi} = \dfrac{m}{n}$ is rational, and the corresponding circle then consists entirely of fixed-points of U^n.

Since the elliptic case is the most interesting one, henceforth we will restrict ourselves to it. Here, as opposed to the hyperbolic case, the derivation of the above result required the normal form, and therefore it is necessary to assume that C, and with it U, converges. Without this it has not been possible to prove existence of a one parameter family of curves invariant under S corresponding to the above family of concentric circles, and in fact, no such family exists in general. In the next section, however, we will be able to make a useful assertion on the problem of fixed-points without drawing on the full normal form. In preparation for this we obtain now, by means of a convergent area-preserving substitution, at least a suitable approximation to the normal form.

With this in mind, we derive a parametric representation of the substitutions in the group Δ, which comes from the considerations in § 3 about canonical transformations. Each two-by-two matrix

$$\mathfrak{M} = \begin{pmatrix} p & q \\ r & s \end{pmatrix}$$

satisfies the relation

$$\mathfrak{M}'\mathfrak{J}\mathfrak{M} = |\mathfrak{M}|\,\mathfrak{J}, \quad \mathfrak{J} = \begin{pmatrix} 0 & 1 \\ -1 & 0 \end{pmatrix},$$

and therefore the Jacobian matrix of each convergent area-preserving substitution is symplectic. Since for the substitution $z = C\zeta$ in (8) the derivative x_ξ at the point $(\xi, \eta) = (0, 0)$ has the value 1, it follows from (3; 4) that C can be obtained via the expression

$$y = \varrho_x, \qquad \xi = \varrho_\eta \tag{40}$$

for a suitable generating function $\varrho(x, \eta)$. The function ϱ will then be analytic in a neighborhood of $x = 0, \eta = 0$ and, if the irrelevant constant term is set equal to 0, it will have a convergent series expansion of the form

$$\varrho = x\eta + \cdots. \tag{41}$$

One would naturally expect that by considering all formal power series ϱ in x, η of the form (41) one can obtain via (40) all substitutions of the form $x=\xi+\cdots$, $y=\eta+\cdots$ in the group Δ, of which Δ_0 is a subgroup. To see this without returning to the developments in § 2 we proceed as follows. If the first equation in (8) is solved for ξ, the corresponding area-preserving substitution $z=C\zeta$ can be represented in the form

(42) $$\xi=P(x,\eta)=x+\cdots, \quad y=Q(x,\eta)=\eta+\cdots$$

with P and Q formal power series in x, η satisfying

$$P(\phi(\xi,\eta),\eta)=\xi, \quad Q(\phi(\xi,\eta),\eta)=\psi(\xi,\eta).$$

From this it follows that

(43) $$P_x\phi_\xi=1, \quad Q_x\phi_\xi=\psi_\xi, \quad Q_x\phi_\eta+Q_\eta=\psi_\eta,$$

and because C is area-preserving, then also

(44) $$1=\phi_\xi\psi_\eta-\phi_\eta\psi_\xi=\phi_\xi Q_\eta, \quad P_x=P_x\phi_\xi Q_\eta=Q_\eta.$$

The last equation, on the other hand, expresses precisely the necessary integrability condition that guarantees the existence of a formal power series $\varrho(x,\eta)$ of the form (41) with the prescribed derivatives $\varrho_x=Q$, $\varrho_\eta=P$. In view of (42), the substitution C is then represented in the form (40). If C is real, all the coefficients in ϱ will also be real. Conversely, if one sets up the expressions (40), (42) for ϱ an arbitrary power series of the form (41), then $P_x=Q_\eta$ and (43) holds, from which also the first equation in (44) readily follows, and the substitution C defined by (40) is again in Δ.

In analogy to (30), we next express the convergent area-preserving transformation T in (7) in real form by introducing the unknowns $\frac{1}{2}(x+y)$, $\frac{1}{2i}(x-y)$ as new variables in place of x, y. Using (28), we can express T as the real convergent area-preserving transformation

$$z_1=T^*z, \quad x_1=x\cos\gamma_0-y\sin\gamma_0+\cdots, \quad y_1=x\sin\gamma_0+y\cos\gamma_0+\cdots$$

and, by what was shown subsequently, this can be transformed by a real area-preserving substitution

$$z=C\zeta, \quad x=\phi(\xi,\eta)=\xi+\cdots, \quad y=\psi(\xi,\eta)=\eta+\cdots$$

into the real normal form (31) with r, s replaced by ξ, η. We now write C in the form (40), whereby the formal power series ϱ has real coefficients. To obtain a real convergent area-preserving substitution, for a given arbitrary integer $l\geqq 0$ we discard the terms beyond degree $2l+2$ in the series $\varrho(x,\eta)$ and in this way obtain a polynomial $\varrho_l(x,\eta)$ of degree $2l+2$.

By means of the expression

(45) $$y = \varrho_{lx}, \quad \xi = \varrho_{l\eta}$$

this polynomial also generates a real area-preserving substitution $z = C_l \zeta$ which, by the theorems on implicit functions, however, is convergent and agrees with C in the terms of degree below $2l+2$. Then $C_l^{-1} T^* C_l$ has the form

(46) $$\xi_1 = \xi \cos w_l - \eta \sin w_l + \cdots, \quad \eta_1 = \xi \sin w_l + \eta \cos w_l + \cdots$$

with

(47) $$w_l = \sum_{k=0}^{l} \gamma_k (\xi^2 + \eta^2)^k,$$

where the terms in (46) not written down explicitly all have degree at least $2l+2$ with the coefficients real. In this way S is carried by a convergent real area-preserving transformation into a form that agrees with the normal form in all terms of degree less than $2l+2$. Once the existence of the polynomial ϱ_l has been established, it can be determined also directly from (45), (46), (47) by comparison of coefficients. It is thereby evident that one need not assume that $\lambda^k \neq 1$ for all $k = 1, 2, \ldots$, but only for $k = 1, \ldots, 2l+2$. In particular, for the case $l = 1$ it is enough to know that $\lambda^3 \neq 1, \lambda^4 \neq 1$.

For the Birkhoff fixed-point theorem in the next section it will be essential that the series w in (31) not reduce to the constant term, i.e. that the normal form not be just a rotation through a fixed angle γ_0. Under this assumption let l be chosen so that $\gamma_1 = \cdots = \gamma_{l-1} = 0$ and $\gamma_l \neq 0$. If the transformation (46) is then once more expressed in complex form, with $\xi + i\eta$, $\xi - i\eta$, $\xi_1 + i\eta_1$, $\xi_1 - i\eta_1$ denoted again by ξ, η, ξ_1, η_1, then

$$\xi_1 = p(\xi, \eta) = u\xi + P, \quad \eta_1 = q(\xi, \eta) = v\eta + Q$$

with

(48) $$u = e^{i\gamma_0 + i\gamma(\xi\eta)^l}, \quad v = u^{-1}, \quad \gamma = \gamma_l \neq 0$$

and $\bar{p}(\xi, \eta) = q(\eta, \xi)$. Here the series P, Q begin with terms of degree at least $2l+2$ and converge in a neighborhood of $\xi = 0$, $\eta = 0$. Finally, interchanging ξ, η if necessary, one may assume $\gamma > 0$ and by the linear substitution

$$\xi = \xi^* \gamma^{-\frac{1}{2l}}, \quad \eta = \eta^* \gamma^{-\frac{1}{2l}}$$

achieve the simplification $\gamma = 1$ in (48).

§ 24. The Birkhoff Fixed-Point Theorem

Our point of departure is again a real area-preserving mapping $z_1 = Sz$ of the form given by (23;1), (23;2), where the power series $f(x, y)$, $g(x, y)$ are assumed to converge in some neighborhood of the origin. We consider the elliptic case, whereby the eigenvalues λ, λ^{-1} of the matrix

$$\mathfrak{S} = \begin{pmatrix} a & b \\ c & d \end{pmatrix}$$

have absolute value 1 but are not ± 1. Assuming in addition that $\lambda^k \neq 1$ $(k = 3, \ldots, 2l+2)$ we compute the invariants $\gamma_1, \ldots, \gamma_l$ in (23;31) and take γ_l to be the first nonvanishing one. By the result at the end of the previous section there exists then a convergent substitution $z = C\zeta$ such that $C^{-1}SC = T$ is area-preserving and has the form

$$(1) \qquad \begin{cases} \xi_1 = p(\xi, \eta) = u\xi + P, & \eta_1 = q(\xi, \eta) = v\eta + Q, \quad uv = 1, \\ u = e^{i(\alpha + r^{2l})}, & r^2 = \xi\eta, \quad \bar{p}(\xi, \eta) = q(\eta, \xi), \end{cases}$$

where P and Q are series that begin with terms of degree at least $2l+2$ and α is a real constant. In order that the original variables x, y be real, one has to take $\eta = \bar{\xi}$, $r = |\xi|$. We will prove that for every sufficiently small neighborhood G of the origin in the (x, y)-plane and for all sufficiently large integers $n > n_0(G)$ there exist fixed-points $z \neq 0$ of S^n such that $S^k z \in G$ $(k = 0, \ldots, n-1)$. This assertion will be referred to as the Birkhoff fixed-point theorem. The following proof differs from that given by Birkhoff in that we make the necessary estimates more precise.

We introduce the polar coordinates r, ϕ $(r > 0)$, so that $\xi = re^{i\phi}$, $\eta = re^{-i\phi}$, and denote by ξ_k, r_k, ϕ_k $(k = 0, 1, \ldots)$ the coordinates ξ, r, ϕ of $\zeta_k = T^k\zeta$. Also, $c_1, \ldots, c_{17}$ will denote suitable positive constants that depend only on the given mapping S, while $\vartheta_0, \vartheta_1, \ldots$ will be certain functions of r, ϕ defined in each case by the equation in which they first appear. When there is no danger of confusion, the symbol ϑ will occasionally be used for several different functions. Let c_1 be chosen so that in the circle $r \leqq c_1^{-1}$ the series P, Q converge absolutely and satisfy the estimate

$$(2) \qquad |P| + |Q| \leqq c_2 r^{2l+2}.$$

By (1), (2) we then have

$$r_1^2 = \xi_1\eta_1 = r^2 + \vartheta r^{2l+3}, \quad |\vartheta| < c_3,$$

$$(3) \qquad r_1 = r(1 + \vartheta r^{2l+1})^{\frac{1}{2}} = r + \vartheta_1 r^{2l+2}, \quad |\vartheta_1| < c_4 \quad (r < c_1^{-1}).$$

Next we prove the following lemma:

If r and a natural number n satisfy

$$(4)\qquad 0<r<\frac{4}{5}c_1^{-1}, \quad nr^{2l+1}<\frac{1}{6l+6}c_4^{-1},$$

then

$$(5)\quad 0<\tfrac{3}{4}r<r_k<\tfrac{5}{4}r<c_1^{-1},\ r_k=r+k\vartheta_k r^{2l+2},\ |\vartheta_k|<3c_4\quad (k=0,\dots,n).$$

The proof is by induction on k. For $k=0$ the assertion is trivial, since $\xi_k=\xi\neq 0$ and one can take $\vartheta_0=0$. Assume now that the assertion has been proved for some $k<n$. Then (3) gives the estimate

$$r_{k+1}=r_k+\vartheta_1 r_k^{2l+2}=r+k\vartheta_k r^{2l+2}+\vartheta_1 r^{2l+2}(1+k\vartheta_k r^{2l+1})^{2l+2},$$

$$(6)\qquad |r_{k+1}-r|\leqq r^{2l+2}\{k|\vartheta_k|+|\vartheta_1|(1+k|\vartheta_k|r^{2l+1})^{2l+2}\},$$

while by (4) and the induction hypothesis (5) we have

$$(7)\qquad k|\vartheta_k|\,r^{2l+1}<\frac{1}{2l+2},$$

so that the expression in the curved bracket in (6) is less than

$$3kc_4+c_4e<(3k+3)\,c_4.$$

This implies the second assertion of (5) with $k+1$ in place of k, while from (4), (7) one also has

$$r_{k+1}\leqq r\{1+(k+1)\,|\vartheta_{k+1}|\,r^{2l+1}\}<r(1+\tfrac{1}{4})=\tfrac{5}{4}r<c_1^{-1},$$
$$r_{k+1}\geqq r\{1-(k+1)\,|\vartheta_{k+1}|\,r^{2l+1}\}>r(1-\tfrac{1}{4})=\tfrac{3}{4}r>0,$$

which completes the induction.

Taking logarithms in (1) we obtain the relation

$$(8)\qquad \log r_1+i\phi_1=\log r+i\phi+i\alpha+ir^{2l}+\log\left(1+\frac{P}{u\xi}\right),$$

and considering the imaginary parts, after suitably adjusting the integral multiple of 2π in the continuous function $\phi_1-\phi$ and using (2), we obtain

$$(9)\quad \phi_1-\phi=\alpha+r^{2l}+\vartheta r^{2l+1},\qquad |\vartheta|<c_5\quad (0<r<c_6^{-1}\leqq c_1^{-1}).$$

If in addition

$$(10)\qquad 0<r<\frac{4}{5}c_6^{-1}, \quad nr^{2l+1}<\frac{1}{6l+6}c_4^{-1},$$

then, by the lemma, $0<r_k<c_6^{-1}$ for $k=0,\dots,n$. We wish r and n to satisfy (10) so as to have (9) apply also to the images ξ_k $(k=0,\dots,n-1)$

in place of ξ and give

$$\phi_{k+1}-\phi_k=\alpha+r_k^{2l}+\vartheta_k r_k^{2l+1}, \quad |\vartheta_k|<c_5.$$

In that case (5) implies

$$\phi_{k+1}-\phi_k=\alpha+r^{2l}+\vartheta_k r^{2l+1}(1+nr^{2l}), \quad |\vartheta_k|<c_7 \quad (k=0,\ldots,n-1)$$

and addition over k gives

$$\phi_n-\phi=n(\alpha+r^{2l})+\tau \tag{11}$$

with

$$\tau=n\vartheta r^{2l+1}(1+nr^{2l}), \quad |\vartheta|<c_8. \tag{12}$$

Let M,δ be any two positive numbers satisfying the inequalities

$$M>4\pi, \quad \delta<\min\left\{\frac{c_4^{-1}}{(6l+6)M}, \frac{4c_6^{-1}}{5}, \frac{\pi c_8^{-1}}{2M(1+M)}\right\}, \tag{13}$$

and choose a natural number

$$n>M\delta^{-2l}. \tag{14}$$

We assign to the number $n\alpha$ an integer g in accordance with the conditions

$$n\alpha=2g\pi+\beta, \quad -\pi\leqq\beta<\pi, \tag{15}$$

which determine g,β uniquely. Furthermore, let h be any natural number in the interval

$$1\leqq h\leqq\frac{M}{2\pi}-1 \tag{16}$$

and define I_h as the set of r satisfying

$$-\frac{\pi}{2}\leqq nr^{2l}-2h\pi+\beta\leqq\frac{\pi}{2}. \tag{17}$$

Since

$$2h\pi-\frac{\pi}{2}-\beta>\frac{\pi}{2}>0, \quad 2h\pi+\frac{\pi}{2}-\beta\leqq 2h\pi+\frac{3\pi}{2}<M, \tag{18}$$

we see by (14) that the interval I_h lies in the interior of $0<r<\delta$. Keeping ϕ fixed, let r increase over the interval I_h. The points $\xi=re^{i\phi}$ then generate a closed interval $I_h(\phi)$ in the complex plane that lies on the ray emanating from the origin and making an angle ϕ with the positive real axis. In view of (13) we also have

$$0<r<\delta<\frac{4}{5}c_6^{-1}, \quad nr^{2l+1}<Mr<M\delta<\frac{1}{6l+6}c_4^{-1}, \tag{19}$$

so that (10) holds, and for the function $\tau = \tau(r, \phi)$ one obtains by (12) the estimate

$$|\tau| \leqq |\vartheta| M r(1+M) < c_8 M \delta(1+M) < \frac{\pi}{2}.$$

Now by (11), (15), (17) the expression

$$F(r, \phi) = \phi_n - \phi - 2(g+h)\pi = nr^{2l} - 2h\pi + \beta + \tau$$

has opposite signs at the two endpoints of $I_h(\phi)$, and as a continuous function of r it must therefore change sign at least once in the interval. On the other hand, for

$$\phi_n - \phi = 2(g+h)\pi \tag{20}$$

the image $\xi_n = r_n e^{i\phi_n}$ of ξ under T^n also lies on the same ray as ξ, and, moreover, we have there $0 < r_n < c_6^{-1}$.

From the analytic dependence of the coordinates ξ_k, η_k on ξ, η it follows that ϕ_n, and with it $F(r, \phi)$, is actually analytic in the variables r, ϕ for $0 < r < \frac{4}{5} c_6^{-1}$. Moreover, we will subsequently show that the partial derivative $F_r = \phi_{nr}$ is positive in the whole interval $I_h(\phi)$ provided in addition to (13) we require that

$$\delta < e^{-c_9 M} \tag{21}$$

for a suitable c_9 to be determined later. All the conditions on δ will then certainly be satisfied if we choose

$$\delta < e^{-c_{10} M}. \tag{22}$$

Under this assumption the equation $F(r, \phi) = 0$ now has exactly one solution $r = r(\phi)$ in $I_h(\phi)$, which according to the existence theorem for implicit functions is differentiable and even analytic in ϕ. As ϕ runs over the interval $0 \leqq \phi \leqq 2\pi$ the point $r(\phi)$ describes a smooth closed curve K that lies in the punctured disk $0 < r < \frac{4}{5} c_6^{-1}$ and surrounds the origin. The image $K_n = T^n K$ of K under the mapping T^n then is a smooth curve in the disk $0 < r < c_6^{-1}$ that likewise surrounds the origin, and indeed, by (20), for each ξ in K the image ξ_n in K_n lies on the ray from 0 through ξ. Now if the two simple closed curves K and K_n did not intersect, then one would have to lie entirely in the interior of the other, which would contradict that T is area-preserving. Consequently, the curves have, in fact, at least two distinct points in common. On the other hand, for each point where K and K_n intersect we have $\xi = \xi_n$. Thus the above assumptions lead to at least two fixed-points $\xi \neq 0$ for the mapping T^n with the images ξ_k $(k = 0, \ldots, n)$ all remaining in the circle $|\xi| < \frac{5}{4}\delta$. Returning to the original coordinates x, y in place of ξ, η and observing that the M in (22) can be taken arbitrarily large, we then obtain the previously mentioned fixed-point theorem of Birkhoff.

According to (17), for n fixed, the intervals I_h corresponding to the different h are disjoint from one another, and hence so are the sets of fixed-points corresponding to the values

$$h=1,2,\ldots,\left[\frac{M}{2\pi}\right]-1$$

permitted by (16). Of course if one also changes n in keeping with (14), fixed-points corresponding to different pairs n, h may agree. However, for $M>c_{11}$ this will not happen if n runs only over pairwise relatively prime integers, or, in particular, only over prime numbers. For if $T^m\zeta=T^n\zeta=\zeta$ and the largest common divisor is $(m,n)=1$, there exist integers p, q such that $pm+qn=1$, so that $(T^m)^p(T^n)^q=T$ and $T\xi=\xi$, while in a sufficiently small neighborhood of the origin the only fixed-point of T is $(0,0)$. It follows that for prime numbers n one of our constructed fixed-points of T^n is also a fixed-point of T^m if and only if m is divisible by n.

It still remains to prove the earlier required estimate that for a suitable choice of c_9 we have $\phi_{nr}>0$ throughout the interval $I_h(\phi)$. For this we consider the total differential in (8) and introduce the abbreviations $\log r=\varrho$, $\log r_k=\varrho_k$ $(k=0,\ldots,n)$ whereupon it follows that

$$d\varrho_1+id\phi_1=d\varrho+id\phi+2ilr^{2l}d\varrho+r^{2l+1}(\vartheta d\varrho+\tilde{\vartheta}d\phi),$$
$$|\vartheta|+|\tilde{\vartheta}|<c_{12}\qquad(0<r<c_6^{-1}).$$

Assuming (10), we have $0<r_k<c_6^{-1}$ for $k=0,\ldots,n$, and therefore

$$\text{(23)}\qquad\begin{cases}d\varrho_{k+1}+id\phi_{k+1}=d\varrho_k+id\phi_k+2ilr_k^{2l}d\varrho_k+r_k^{2l+1}(\vartheta_k d\varrho_k+\tilde{\vartheta}_k d\phi_k),\\ |\vartheta_k|+|\tilde{\vartheta}_k|<c_{12}\qquad(k=0,\ldots,n-1).\end{cases}$$

Now let

$$\text{(24)}\qquad\mathfrak{A}_k=\begin{pmatrix}1&0\\2lr_k^{2l}&1\end{pmatrix},\quad\mathfrak{B}_k=r_k^{2l+1}\begin{pmatrix}\vartheta_{1k}&\vartheta_{2k}\\\vartheta_{3k}&\vartheta_{4k}\end{pmatrix},$$

so that (23) can be expressed in real vector form as

$$\begin{pmatrix}d\varrho_{k+1}\\d\phi_{k+1}\end{pmatrix}=\mathfrak{M}_k\begin{pmatrix}d\varrho_k\\d\phi_k\end{pmatrix},\quad\mathfrak{M}_k=\mathfrak{A}_k+\mathfrak{B}_k$$

with

$$|\vartheta_{1k}|+|\vartheta_{2k}|+|\vartheta_{3k}|+|\vartheta_{4k}|<c_{13}.$$

Thus, if

$$(25) \qquad \mathfrak{M}_{n-1} \dots \mathfrak{M}_1 \mathfrak{M}_0 = \begin{pmatrix} \varkappa & \lambda \\ \mu & \nu \end{pmatrix},$$

the partial derivative in question appears as $\phi_{nr} = r^{-1}\mu$, and it remains to show that $\mu > 0$.

In what follows, for two real matrices $\mathfrak{X}$ and $\mathfrak{Y}$, let the formula $\mathfrak{X} \prec \mathfrak{Y}$ mean that the absolute values of the elements in $\mathfrak{X}$ do not exceed the corresponding elements in $\mathfrak{Y}$; in particular, the elements of $\mathfrak{Y}$ must then all be $\geqq 0$. If in addition we set

$$\mathfrak{B} = \frac{1}{2}\begin{pmatrix} 1 & 1 \\ 1 & 1 \end{pmatrix},$$

then $\mathfrak{B}^2 = \mathfrak{B}$ and

$$\mathfrak{A}_k \prec \mathfrak{E} + c_{14} r^{2l} \mathfrak{B} = \mathfrak{A}, \quad \mathfrak{B}_k \prec c_{14} r^{2l+1} \mathfrak{B} = \mathfrak{B} \qquad (k = 0, \dots, n-1).$$

Because $\mathfrak{A}$ and $\mathfrak{B}$ commute, it follows that

$$(26) \qquad \mathfrak{M}_{n-1} \dots \mathfrak{M}_1 \mathfrak{M}_0 - \mathfrak{A}_{n-1} \dots \mathfrak{A}_1 \mathfrak{A}_0 \prec (\mathfrak{A} + \mathfrak{B})^n - \mathfrak{A}^n,$$

$$(27) \qquad \begin{cases} (\mathfrak{A} + \mathfrak{B})^n - \mathfrak{A}^n = \mathfrak{B} \displaystyle\sum_{k=0}^{n-1} (\mathfrak{A} + \mathfrak{B})^{n-k-1} \mathfrak{A}^k \prec n\mathfrak{B}(\mathfrak{A} + \mathfrak{B})^{n-1} \\ = c_{14} n r^{2l+1} (\mathfrak{A} + \mathfrak{B})^{n-1} \mathfrak{B} \\ = c_{14} n r^{2l+1} (1 + c_{14} r^{2l} + c_{14} r^{2l+1})^{n-1} \mathfrak{B} \prec c_{15} n r^{2l+1} e^{c_{16} n r^{2l}} \mathfrak{B}, \end{cases}$$

where (10) has been kept in mind throughout. Furthermore (24) leads to the estimate

$$\mathfrak{A}_{n-1} \dots \mathfrak{A}_1 \mathfrak{A}_0 = \begin{pmatrix} 1 & 0 \\ \sigma & 1 \end{pmatrix}, \quad \sigma = 2l \sum_{k=0}^{n-1} r_k^{2l} > 2l \left(\frac{3}{4}\right)^{2l} n r^{2l} > c_{17}^{-1} n r^{2l},$$

and this in conjunction with (25), (26), (27) gives

$$\mu > n r^{2l} (c_{17}^{-1} - c_{15} r e^{c_{16} n r^{2l}}).$$

Thus indeed $\mu > 0$ provided that

$$r < (c_{15} c_{17})^{-1} e^{-c_{16} n r^{2l}},$$

and this will be guaranteed in $I_n(\phi)$ by (19), (21) if c_9 is chosen sufficiently large. We note that it is only at this point that c_9, c_{10}, c_{11} are determined.

This completes the proof of the Birkhoff fixed-point theorem [1]. The proof of Poincaré's fixed-point theorem as given by Birkhoff is based on the same idea, namely, constructing a curve that surrounds the origin and whose points are translated by S^n radially.

We apply Birkhoff's theorem to a Hamiltonian system

$$\dot{x}_k = E_{y_k}, \quad \dot{y}_k = -E_{x_k} \quad (k=1,2) \tag{28}$$

for which a periodic solution that is not an equilibrium solution is assumed to be known. With $x_k(t,\xi,\eta)$, $y_k(t,\xi,\eta)$ denoting the solution with initial values $x_k=\xi_k$, $y_k=\eta_k$ at $t=0$, let $\xi_k=\xi_k^*$, $\eta_k=\eta_k^*$ correspond to the known periodic solution, and assume in addition that this closed orbit in the (x,y)-space is not tangent to the plane $y_2=\eta_2^*$, or, equivalently, that $E_{x_2}(\xi^*,\eta^*)\neq 0$. Keeping $\eta_2=\eta_2^*$ and the value $E(\xi,\eta)=E(\xi^*,\eta^*)$ fixed, we consider ξ_1,η_1 as independent variables in a small neighborhood of $\xi_1=\xi_1^*$, $\eta_1=\eta_1^*$. From the considerations at the end of § 22 we see that in tracing the corresponding solutions for increasing t until their next intersection with the plane $y=\eta_2^*$, we are led to an analytic area-preserving mapping S in the two-dimensional (x_1,y_1)-plane with $x_1=\xi_1^*$, $y_1=\eta_1^*$ as a fixed-point. Let us assume that we are in the elliptic case, whereby if there exists a natural number l such that $\lambda^k\neq 1$ $(k=1,\dots,2l+2)$ and $\gamma_1=\dots=\gamma_{l-1}=0$, $\gamma_l\neq 0$, then the mapping S can be transformed into the form (1) and Birkhoff's fixed-point theorem applies. The latter then leads to the existence of infinitely many periodic solutions arbitrarily near our original solution, all having the same value $E(\xi^*,\eta^*)$ for the Hamiltonian function, and indeed for each sufficiently large prime number n there is an orbit that closes in on itself only after circling n times. If at the point ξ^*,η^* we have $E_{x_2}=0$ but $E_{y_2}\neq 0$, then replacement of x,y by $y,-x$ leads again to the above case.

As an example, let us consider the restricted three-body problem. As before: let the points P_1,P_2,P_3 have masses μ, $1-\mu$, 0 with $0<\mu<1$; let the particles P_1,P_2 rotate with angular velocity 1 about their center of mass; and, in the corresponding rotating coordinate system, let the coordinates of the three points be $(1-\mu,0)$, $(-\mu,0)$, (x_1,x_2). The equations of motion (21; 28) can be readily expressed in canonical form by introducing $y_1=x_3-x_2$, $y_2=x_1+x_4$ in place of x_3,x_4 and setting

$$E=\tfrac{1}{2}(y_1^2+y_2^2)+x_2y_1-x_1y_2-F(x_1,x_2) \tag{29}$$

with F given by (21; 29), whereby (21; 28) is transformed into (28). In § 21 we applied Poincaré's continuation method, starting for $\mu=0$ from the periodic solution (21; 30) with $r^3(\omega+1)^2=1$ subject to certain restrictions on ω. For $\mu>0$ sufficiently small we obtained periodic solutions to the restricted three-body problem in the vicinity of that particular solution. Such a solution will now be used as the initial solution in the application of Birkhoff's fixed-point theorem, whereby $\mu=\mu_0>0$. For $\mu=0$ the periodic solution (21; 30) has initial values $\xi_1^*=r$, $\eta_1^*=0$, $\xi_2^*=0$, $\eta_2^*=r(\omega+1)$ at which $E_{y_2}(\xi^*,\eta^*)=\eta_2^*-\xi_1^*=r\omega\neq 0$, so that the fixed-point method can be applied. On the other hand, the

Hamiltonian function E is analytic in the parameter μ and therefore, by the existence theorems, the same is true for the solution $x(t,\xi,\eta)$, $y(t,\xi,\eta)$. In particular it follows from this that the fixed-point method applies also to the periodic solutions corresponding to sufficiently small values of μ_0. Moreover, if for $\mu=0$ we are in the elliptic case and if the earlier conditions $\lambda^k \neq 1$ $(k=1,\ldots,2l+2)$, $\gamma_1=\cdots=\gamma_{l-1}=0$, $\gamma_l \neq 0$ are satisfied by the eigenvalue λ for some natural number l, then in view of the analytic dependence on μ the same is true for $\mu=\mu_0$ sufficiently small, and indeed with the same or a smaller value of l. Consequently one has to compute the mapping S only for $\mu=0$. For this, however, the equation of variation (21; 7) can be solved explicitly, as was already mentioned in § 21, whereupon after an elementary computation one obtains for S the series expansion

$$(30)\quad \begin{cases} x_1 = c\xi_1 + (\omega+1)^{-1} s\eta_1 + \cdots, \quad y_1 = -(\omega+1)s\xi_1 + c\eta_1 + \cdots, \\ c = \cos\dfrac{2\pi}{\omega}, \quad s = \sin\dfrac{2\pi}{\omega} \end{cases}$$

with ξ_1, η_1 the initial values of x_1, y_1 at $t=0$. The eigenvalues of the matrix corresponding to the linear terms are found to be λ, λ^{-1} with $\lambda = e^{\frac{2\pi i}{\omega}}$, where the assumptions $\omega \neq 0, -1, -2$ and (21; 32) were made already in § 21. If in addition one assumes that

$$(31)\qquad \omega \neq 3g^{-1}, 4g^{-1} \quad (g = \pm 1, \pm 2, \ldots),$$

then we are in the elliptic case and $\lambda^k \neq 1$ for $k=1,\ldots,4$. By actually computing also the second and third order terms in the series expansion (30) one can obtain the invariant γ_1 explicitly [2], and one finds that $\gamma_1 = -3\pi r^{-2}\omega^{-3} \neq 0$, so that $l=1$. All the assumptions for $\omega \neq 0$ are contained in (31), whereupon for $\mu>0$ sufficiently small we are assured of the existence of infinitely many periodic solutions to the restricted three-body problem in the vicinity of the corresponding initial solution, and indeed of solutions which close in on themselves only after circling many times and which have the same value for the Jacobian integral E as the corresponding initial solution.

One might suppose that these solutions can also be obtained by means of the continuation method as follows. For $\mu=0$ all the solutions of (21; 28) in the (x_1, x_2)-plane represent conic sections rotating with angular velocity -1 about a focus that lies at the origin. In the neighborhood of the circular solution (21; 30) with period $\dfrac{2\pi}{|\omega|}$ there are orbits corresponding to rotating ellipses. Such an orbit is closed in the rotating coordinate system if and only if the time required to traverse the ellipse is commensurable with 2π, that is, if this time is $\tau = 2\pi \frac{k}{l}$ with l/k a rational number near ω. For l/k expressed as an irreducible fraction the corresponding period is then $2\pi k$, and the orbit first closes after circling $|l|$ times. If the earlier assumptions for the application of the continuation method were satisfied, we could deduce the existence of such periodic solutions for sufficiently small positive values μ. It turns out, however, that the continuation method is not applicable here in its usual form, in that the assumption about the rank of the Jacobian matrix (21; 27) is not satisfied. The difficulty is related to the fact that the differential equations of the restricted three-body problem possess for $\mu = 0$ the angular momentum integral and the energy integral, while for $\mu > 0$ we have only the Jacobian integral (29) at our disposal.

The long-periodic solutions of the restricted three-body problem constructed here lie close to circular orbits about the heavier particle of mass $1 - \mu$. One may inquire for similar solutions circling about the particle of smaller mass μ, as it would be natural for the lunar problem. In this case one has to restrict the Jacobian integral (29) to large negative values. Of course, for $\mu \to 0$ such solutions loose their meaning as the particle of small mass μ disappears. Nevertheless, using another version of the Birkhoff fixed-point theorem, Conley [3] succeeded in establishing the existence of such long-periodic solutions. Incidentally, this paper also contains a series expansion for the periodic solutions of the first kind about the small mass point which correspond to the solutions discussed in § 19, § 20. This derivation is of interest since it shows clearly the connection between this problem and the existence theorem of § 16, § 17.

Chapter Three

Stability

§ 25. The Function-Theoretic Center Problem

We begin with the definition of stability and instability. Let $\mathfrak{R}$ be a topological space whose points we denote by $\mathfrak{p}$, and let $\mathfrak{a}$ be a certain point in $\mathfrak{R}$. By a neighborhood here we will always mean a neighborhood of $\mathfrak{a}$ in $\mathfrak{R}$. Let $\mathfrak{p}_1 = S\mathfrak{p}$ be a topological mapping of a neighborhood $\mathfrak{U}_1$ onto a neighborhood $\mathfrak{B}_1$, whereby $\mathfrak{a} = S\mathfrak{a}$ is mapped onto itself. The inverse mapping $\mathfrak{p}_{-1} = S^{-1}\mathfrak{p}$ then carries $\mathfrak{B}_1$ onto $\mathfrak{U}_1$, and in general $\mathfrak{p}_n = S^n\mathfrak{p}$ $(n = 0, \pm 1, \pm 2, \ldots)$ is a topological mapping of a neighborhood $\mathfrak{U}_n$ onto a neighborhood $\mathfrak{B}_n$, having $\mathfrak{a}$ as a fixed-point. For each point $\mathfrak{p} = \mathfrak{p}_0$ in the intersection $\mathfrak{U}_1 \cap \mathfrak{B}_1 = \mathfrak{W}$ we construct the successive images $\mathfrak{p}_{k+1} = S\mathfrak{p}_k$ $(k = 0, 1, \ldots)$, as long as $\mathfrak{p}_k$ lies in $\mathfrak{U}_1$, and similarly $\mathfrak{p}_{-k-1} = S^{-1}\mathfrak{p}_{-k}$, as long as $\mathfrak{p}_{-k}$ lies in $\mathfrak{B}_1$. If the process terminates with a largest $k + 1 = n$, then $\mathfrak{p}_0, \ldots, \mathfrak{p}_{n-1}$ all still lie in $\mathfrak{U}_1$, but $\mathfrak{p}_n$ no longer does; similarly for the negative indices. In this way, to each $\mathfrak{p}$ in $\mathfrak{W}$ there is associated a sequence of image points $\mathfrak{p}_k$ $(k = \ldots, -1, 0, 1, \ldots)$, which is finite, infinite on one side, or infinite on both sides.

The mapping S is said to be stable at the fixed-point $\mathfrak{a}$ if for each neighborhood $\mathfrak{U} \subset \mathfrak{W}$ there exists a neighborhood $\mathfrak{B} \subset \mathfrak{U}$ of $\mathfrak{a}$ whose images $S^n\mathfrak{B}$ $(n = \pm 1, \pm 2, \ldots)$ all lie in $\mathfrak{U}$. Instability on the other hand, is defined not as the logical negation of stability, but in terms of the following stronger requirement. The mapping S is said to be unstable at the fixed-point $\mathfrak{a}$ if there exists a neighborhood $\mathfrak{U} \subset \mathfrak{W}$ such that for each point $\mathfrak{p} \neq \mathfrak{a}$ in $\mathfrak{U}$ at least one image point $\mathfrak{p}_n$ lies outside $\mathfrak{U}$.

Let us restate the above definition in another form. A point set $\mathfrak{M} \subset \mathfrak{W}$ is said to be invariant under the mapping S if $\mathfrak{M} = S\mathfrak{M}$. The fixed-point $\mathfrak{a}$ is, of course, trivially an invariant point set. We now show that S is stable if and only if each neighborhood $\mathfrak{U}$ contains an invariant neighborhood $\mathfrak{B}$. If for each neighborhood $\mathfrak{U}$ there exists a neighborhood $\mathfrak{B} = S\mathfrak{B} \subset \mathfrak{U}$, then certainly $\mathfrak{B}$ has the necessary property required in the definition of stability, and consequently S is stable. Conversely, under the assumption that S is stable, for each neighborhood $\mathfrak{U} \subset \mathfrak{W}$ there exists a neighborhood $\mathfrak{Q} \subset \mathfrak{U}$ such that $S^n\mathfrak{Q} \subset \mathfrak{U}$ $(n = 0, \pm 1, \pm 2, \ldots)$. The union

$\mathfrak{B}=\bigcup_n(S^n\mathfrak{Q})$ of all the $S^n\mathfrak{Q}$ is then invariant under S and is again a neighborhood, whereby the assertion is proved. Correspondingly, let us show that S is unstable if and only if there exists a neighborhood $\mathfrak{U}$ that contains no invariant subsets other than the fixed-point $\mathfrak{a}$. Indeed, if such a neighborhood $\mathfrak{U}$ exists then certainly the intersection $\mathfrak{U}\cap\mathfrak{W}$ has the same property, and we may therefore assume that $\mathfrak{U}\subset\mathfrak{W}$. If $\mathfrak{p}$ is then any point $\neq\mathfrak{a}$ in $\mathfrak{U}$, the images $\mathfrak{p}_n$ cannot all lie in $\mathfrak{U}$ or else $\mathfrak{M}=\bigcup_n\mathfrak{p}_n$ would be an invariant subset of $\mathfrak{U}$ that contains a point $\neq\mathfrak{a}$. Consequently S is unstable. Conversely, if S is unstable, there exists a neighborhood $\mathfrak{U}\subset\mathfrak{W}$ such that for each $\mathfrak{p}\neq\mathfrak{a}$ in $\mathfrak{U}$ at least one image $\mathfrak{p}_n$ does not lie in $\mathfrak{U}$. If $\mathfrak{p}$ is now any point of an invariant subset $\mathfrak{M}=S\mathfrak{M}$ of $\mathfrak{U}$, all the images $\mathfrak{p}_n$ of $\mathfrak{p}$ must lie in $\mathfrak{M}$ and therefore certainly in $\mathfrak{U}$, from which it follows that $\mathfrak{p}=\mathfrak{a}$. This again proves the assertion.

A mapping S that is not unstable thus has the property that each neighborhood contains an invariant point set with $\mathfrak{a}$ as a proper subset, while for a stable mapping S each neighborhood actually contains an invariant neighborhood. Consequently a stable mapping is necessarily not unstable, but a mapping that is not stable need not be unstable. A mapping S is said to be mixed at a fixed-point $\mathfrak{a}$ if it is neither stable nor unstable there. That there actually exist mixed mappings is seen by the simple example of the affine mapping $x_1=x+y$, $y_1=y$ in the (x,y)-plane, which has each point of the abscissa axis as a fixed-point. A bounded set is invariant under this mapping if and only if it lies on the abscissa axis. Since for arbitrary $r>0$ the disk $x^2+y^2<r^2$ contains no invariant neighborhood of $(x,y)=(0,0)$ but contains the invariant interval $-r<x<r$, $y=0$, at the origin this mapping is neither stable nor unstable.

We carry over the definition of stability and instability to systems of differential equations

$$\dot{x}_k=f_k(x)\quad(k=1,\ldots,m)\,. \tag{1}$$

Let $x=\xi^*$ be an equilibrium solution, so that $f_k(\xi^*)=0$, and assume that a Lipschitz condition holds in a neighborhood of $x=\xi^*$. We again denote by $x(t,\xi)$ the solution to (1) with initial values $x_k=\xi_k$ at $t=0$. Passage from ξ to $x(t,\xi)$ then defines for each fixed t a topological mapping S_t in a neighborhood of the fixed-point $x=\xi^*$. The definitions of stability and instability of the system (1) at the given equilibrium point are then obtained by taking for $\mathfrak{a}$, $\mathfrak{p}$, S^n, and $\mathfrak{p}_n=S^n\mathfrak{p}$ $(n=0,\pm1,\ldots)$ in the previous definitions the corresponding quantities ξ^*, ξ, S_t, and $\xi_t=x(t,\xi)$ as t varies over the reals. By introducing the modification that only positive values of t are permitted, one may speak also of stability or instability with respect to future time. This notion has, of course, significance in problems dealing with mechanics. Also the definition of the mixed case carries over in an obvious way.

Before turning to problems relating to stability for differential equations, we will look at the particular case when S is a conformal mapping in the plane. Already here some of the characteristic difficulties show up, although they can still be overcome by the available methods of analysis. Without loss of generality the fixed-point may be taken as the origin of the complex z-plane. The conformal mapping is then given by a power series

$$z_1 = f(z) = \lambda z + a_2 z^2 + a_3 z^3 + \cdots \qquad (\lambda \neq 0) \tag{2}$$

with complex coefficients, which converges in a neighborhood of $z = 0$. We wish to investigate when this mapping is stable, unstable, or mixed at $z = 0$. Assume first that S is stable. The circle of convergence $\mathfrak{K}$ for the series (2) then contains an invariant neighborhood $\mathfrak{B} = S\mathfrak{B}$ of the origin. This neighborhood may not be connected, but it does contain a connected invariant neighborhood; indeed, if $\mathfrak{L}$ is an open disk in $\mathfrak{B}$ containing the origin, the union of all the images $S^n \mathfrak{L}$ $(n = 0, \pm 1, \ldots)$ has the desired property. We may therefore assume that $\mathfrak{B}$ is already connected. Our aim here is to find an invariant neighborhood in $\mathfrak{K}$ that can be mapped conformally onto the unit disk. This can be achieved in, say, one of the following two ways. Perhaps $\mathfrak{B}$ is not simply connected. Then one adds to $\mathfrak{B}$ all points that lie in the interior of any simple closed curve $\mathfrak{C}$ contained in $\mathfrak{B}$. The resulting set $\mathfrak{U}$ is then again a connected neighborhood within $\mathfrak{K}$, and is easily seen to be simply connected. Because of the invariance of $\mathfrak{B}$, together with $\mathfrak{C}$ also $S\mathfrak{C}$ belongs to $\mathfrak{B}$, from which it follows that $\mathfrak{U}$ is invariant. Now, by the Riemann mapping theorem, $\mathfrak{U}$ can be mapped conformally onto a disk $|\zeta| < \varrho$ so that $z = 0$ goes into $\zeta = 0$ and the derivative z_ζ at $\zeta = 0$ has the value 1. Let

$$z = \phi(\zeta) = \zeta + b_2 \zeta^2 + \cdots \qquad (|\zeta| < \varrho) \tag{3}$$

be the inverse conformal mapping, whereby the series converges certainly in the circle $|\zeta| < \varrho$. We denote the mapping (3) by C and consider $T = C^{-1} S C$. Since the region $\mathfrak{U}$ was invariant under S, the disk $|\zeta| < \varrho$ is evidently invariant under the conformal mapping T, which has the center $\zeta = 0$ as a fixed-point. It follows from a well-known theorem in function theory that T is a linear mapping of the form

$$\zeta_1 = \mu \zeta \qquad (|\mu| = 1), \tag{4}$$

that is, a rotation about the origin. One can also arrive at (4) as follows, without constructing the set $\mathfrak{U}$. One constructs for $\mathfrak{B}$ the universal covering surface $\tilde{\mathfrak{B}}$, which by definition is simply connected. It has more than one boundary point, since this is already true of $\mathfrak{B}$. The conformal mapping S can now be extended to $\tilde{\mathfrak{B}}$ so that $S\tilde{\mathfrak{B}} = \tilde{\mathfrak{B}}$ and a fixed-point lies over the point $z=0$ of $\mathfrak{B}$. By the uniformization theorem one can then again map $\tilde{\mathfrak{B}}$ conformally onto a circle in the ζ-plane and set up the expression (3), where z now varies over the covering surface as ζ runs over $|\zeta| < \varrho$. The subsequent conclusion then follows as before.

The relation $T = C^{-1}SC$ can be expressed in the form $CT = SC$, whereupon (2), (3), (4) combine into the identity $\phi(\mu\zeta) = f(\phi(\zeta))$. This is the Schröder functional equation [1]. By comparison of the linear terms it follows that $\lambda = \mu$. Denoting the two conformal mappings determined in the previous paragraph by C_1 and C_2, one sees from $C_1^{-1}SC_1 = T = C_2^{-1}SC_2$ that $C_1^{-1}C_2 = C_0$ commutes with T. If λ is not a root of unity, by inserting the respective power series into the relation $C_0T = TC_0$ one finds that C_0 is the identity mapping, so that $C_1 = C_2$. In particular, this implies that $\mathfrak{B} = \tilde{\mathfrak{B}} = \mathfrak{U}$ is simply connected, although this will not be used subsequently.

In view of (4) one has $|\lambda| = 1$, which therefore is a necessary condition for stability of S. We will now show that S is stable if and only if $|\lambda| = 1$ and the Schröder functional equation

$$\phi(\lambda\zeta) = f(\phi(\zeta)) \tag{5}$$

has a convergent power series solution $\phi(\zeta) = \zeta + \cdots$. The necessity of this condition is precisely what was shown in the preceding argument. Conversely, if there exists a convergent solution $\phi(\zeta)$ to (5) with $|\lambda| = 1$, then the substitution $z = \phi(\zeta)$, $z_1 = \phi(\zeta_1)$ transforms the given mapping $z_1 = f(z)$ into the rotation $\zeta_1 = \lambda\zeta$ which trivially is stable, since as invariant neighborhoods one can take all circles in the ζ-plane with center at the origin $\zeta = 0$. Because $\phi(\zeta)$, as well as its inverse series, converges in a sufficiently small neighborhood of the origin, it follows that also the given mapping $S = CTC^{-1}$ is stable. This proves the assertion. The name "center problem" is derived from the fact that in case of stability the family of concentric circles about the origin in the ζ-plane gives rise to the invariant neighborhoods of $z = 0$.

To investigate whether the mapping S is stable it is therefore enough to discuss whether Schröder's functional equation can be solved by a convergent power series $\phi(\zeta) = \zeta + \cdots$. Setting up $\phi(\zeta)$ as a series with undetermined coefficients, we first seek a solution to (5) in terms of a formal power series. Under the assumption that λ is not a root of unity, comparison of coefficients will give rise to exactly one solution, which we will call the Schröder series. Let $n \geqq 2$, and assume that the coefficients b_k $(1 < k < n)$ in (3) have already been determined so that both sides of (5) agree in terms of order $k < n$. For $n = 2$ the assumption is valid.

Expressing (5) in the form

$$\phi(\lambda\zeta) - \lambda\phi(\zeta) = f(\phi(\zeta)) - \lambda\phi(\zeta)$$

we have

$$\sum_{l=2}^{\infty} (\lambda^l - \lambda)\, b_l \zeta^l = \sum_{l=2}^{\infty} a_l\, \phi^l(\zeta), \tag{6}$$

and consequently the coefficient of ζ^n on the right is a polynomial in the a_l $(l=2, \ldots, n)$ and the already known b_k $(k=2, \ldots, n-1)$ with integral coefficients, while the corresponding coefficient on the left side of (6) is equal to $(\lambda^n - \lambda)\, b_n$. Since λ is not a root of unity and $\neq 0$ we have $\lambda^n - \lambda \neq 0$ $(n=2, 3, \ldots)$, and consequently b_n is uniquely determined. In this way one obtains recursively the coefficients of the Schröder series $\phi(\zeta) = \zeta + b_2\zeta^2 + \cdots$ which formally satisfies the Schröder functional equation (5).

Before investigating convergence of the above series $\phi(\zeta)$, we will consider the case when λ is a root of unity. Let $\lambda^n = 1$ $(n > 0)$, where also $n=1$ is admitted. If S is stable, then $T = C^{-1}SC$ again has the normal form $\zeta_1 = \lambda\zeta$, and $T^k = C^{-1}S^kC$ is the mapping $\zeta_1 = \lambda^k\zeta$. Consequently T^n is the identity mapping E, and therefore also $S^n = E$. Conversely, if $S^n = E$ and $\mathfrak{U}$ is a neighborhood of $z=0$ within the circle of convergence $\mathfrak{K}$ of $f(z)$, one selects any sufficiently small neighborhood $\mathfrak{B}$ of $z=0$ for which the n images $S^k\mathfrak{B}$ $(k=0, \ldots, n-1)$ are still completely contained in $\mathfrak{U}$. Because $S^n\mathfrak{B} = \mathfrak{B}$, the union of the $S^k\mathfrak{B}$ is an invariant neighborhood within $\mathfrak{U}$, and it follows that S is stable. Thus, in the case $\lambda^n = 1$ $(n > 0)$ the mapping S is stable if and only if $S^n = E$. As an example we consider the mapping

$$z_1 = \frac{z}{1-z} = z + z^2 + \cdots, \quad \lambda = 1,$$

for which S^n is given by

$$z_n = \frac{z}{1-nz} \quad (n = \pm 1, \pm 2, \ldots),$$

and consequently is never the identity. Because $S \neq E$ and $\lambda = 1$, this mapping is necessarily not stable. This can also be seen directly by setting $z = 1/n$, where the natural number n can be arbitrarily large. On the other hand, if one sets $z = ir, 0 < r < 1$, then $|z_n| < r$ and the totality of images of z together with z form an invariant set within the circle $|z| \leqq r$. This shows that S is not unstable, and is therefore mixed. It is not known, however, whether it may happen that λ is a root of unity and S is unstable. From now on we will assume throughout that λ is not a root of unity.

We next investigate convergence of the formally constructed Schröder series $\phi(z)$ for the case $|\lambda| \neq 1$. This can be readily accomplished by the usual method of majorants. From the convergence of the series (2) there exists a positive number a such that $|a_{n+1}| < a^n$ $(n=1,2,\ldots)$. If az_1, az are introduced as variables in place of z_1, z in the transformation (2), one obtains again a conformal mapping of the form (2) with the same value for λ, but for which now

(7) $$|a_{n+1}| < 1 \quad (n=1,2,\ldots).$$

For the investigation of convergence we may therefore assume (7) at the outset. Moreover, since $|\lambda| \neq 1$, there exists a positive constant c such that

(8) $$|\lambda^{n+1} - \lambda| > c > 0 \quad (n=1,2,\ldots).$$

If the coefficients b_{n+1} in the Schröder series are determined by the recursive procedure associated with (6), it follows from (7), (8) that the formal solution $\Phi(\zeta) = \zeta + c_2\zeta^2 + \cdots$ to the functional equation

(9) $$c(\Phi - \zeta) = \sum_{l=2}^{\infty} \Phi^l$$

is a majorant for $\phi(\zeta)$. On the other hand, the series

$$\zeta = \Phi - c^{-1} \sum_{l=2}^{\infty} \Phi^l,$$

which converges for $|\Phi| < 1$, has an inverse that converges in a neighborhood of $\zeta = 0$. This completes the convergence proof. As in § 17, one can also readily obtain from this a lower bound for the radius of convergence. We already know from $|\lambda| \neq 1$ that the mapping S is not stable. Because of the convergence of C just proved, we can construct the normal form $C^{-1}SC = T$, and it is immediately evident that the mapping $\zeta_1 = \lambda\zeta$ is actually unstable. Indeed, if one considers any point $\zeta \neq 0$ in an arbitrary bounded neighborhood $\mathfrak{U}$ of $\zeta = 0$, then because $|\lambda| \neq 1$, for n sufficiently large, positive or negative, the point $\zeta_n = \lambda^n\zeta$ will no longer lie in $\mathfrak{U}$. The instability of T implies that of $S = CTC^{-1}$, and consequently for $|\lambda| \neq 1$ the mapping S is necessarily unstable. This can also be shown directly without use of the normal form T.

For future discussion we may restrict ourselves to the case where λ is in absolute value 1, and is not a root of unity. In this case the investigation of convergence of the Schröder series requires finer estimates, to which we now turn. We first show that the set of λ for which there exists a convergent power series $f(z) = \lambda z + \cdots$ whose Schröder series diverges form a dense set on the unit circle $|\lambda| = 1$ [2]. For this divergence proof it will be enough to consider only series $f(z)$ whose coefficients a_n $(n=2,3,\ldots)$ are always equal to $\pm\frac{1}{n!}$, with the choice in sign determined recursively. In particular, such $f(z)$ converge everywhere. We turn once more to the determination of the b_n from equation (6). By comparison of coefficients one obtains for each $n>1$ the expression $(\lambda^n-\lambda)b_n-a_n$ as a polynomial in the a_k, b_k with $1<k<n$, and it is therefore obviously possible to choose $a_n=\pm\frac{1}{n!}$ recursively in such a way that

$$(10)\qquad |b_n|\geqq\frac{1}{n!}|\lambda^n-\lambda|^{-1}=\frac{1}{n!}|\lambda^{n-1}-1|^{-1}\qquad(n=2,3,\ldots).$$

Suppose now that for a given λ the inequality

$$(11)\qquad |\lambda^n-1|<(n!)^{-2}$$

is satisfied for infinitely many natural numbers n, and let $f(z)$ be a power series whose coefficients $a_2, a_3, \ldots$ have been determined in the above manner. Then on the one hand the series in z is everywhere convergent, while on the other hand the corresponding Schröder series $\phi(\zeta)$ diverges for each $\zeta\neq 0$, since by (10), (11) the general term $b_n\zeta^n$ does not even tend to 0. The mapping $z_1=f(z)=\lambda z+\cdots$ therefore is not stable. However, it is not known whether it is mixed or unstable.

It remains to show that there is a dense set of values λ on the unit circle that are not roots of unity and that satisfy the inequality (11) for infinitely many n. If one sets $\lambda=e^{2\pi i\alpha}$ $(0\leqq\alpha<1)$ and for each natural number n chooses the integer m so that

$$(12)\qquad -\tfrac{1}{2}\leqq n\alpha-m<\tfrac{1}{2},$$

then

$$|\lambda^n-1|=|e^{2\pi i n\alpha}-1|=|e^{\pi i n\alpha}-e^{-\pi i n\alpha}|$$
$$=2|\sin(\pi n\alpha)|=2\sin(\pi|n\alpha-m|).$$

Because $|n\alpha-m|=\vartheta\leqq\frac{1}{2}$, it follows that $2\vartheta\leqq\sin(\pi\vartheta)\leqq\pi\vartheta$ and therefore

$$(13)\qquad 4\vartheta\leqq|\lambda^n-1|\leqq 2\pi\vartheta\leqq 7\vartheta.$$

Consequently it is enough to construct a set of irrational numbers α, dense in the interval $0\leqq\alpha<1$, for which the inequalities

$$(14)\qquad |n\alpha-m|<\frac{1}{7(n!)^2},\qquad n>0$$

have infinitely many integral solutions n, m. This can be readily accomplished as follows, using the representation of real numbers in terms of simple continued fractions. As is well known, to each irrational number α in the interval $0<\alpha<1$ one can associate a sequence of natural numbers $r_1, r_2, \ldots$ so that the sequence of fractions $\frac{p_k}{q_k}$ $(k=0,1,\ldots)$, recursively defined according to the prescription

$$(15)\quad \begin{cases} p_0=0, \quad q_0=1, \quad p_1=1, \quad q_1=r_1, \\ p_k=r_k p_{k-1}+p_{k-2}, \quad q_k=r_k q_{k-1}+q_{k-2} \quad (k=2,3,\ldots), \end{cases}$$

converges to α. The numbers $r_1, r_2, \ldots$ are uniquely determined by α and are known as the partial quotients of α. Moreover from the theory of continued fractions one obtains the inequality

$$(16)\qquad |q_k\alpha-p_k|<\frac{1}{q_{k+1}}<\frac{1}{r_{k+1}q_k}\leqq\frac{1}{r_{k+1}} \qquad (k=1,2,\ldots).$$

Conversely, corresponding to each such sequence $r_1, r_2, \ldots$ there is an irrational number α in the interval $0<\alpha<1$ with these prescribed partial quotients in its continued fraction representation.

Let β now be an arbitrary irrational number in the interval $0<\beta<1$ with $s_1, s_2, \ldots$ the partial quotients in its continued fraction expansion. For l an arbitrary fixed natural number one defines

$$(17)\qquad r_k=s_k \quad (0<k\leqq l), \qquad r_{k+1}=7(q_k!)^2 \quad (k\geqq l),$$

where $q_0, q_1, \ldots, q_k$ are again recursively determined in accordance with (15). For the continued fraction α with partial quotients $r_1, r_2, \ldots$ we have inequality (16), and since the first l partial quotients in the continued fractions of α and β agree, also $|q_l\beta-p_l|<q_l^{-1}$. It follows that

$$|\alpha-\beta|\leqq\left|\alpha-\frac{p_l}{q_l}\right|+\left|\beta-\frac{p_l}{q_l}\right|<2q_l^{-2}\leqq 2l^{-2},$$

while on the other hand, by (16), (17), the infinitely many pairs $n=q_k$, $m=p_k$ $(k=l, l+1, \ldots)$ satisfy (14). Since l can be chosen arbitrarily large, the corresponding numbers $\alpha=\alpha_l$ accumulate at β, and β being arbitrary, we have shown that the set of values α in question form a dense set in the unit interval.

Let Λ denote the set of values $\lambda=e^{2\pi i\alpha}$ on the unit circle such that for each power series $f(z)=\lambda z+a_2z^2+\cdots$ convergent in a neighborhood of $z=0$ the corresponding solution $\phi(\zeta)=\zeta+b_2\zeta^2+\cdots$ to the Schröder functional equation converges in a neighborhood of $\zeta=0$. We will now prove that Λ, as a subset of the unit circle, has linear Lebesgue measure 2π, or, equivalently, that the set A of the corresponding values α in the unit interval has Lebesgue measure 1. This will show that the set of irrational α for which there is at least one convergent series $f(z)$, with leading coefficient $\lambda = e^{2\pi i\alpha}$, whose Schröder series $\phi(\zeta)$ diverges, is a set of measure zero. In particular, this says that generally the mapping S is stable, provided that the necessary condition $|\lambda| = 1$ is satisfied.

For two given positive numbers ε, μ we consider the set $B(\varepsilon, \mu)$ of all numbers α in the unit interval E for which the inequalities

(18) $$|n\alpha - m| < \varepsilon n^{-\mu}, \quad n > 0$$

have at least one integral solution n, m. Obviously

$$B(\varepsilon', \mu') \subset B(\varepsilon, \mu) \quad (\varepsilon' \leqq \varepsilon, \mu \leqq \mu').$$

If one permits k to range over all natural numbers and considers the intersection

(19) $$B = \bigcap_k B(k^{-1}, 2)$$

of all the $B(k^{-1}, 2)$, then certainly

(20) $$B \subset B(\varepsilon, 2)$$

for each ε. We denote the Lebesgue measure of a measurable set Γ by $m(\Gamma)$ and estimate the measure of $B(\varepsilon, 2)$ from above. By (18), this set is a countable union of intervals, and therefore measurable, while by (19) then also B is measurable. For each solution n, m of (18) we have

(21) $$-\varepsilon < m < n + \varepsilon$$

whenever α is in E, while on the other hand for given n, m the interval for α defined by (18) has length $2\varepsilon n^{-\mu-1}$. Noting that for each fixed natural number n the number of integers m satisfying (21) is smaller than $n + 2\varepsilon + 1$, and keeping in mind (20), we finally have

$$m(B(\varepsilon, 2)) \leqq \sum_{n=1}^{\infty} 2\varepsilon(n + 2\varepsilon + 1)n^{-3} < 4\varepsilon(\varepsilon + 1) \sum_{n=1}^{\infty} n^{-2}$$

$$m(B) < \frac{2\pi^2}{3}\varepsilon(\varepsilon + 1),$$

and since ε can be arbitrarily small, it follows that $m(B) = 0$. If Δ denotes the set of all α in E for which (18) has a solution for each choice of ε, μ, then by (19) the set Δ is contained in B, so that certainly $m(\Delta) = 0$. The complementary set $\Gamma = E - \Delta$ thus has $m(\Gamma) = 1$, and Γ is characterized by the property that for each number α in Γ there exist two positive numbers ε, μ such that for all natural numbers n and integers m we have

(22) $$|n\alpha - m| > \varepsilon n^{-\mu}.$$

We will show in the next section that for all α in Γ the Schröder series of any convergent series $f(z)=\lambda z+\cdots$, with $\lambda=e^{2\pi i\alpha}$, is also convergent. In that case, by definition, we have $A\supset\Gamma$, so that also $m(A)=1$, as asserted.

§ 26. The Convergence Proof

In the previous section we saw that Schröder's functional equation has a unique formal power series solution $\phi(\zeta)=\zeta+\cdots$. To prove convergence of this series we will begin with a new construction of ϕ using an iteration process that converges rapidly enough to be insensitive to the effect of the small divisors λ^n-1.

Using the notation of the previous section, we consider an α in Γ, which by (25; 22) is irrational. Consequently $\lambda=e^{2\pi i\alpha}$ is not a root of unity, and we set

$$\varrho_n=|\lambda^n-1|^{-1} \qquad (n=1,2,\ldots).$$

With m as in (25; 12), we obtain from (25; 13), (25; 22) the estimate

$$\varrho_n\leqq\frac{1}{4}|n\alpha-m|^{-1}\leqq\frac{n^\mu}{4\varepsilon}=\frac{c_0 n^\mu}{\mu!}, \tag{1}$$

where μ is a natural number and ε,μ may depend on α. Here $c_0=\mu!/4\varepsilon$ and c_1,c_2,c_3 will denote positive constants that depend only on α.

In the construction to follow we will obtain ϕ not by comparison of coefficients but by an iteration process consisting of a repeated substitution of variables. To describe this process, we denote the given transformation (25; 2) symbolically by S_0 and recall that our aim is to find a substitution C such that $T=C^{-1}S_0C$ is the linear transformation $\zeta_1=\lambda\zeta$. Rather than do this directly, we will first construct a substitution C_0 such that $C_0^{-1}S_0C_0=S_1$ is merely closer to the linear transformation T than was S_0. Then, starting with S_1, we will repeat this process to construct a substitution C_1 leading to a transformation

$$S_2=C_1^{-1}S_1C_1=C_1^{-1}C_0^{-1}S_0C_0C_1$$

that approximates T even more closely. Inductively, this process will lead to substitutions C_ν $(\nu=0,1,2,\ldots)$ and transformations

$$S_{\nu+1}=C_\nu^{-1}S_\nu C_\nu=B_\nu^{-1}S_0B_\nu,$$

with

$$B_\nu=C_0C_1\ldots C_\nu \tag{2}$$

converging to the desired substitution. For the success of this method it will be important that the composition of $C_0, C_1, \ldots, C_\nu$ is well defined, i.e. that the range of C_ν lies in the domain of definition of $C_{\nu-1}$, and that the sequence B_ν converges in some fixed neighborhood of the origin to an invertible substitution B of the form $z=\phi(\zeta)=\zeta+\cdots$, while $S_\nu \to T$ as $\nu \to \infty$.

If we assume the above statements to be true, it follows that

$$T=B^{-1}S_0B,$$

so that B will be the desired substitution. The uniqueness of the formal power series for ϕ then assures us that B is represented by the Schröder series, which therefore must converge, since B is analytic near $\zeta=0$.

To carry out the above procedure, we begin with a transformation S expressed in the form

$$z_1=f(z)=\lambda z+\hat{f}(z)$$

where $\hat{f}$ is a convergent power series starting with the quadratic term. The derivative $\hat{f}'$ is also analytic in some disk about the origin, and given $\delta>0$ we can find $r>0$ such that

(3) $$|\hat{f}'|<\delta \quad \text{in} \quad |z|<r.$$

The substitution $z=\phi(\zeta)=\zeta+\hat{\phi}$ that linearizes the above transformation satisfies Schröder's functional equation (25; 5), which can be expressed in the form

$$\hat{\phi}(\lambda\zeta)-\lambda\hat{\phi}(\zeta)=\hat{f}(\phi(\zeta)).$$

Rather than solve this equation, we define the substitution $z=\zeta+\psi(\zeta)$, with ψ a power series beginning with the quadratic term, as the solution to the linear equation

(4) $$\psi(\lambda\zeta)-\lambda\psi(\zeta)=\hat{f}(\zeta)=\sum_{k=2}^{\infty} a_k\zeta^k.$$

This substitution, which we denote by C, forms the basic step in the iteration process. Defining $S_+=C^{-1}SC$ and expressing it as

(5) $$\zeta_1=g(\zeta)=\lambda\zeta+\hat{g}(\zeta),$$

we will show that, with δ, r suitably chosen, the function $\hat{g}$, which measures the deviation of S_+ from the linear transformation, is indeed smaller than the previous function $\hat{f}$.

To this end we choose constants δ, θ so that

(6) $$0<\theta<\tfrac{1}{5}, \quad c_0\delta<\theta^{\mu+2}, \quad 0<\delta<\theta,$$

and then take $r>0$ sufficiently small so as to satisfy (3). First we estimate the solution ψ of (4) given by the series

$$\psi=\sum_{k=2}^{\infty}\frac{a_k}{\lambda^k-\lambda}\zeta^k.$$

Since f is analytic in $|z|<r$, by Cauchy's estimate we have

$$k|a_k|\leqq\frac{\delta}{r^{k-1}},$$

and in view of (1) the series ψ is seen to converge in $|\zeta|<r$. Indeed, in the somewhat smaller domain $|\zeta|<r(1-\theta)$ one obtains, using (1), the estimate

$$|\psi'|\leqq\sum_{k=2}^{\infty}\frac{k|a_k|}{|\lambda^k-\lambda|}|\zeta|^{k-1}\leqq\frac{c_0\delta}{\mu!}\sum_{k=1}^{\infty}k^{\mu}(1-\theta)^k$$

$$<c_0\delta\sum_{k=0}^{\infty}\binom{k+\mu}{\mu}(1-\theta)^k=\frac{c_0\delta}{\theta^{\mu+1}},$$

which together with (6) gives

(7) $$|\psi'|<\theta \quad\text{in}\quad |\zeta|<r(1-\theta),$$

whereupon integration yields

(8) $$|\psi|<\frac{c_0\delta}{\theta^{\mu+1}}r<\theta r \quad\text{in}\quad |\zeta|<r(1-\theta).$$

This inequality shows that the substitution C maps the disk $|\zeta|<r(1-4\theta)$ into $|z|<r(1-3\theta)$, for by (8) we have

$$|z|\leqq|\zeta|+|\psi|<r(1-4\theta)+r\theta=r(1-3\theta).$$

Moreover, we claim that C^{-1} is defined in $|z|<r(1-2\theta)$ and maps this disk into $|\zeta|<r(1-\theta)$. To prove this, we have to show that for z in $|z|<r(1-2\theta)$ the equation

$$\zeta+\psi(\zeta)=z$$

has a unique solution in $|\zeta|<r(1-\theta)$. This follows, for example, from the explicit construction of ζ as $\lim_{n\to\infty}\zeta_n$, where $\zeta_0=0$ and

$$\zeta_{n+1}+\psi(\zeta_n)=z \quad (n=0,1,2,\ldots).$$

Indeed, the above defines a sequence ζ_n of analytic functions of z which, by (7), satisfies

$$|\zeta_{n+1}-\zeta_n|=|\psi(\zeta_n)-\psi(\zeta_{n-1})|\leqq\theta|\zeta_n-\zeta_{n-1}|,$$

and hence

$$|\zeta_{n+1}-\zeta_n| \leqq \theta^n |\zeta_1-\zeta_0| = \theta^n |z| ,$$

provided that $|\zeta_k| < r(1-\theta)$ $(k=0, 1, \ldots, n)$. Thus, for $|z| < r(1-2\theta)$ we have

$$|\zeta_{n+1}| \leqq \sum_{k=1}^{n+1} |\zeta_k - \zeta_{k-1}| < (1-\theta)^{-1}|z| < \frac{1-2\theta}{1-\theta} r < (1-\theta)r ,$$

which shows that the functions ζ_n are defined and analytic for $|z| < (1-2\theta)r$ $(n=0, 1, 2, \ldots)$. Since $\theta < 1$, the sequence ζ_n converges for $|z| < (1-2\theta)r$ to $\zeta = \zeta(z)$, the desired inverse function of $\zeta + \psi(\zeta)$.

This allows us to define the transformation $S_+ = C^{-1}SC$ in the disk $|\zeta| < r(1-4\theta)$, since C maps this disk into $|z| < r(1-3\theta)$ which, in view of (3), (6), is mapped by S into

$$|z_1| \leqq |z| + |\hat{f}| < |z| + \delta r < r(1-2\theta)$$

and, finally, the latter is mapped by C^{-1} into the disk $|\zeta_1| < r(1-\theta)$. Consequently the function $\hat{g}$ in (5) is certainly analytic for $|\zeta| < r(1-4\theta)$, and we will estimate it in this region. To this end we express the relation $CS_+ = SC$ in the form

$$g(\zeta) + \psi(g) = f(\zeta + \psi)$$

or

$$\hat{g}(\zeta) + \psi(\lambda\zeta + \hat{g}) = \lambda\psi + \hat{f}(\zeta + \psi) ,$$

whereupon subtracting the defining relation (4) for ψ we obtain

$$\hat{g}(\zeta) = \psi(\lambda\zeta) - \psi(\lambda\zeta + \hat{g}) + \hat{f}(\zeta + \psi) - \hat{f}(\zeta) .$$

We now use the mean value theorem to estimate $\gamma = \sup|\hat{g}(\zeta)|$ over $|\zeta| < r(1-4\theta)$, obtaining

$$\begin{aligned} \gamma &\leqq \sup|\psi'|\gamma + \sup|\hat{f}(\zeta+\psi) - \hat{f}(\zeta)| \\ &\leqq \theta\gamma + \sup|\hat{f}(\zeta+\psi) - \hat{f}(\zeta)| , \end{aligned}$$

and since $\theta < \frac{1}{5}$, from (3), (8) we have

$$\begin{aligned} \gamma &\leqq \frac{5}{4} \sup|\hat{f}(\zeta+\psi) - \hat{f}(\zeta)| \\ &\leqq \frac{5}{4} \delta \sup|\psi| < c_1 \frac{\delta^2}{\theta^{\mu+1}} r . \end{aligned}$$

Finally, applying Cauchy's estimate to $|\hat{g}'|$ in a somewhat smaller domain, we get

$$(9) \qquad |\hat{g}'| < c_1 \frac{\delta^2}{\theta^{\mu+2}} \quad \text{for} \quad |\zeta| < r_+ = r(1-5\theta).$$

The essential feature of this estimate is the quadratic dependence of $|\hat{g}'|$ on the previous deviation δ, which gives rise to very fast convergence. We now iterate the above construction with $S_\nu, C_\nu, S_{\nu+1}$ in place of S, C, S_+, taking care that the quantities $r_\nu, \theta_\nu, \delta_\nu$ appearing in place of r, θ, δ, which now depend on ν ($\nu = 0, 1, 2, \ldots$), are chosen so that (6), (9) hold at each step, and that the successive domains do not shrink to a point. This is achieved by making the initial choice of δ_0 sufficiently small and then setting

$$r_\nu = \frac{r}{2}(1+2^{-\nu}) \qquad (\nu = 0, 1, \ldots),$$

while defining θ_ν by the relation

$$\frac{r_{\nu+1}}{r_\nu} = 1 - 5\theta_\nu.$$

This last choice, which gives

$$5\theta_\nu = \frac{1}{2(2^\nu+1)},$$

is motivated by the need to replace $|z| < r$ by $|\zeta| < r(1-5\theta)$ when passing from the estimate of S to that of S_+ in (9). The quantity $\delta_{\nu+1}$, which will estimate the deviation of $S_{\nu+1}$ from T, is now defined by

$$(10) \qquad \delta_{\nu+1} = \frac{c_1 \delta_\nu^2}{\theta_\nu^{\mu+2}} < c_2^{\nu+1} \delta_\nu^2 \qquad (\nu = 0, 1, \ldots).$$

Observing that the sequence $\eta_\nu = c_2^{\nu+2} \delta_\nu$ satisfies

$$0 < \eta_{\nu+1} < \eta_\nu^2$$

and therefore tends to zero faster than exponentially, provided that $\eta_0 < 1$, we see that for

$$\delta_0 < c_2^{-2}$$

the sequence δ_ν tends to zero as $\nu \to \infty$. One readily verifies that also (6) holds for $\delta = \delta_\nu, \theta = \theta_\nu$ ($\nu = 0, 1, 2, \ldots$), provided that δ_0 is chosen sufficiently small.

With the above choice of $r_\nu, \theta_\nu, \delta_\nu$, we proceed with our construction. We assume that S_ν is represented in the form

$$z_1 = f(z) = \lambda z + \hat{f}(z)$$

with

$$|\hat{f}'| < \delta_\nu \quad \text{in} \quad |z| < r_\nu ,$$

whereupon the previous considerations lead to a substitution $C = C_\nu$ transforming S_ν into $S_{\nu+1} = C_\nu^{-1} S_\nu C_\nu$, which we express in the form (5). By (9), (10) we have

$$|\hat{g}'| < \delta_{\nu+1} \quad \text{in} \quad |\zeta| < r_{\nu+1} ,$$

and we may therefore proceed inductively.

It is now easy to show that the sequence B_ν in (2) converges to the desired substitution in $|\zeta| < r/2$. First we recall that $C_{\nu+1}$ maps the disk of radius $r_{\nu+1}(1 - 4\theta_{\nu+1})$ into that of radius

$$r_{\nu+1}(1 - 3\theta_{\nu+1}) < r_{\nu+1} = r_\nu(1 - 5\theta_\nu) < r_\nu(1 - 4\theta_\nu) ,$$

so that $C_\nu C_{\nu+1}$ is defined for $|\zeta| < r_{\nu+1}(1 - 4\theta_{\nu+1})$. It follows by induction that B_ν is defined in $|\zeta| < r_\nu(1 - 4\theta_\nu)$, and since this radius is larger than

$$r_\nu(1 - 5\theta_\nu) = r_{\nu+1} > \frac{r}{2} ,$$

each substitution B_ν maps $|\zeta| < r/2$ into $|z| < r$. Moreover, since each of the factors $C_\varkappa$ in B_ν is a substitution of the form $z = \chi_\varkappa(\zeta) = \zeta + \psi_\varkappa$ with the identity as its linearized part, the same is true of each B_ν.

To show convergence of the sequence B_ν in $|\zeta| < r/2$, we express B_ν in the form $z = \beta_\nu(\zeta)$, with β_ν defined inductively by $\beta_0(\zeta) = \chi_0(\zeta)$ and

$$\beta_\nu(\zeta) = \beta_{\nu-1}(\chi_\nu(\zeta)) \quad (\nu = 1, 2, \ldots) .$$

Using $\|\beta'_\nu\|$ to denote the maximum of $|\beta'_\nu|$ in $|\zeta| \leqq r_\nu(1 - \theta_\nu)$, we have in this domain

$$|\beta'_\nu| \leqq \|\beta'_{\nu-1}\| \, (1 + |\psi'_\nu|) ,$$

which in conjunction with (7) gives

$$\begin{aligned} \|\beta'_\nu\| &\leqq \|\beta'_{\nu-1}\| \, (1 + \theta_\nu) \\ &\leqq \prod_{\varkappa=0}^{\nu} (1 + \theta_\varkappa) \leqq \prod_{\varkappa=0}^{\infty} (1 + \theta_\varkappa) = c_3 , \end{aligned}$$

where the infinite product is readily seen to converge. This together with (8) shows that

$$\begin{aligned} |\beta_{\nu+1}(\zeta) - \beta_\nu(\zeta)| &= |\beta_\nu(\chi_{\nu+1}(\zeta)) - \beta_\nu(\zeta)| \\ &\leqq c_3 |\chi_{\nu+1} - \zeta| = c_3 |\psi_{\nu+1}| < c_3 \theta_{\nu+1} r , \end{aligned}$$

from which it follows that as $\nu \to \infty$ the sequence β_ν converges in $|\zeta| < r/2$ to an analytic function $\beta(\zeta)$. Since β_ν transforms S_0 into $S_{\nu+1}$, we have $B_\nu S_{\nu+1} = S_0 B_\nu$ or, writing $S_\varkappa$ in the form $z = f_\varkappa(\zeta)$ $(\varkappa = 0, 1, \ldots)$,

$$\beta_\nu(f_{\nu+1}(\zeta)) = f_0(\beta_\nu(\zeta))$$

with $|f_{\nu+1}(\zeta) - \lambda\zeta| < \delta_{\mu+1} r \to 0$ as $\nu \to \infty$. Consequently, letting $\nu \to \infty$, we obtain

$$\beta(\lambda\zeta) = f_0(\beta(\zeta)),$$

so that β is a solution to Schröder's functional equation. Moreover, since uniform convergence of analytic functions implies convergence of derivatives in the interior, we have $\beta'(0) = 1, \beta(0) = 0$. Thus, the power series for β must agree with the unique formal expansion of ϕ. This completes the convergence proof.

The basic idea of using such a sequence of substitutions to prove convergence was introduced by A. N. Kolmogorov [1, 2] in a different context. The crucial point in this method is to find an iteration scheme in which the new error depends quadratically on the previous one, as in the case of Newton's method for finding roots of a function. This suffices to counteract efficiently the growth factors due to the small divisors $\lambda^n - 1$, as seen in (10), where the effect of the small divisors is reflected in the coefficient $c_2^{\nu+1}$. For a discussion of this method in the problem of transforming mappings into normal form we refer to [3].

In his original proof of this theorem, Siegel [4] actually succeeded by direct estimates of the coefficients, as in Cauchy's method of majorants. This, however, required more delicate estimates on the small divisors than (1) and, in particular, it was necessary to use the fact that the expressions $\lambda^n - 1$ are small for only relatively few integers n. On the other hand, for the theory of stability, to be developed in this chapter, it will be essential to have a quadratically convergent scheme, for which cruder estimates of the small divisors suffice.

§ 27. The Poincaré Center Problem

We consider a system of differential equations

$$\dot{x}_k = f_k(x) \quad (k = 1, \ldots, m) \tag{1}$$

for which $x = 0$ is an equilibrium solution, whereby the functions $f_k(x)$ are convergent power series in a neighborhood of $x = 0$ with real coefficients and without constant term. If $x(t, \xi)$ denotes the solution to (1) with initial condition $x(0, \xi) = \xi$, the association of $x(t, \xi)$ to ξ for each fixed t defines in a sufficiently small neighborhood of the origin a mapping S_t that evidently has $\xi=0$ as a fixed-point. We wish to investigate now when the equilibrium solution $x=0$ is stable, and therefore, in accordance with the definitions given in the beginning of § 25, we have to consider the mapping S_t in a neighborhood of the origin for all real t. To facilitate this investigation we will bring the system (1) into as simple a form as possible by a suitable change of variables

$$x_k=\phi_k(u) \quad (k=1,\ldots,m). \tag{2}$$

The ϕ_k shall thereby again be power series in m new variables $u_1,\ldots,u_m$ which also have no constant term, so that the origin remains fixed under the substitution (2). The analysis to be presently carried out is to a large extent analogous to that in § 23, where we constructed normal forms for analytic mappings in the plane, and therefore we will again first work with formal power series. As in § 16, we will also treat differentiation formally by defining

$$\dot{x}_k=\sum_{l=1}^{m}\phi_{ku_l}\dot{u}_l \quad (k=1,\ldots,m), \tag{3}$$

or in vector form

$$\dot{x}=\phi_u\dot{u}, \qquad \phi_u=(\phi_{ku_l}).$$

It is assumed that the substitution (2) is invertible, which means that the Jacobian determinant $|\phi_u|$ as a power series has a nonzero constant term, or in other words, that the coefficients of the linear parts of the ϕ_k yield a nonvanishing determinant. The substitution (2), (3) takes (1) into

$$\dot{u}=\phi_u^{-1}f(\phi(u)), \tag{4}$$

and conversely the inverse substitution carries (4) into (1). We turn to the problem of determining the substitution (2) so that (4) assumes a certain normal form. This is connected with the following question: If, in addition to (1), we consider a second system

$$\dot{u}_k=h_k(u) \quad (k=1,\ldots,m), \tag{5}$$

where the h_k are power series in $u_1,\ldots,u_m$ without constant term, under what conditions does there exist an invertible substitution (2) that takes (1) into (5)? This problem evidently leads to the system of first order partial differential equations

$$f(\phi(u))=\phi_u h(u) \tag{6}$$

for the unknown series ϕ_k, and a necessary condition for the existence of a solution to (6) is seen immediately from comparison of the linear terms. Namely, if $\mathfrak{F}, \mathfrak{H}, \mathfrak{C}$ are the matrices corresponding to the linear parts of $f(x), h(u), \phi(u)$, then it follows that $\mathfrak{F}\mathfrak{C} = \mathfrak{C}\mathfrak{H}$, and consequently $\mathfrak{F}$ and $\mathfrak{H}$ must have the same elementary divisors.

In the rest of this section we will restrict ourselves to the case $m = 2$. The eigenvalues λ and μ of $\mathfrak{F}$ will be required to be distinct. Writing x, y in place of x_1, x_2, after a preliminary linear substitution of variables we may assume the system (1) to be of the form

$$\dot{x} = f(x, y) = \lambda x + \cdots, \quad \dot{y} = g(x, y) = \mu y + \cdots, \tag{7}$$

where in the real case $\lambda = \bar{\lambda}$, $\mu = \bar{\mu}$ and we may assume that $f(x, y) = \bar{f}(x, y)$, $g(x, y) = \bar{g}(x, y)$, while in the imaginary case $\lambda = \bar{\mu}$ and we have $f(x, y) = \bar{g}(y, x)$. Consideration of the linear system $\dot{x} = \lambda x$, $\dot{y} = \mu y$ suggests that one can have stability at the equilibrium solution $x = y = 0$ of (7) only if λ and μ are both purely imaginary, as will subsequently be shown. Therefore we will first treat the purely imaginary case $\mu = \bar{\lambda} = -\lambda$. It will be shown that by means of a substitution of the form

$$x = \phi(u, v) = u + \phi_2 + \phi_3 + \cdots, \quad y = \psi(u, v) = v + \psi_2 + \psi_3 + \cdots \tag{8}$$

we can then bring the system (7) into the normal form

$$\dot{u} = pu, \quad \dot{v} = qv,$$

where p and q are power series in the product $w = uv$ only. In addition, we will require that the series $\phi(u, v)$, $\psi(u, v)$ not contain terms of the form uw^k, $w^k v$ $(k > 0)$ respectively, and we will show that in that case there is exactly one such substitution (8). In case of convergence of f, g we will also have convergence of ϕ, ψ provided that $p + q = 0$.

Thus, in accordance with (6), we have to solve the corresponding partial differential equations

$$\begin{cases} \phi_u pu + \phi_v qv = f(\phi, \psi) = \lambda\phi + \cdots, \\ \psi_u pu + \psi_v qv = g(\phi, \psi) = -\lambda\psi + \cdots \end{cases} \tag{9}$$

by power series of the form (8), where for p, q one has the expressions

$$p = \sum_{r=0}^{\infty} a_{2r} w^r, \quad q = \sum_{r=0}^{\infty} b_{2r} w^r.$$

In addition we define $a_{2r+1} = 0$, $b_{2r+1} = 0$ $(r = 0, 1, \ldots)$ and compare coefficients in (9). Comparison of the linear terms gives $a_0 = \lambda$, $b_0 = -\lambda$. Let us now make the inductional hypothesis that both sides of (9) agree up to terms of order $k - 1$ $(k > 1)$, and that $\phi_\varkappa, \psi_\varkappa$ $(\varkappa < k)$, $a_\varkappa, b_\varkappa$ $(\varkappa < k - 1)$ have already been uniquely determined in this way. For the terms of degree k in (9), comparison of coefficients then leads to the relations

$$(10)\qquad \begin{cases} \lambda(\phi_{ku}u-\phi_{kv}v-\phi_k)+a_{k-1}w^{\frac{k-1}{2}}u=P_k\,,\\ \lambda(\psi_{ku}u-\psi_{kv}v+\psi_k)+b_{k-1}w^{\frac{k-1}{2}}v=Q_k\,,\end{cases}$$

where P_k, Q_k are homogeneous polynomials in u, v of degree k whose coefficients are expressed in terms of the coefficients of the already known $\phi_\varkappa, \psi_\varkappa$ $(\varkappa<k)$ as well as $a_\varkappa, b_\varkappa$ $(\varkappa<k-1)$. Next we determine a_{k-1}, b_{k-1}. For k even $a_{k-1}=0$, $b_{k-1}=0$ by definition, so let $k=2r+1$ be odd. By assumption, ϕ_k, ψ_k do not contain terms of the form uw^r, w^rv respectively, and it follows from (10) that a_{k-1}, b_{k-1} are uniquely determined. With this choice of a_{k-1}, b_{k-1}, the coefficients of $u^{r+1}v^r$ and u^rv^{r+1}, respectively, agree on both sides of (10). Now we will determine ϕ_k, ψ_k, whereby k may be odd or even. If αu^gv^h, βu^gv^h are the terms of the form u^gv^h $(g+h=k)$ in ϕ_k, ψ_k, then the coefficients of the corresponding terms on the left side of (10) are equal to $\lambda(g-h-1)\,\alpha$, $\lambda(g-h+1)\,\beta$. Since it may be assumed that $g\neq h+1$, $g\neq h-1$ respectively, it follows that ϕ_k, ψ_k are uniquely determined. This completes the induction and shows that the equations (9) can be solved by formal power series ϕ, ψ, p, q.

The comparison of coefficients carried out above arises as the special case $m=2$ in the corresponding development in § 16, and from the result obtained there it follows that the uniquely determined power series satisfy the reality conditions

$$(11)\qquad \phi(u,v)=\bar{\psi}(v,u)\,,\qquad p(uv)=\bar{q}(uv)\,.$$

The normal form constructed above allows us to readily discuss the question of stability of the equilibrium solution. It will be shown that the equilibrium solution is stable if and only if

$$p+q=\sum_{r=1}^{\infty}(a_{2r}+b_{2r})\,w^r=0\,,$$

or

$$(12)\qquad a_{2r}+b_{2r}=0\qquad (r=1,2,\ldots)\,,$$

and that otherwise we have instability. First let us assume that (12) is not satisfied for all r, so that

$$p+q=cw^{n-1}+\cdots,\qquad c\neq 0,\qquad n>1\,,$$

where by (11) the constant c is real. By using $2c^{-1}t$ in place of t we may normalize so that $c=2$. To bypass the question of convergence we break off the series ϕ, ψ and p, q after the terms of order $2n-1$ and $2n-2$

respectively and denote the resulting polynomials by $\tilde{\phi}, \tilde{\psi}, \tilde{p}, \tilde{q}$. Introducing the convergent substitution $x = \tilde{\phi}(u, v)$, $y = \tilde{\psi}(u, v)$ we obtain by (9) for the solutions of (7) the relations

$$\tilde{\phi}_u(u\tilde{p} - \dot{u}) + \tilde{\phi}_v(v\tilde{q} - \dot{v}) = \cdots, \qquad \tilde{\psi}_u(u\tilde{p} - \dot{u}) + \tilde{\psi}_v(v\tilde{q} - \dot{v}) = \cdots,$$

or

$$\dot{u} - u\tilde{p} = \cdots, \qquad \dot{v} - v\tilde{q} = \cdots,$$

where the right sides are convergent power series in u, v that contain no terms of degree less than $2n$. From this one obtains the differential equation

$$\dot{w} - 2w^n = \cdots, \tag{13}$$

with the right side containing no terms of degree less than $2n+1$. For a real solution of (7) we have $v = \bar{u}$ and $w = uv \geqq 0$. Let a positive number ϱ be chosen so that for $w < \varrho$ one has convergence, and also, as a consequence of (13), the inequality

$$\dot{w} \geqq w^n. \tag{14}$$

Then w is a monotonically increasing function of t as long as $w < \varrho$. For $t = 0$ let $0 < w = w_0 < \varrho$. By (14), then

$$w - w_0 \geqq w_0^n t \qquad (t > 0),$$

which for $t = \varrho w_0^{-n}$ is in contradiction with the assumption $w < \varrho$. Consequently, we have instability.

Now suppose that (12) holds for all r. In that case $q = -p$ and the convergence proof in § 17 applies. From

$$\dot{u} = pu, \qquad \dot{v} = qv, \qquad p + q = 0$$

it follows that $w = uv, p, q$ are constant in time, and integration gives

$$u = u_0 e^{pt}, \qquad v = v_0 e^{qt}. \tag{15}$$

According to (11), for real solutions one has to choose $v_0 = \bar{u}_0$, in which case p is purely imaginary. Setting $u = r + is$ with r and s real, one obtains concentric circles in the (r, s)-plane which are uniformly traversed in the time $2\pi|p^{-1}|$. This shows stability and motivates the name "center problem" [1]. Finally, by (8), (15) the original coordinates x, y are given as convergent Fourier series in the variable $|p|\,t$.

We have thus established the following method for deciding stability at the equilibrium solution of the given system (7) in case $\lambda = -\mu$ is purely imaginary and $\neq 0$: Compute recursively the coefficients a_{2r}, b_{2r} $(r = 1, 2, \ldots)$ of p, q and then determine whether the sums $c_r = a_{2r} + b_{2r}$ are all 0 or not. By a suitable choice of the unit of time one may assume that $\lambda = i$. If then

$$f(x,y) = ix + \sum_{g+h>1} \alpha_{gh} x^g y^h, \quad g(x,y) = -iy + \sum_{g+h>1} \beta_{gh} x^g y^h$$

with $\beta_{gh} = \bar{\alpha}_{hg}$ are the power series for f and g, then c_r is given as a polynomial in the α_{gh} and β_{gh} $(g+h<2r+2)$. In particular, suppose that f and g are polynomials of a fixed degree l, in which case all the c_r $(r=1,2,\ldots)$ are polynomials in the finitely many α_{gh} and β_{gh} $(g+h \leq l)$. By the Hilbert basis theorem for polynomial ideals there exists then a natural number $m = m(l)$ such that all the c_r can be expressed in the form

$$c_r = \sum_{k=1}^{m} \gamma_{rk} c_k \quad (r=1,2,\ldots),$$

where the coefficients γ_{rk} are polynomials in the α_{gh}, β_{gh}. Consequently, to check whether the c_r are all 0, which is the necessary and sufficient condition for stability, one has only to verify the finitely many equations $c_k = 0$ for $k = 1, \ldots, m$. However, the proof of the Hilbert basis theorem does not give in our case an upper bound for m as a function of l. For $l = 2$, it is known that $m(2) = 7$ [2–4], but for $l > 2$ the actual determination of such a bound for $m(l)$ remains an interesting unsolved problem.

It should be noted that in the case of instability, that is for $p + q \neq 0$, the investigation of convergence of the series ϕ, ψ, p, q likewise remains an open problem.

§ 28. The Theorem of Liapunov

In the previous section we discussed the stability of an equilibrium solution of the system (27;1) only for the case $m = 2$, and then only for purely imaginary eigenvalues. Now we turn to investigating the general case. Let $\lambda_1, \ldots, \lambda_m$ denote the eigenvalues of the matrix $\mathfrak{F}$ corresponding to the linear parts of $f_1(x), \ldots, f_m(x)$. The theorem of Liapunov [1] states:

If the real parts of $\lambda_1, \ldots, \lambda_m$ are all different from 0, then the equilibrium solution is unstable. If the equilibrium solution is stable, then the real parts of $\lambda_1, \ldots, \lambda_m$ are all 0.

We will prove this theorem only under the restriction that $\lambda_1, \ldots, \lambda_m$ are all distinct from one another, and in the course of the proof we will make still another restrictive assumption. For the case of multiple eigenvalues, not treated here, our method of proof requires a modification that makes the formulas somewhat more intricate, but causes no real conceptual difficulties. After a preliminary linear substitution one may begin with the system in the form

(1) $$\dot{x}_k = f_k(x) = \lambda_k x_k + \chi_k(x) \quad (k = 1, \ldots, m),$$

where the power series χ_k begin with quadratic terms. For $\bar{\lambda}_k = \lambda_l$ with $l = l_k$ $(k = 1, \ldots, m)$, we define $\underline{x_k} = x_l$ and assume the reality conditions

(2) $$f_k(x) = \bar{f}_l(\underline{x}) \quad (l = l_k;\ k = 1, \ldots, m).$$

Let ϱ_k denote the real part of λ_k and assume the indices to be chosen so that $\varrho_1 \leqq \varrho_2 \leqq \cdots \leqq \varrho_m$ with $\varrho_p < 0$ but $\varrho_{p+1} \geqq 0$, where of course one has to admit the possibilities $p = 0$ or $p = m$. First suppose that $p > 0$. We will carry out substitutions of the special form

(3) $$u_k = x_k - \phi_k(x_1, \ldots, x_p) \quad (k = 1, \ldots, m),$$

where the ϕ_k are formal power series in the first p variables $x_1, \ldots, x_p$ only and begin with quadratic terms. It is easily seen that such substitutions form a group. If one sets

$$j_k(u) = j_k(u_1, \ldots, u_m) = \chi_k + \lambda_k \phi_k - \sum_{l=1}^{p} \phi_{k x_l} f_l,$$

where on the right x is to be expressed as a function of u by means of the substitution inverse to (3), then (1) becomes

(4) $$\dot{u}_k = \lambda_k u_k + j_k(u) \quad (k = 1, \ldots, m),$$

with the power series j_k beginning again with quadratic terms. We will now determine the coefficients of the ϕ_k so that none of the series $j_1, \ldots, j_m$ contain products of powers of $u_1, \ldots, u_p$ alone. In other words, the equations

(5) $$j_k(u_1, \ldots, u_p, 0, \ldots, 0) = 0 \quad (k = 1, \ldots, m)$$

are to hold identically.

By (3) the $x_1, \ldots, x_p$ are invertible power series in the p indeterminates $u_1, \ldots, u_p$ only, and moreover, for $u_{p+1} = 0, \ldots, u_m = 0$ we have

(6) $$x_k = \phi_k(x_1, \ldots, x_p) \quad (k = p+1, \ldots, m).$$

Consequently (5) reduces to the requirement that the equations

(7) $$-\lambda_k \phi_k + \sum_{l=1}^{p} \phi_{k x_l} \lambda_l x_l = \chi_k - \sum_{l=1}^{p} \phi_{k x_l} \chi_l \quad (k = 1, \ldots, m)$$

be satisfied identically in $x_1, \ldots, x_p$, where $x_{p+1}, \ldots, x_m$ are defined by (6). Conversely, from (3), (6), (7) we again obtain (5). We now undertake comparison of coefficients in (7). If $\sigma x_1^{g_1} \ldots x_p^{g_p}$ is a term of ϕ_k with $g_1 + \cdots + g_p = h > 1$, the comparison gives

$$\left(-\lambda_k + \sum_{l=1}^{p} g_l \lambda_l\right) \sigma = \gamma,$$

where γ is a polynomial in the coefficients of the terms in $\phi_1, \ldots, \phi_m$ of degree less than h. From now on we make the additional restriction that for all systems of nonnegative integers $g_1, \ldots, g_p$ with $g_1 + \cdots + g_p > 1$ we always have

$$\sum_{l=1}^{p} g_l \lambda_l \neq \lambda_k \quad (k=1, \ldots, p). \tag{8}$$

These actually constitute only finitely many conditions, and (8) is trivially true also for $k=p+1, \ldots, m$. In that case induction shows that (5) has exactly one solution in power series $\phi_1, \ldots, \phi_m$. Moreover, because of (2) we also have

$$\phi_k(x) = \bar{\phi}_l(\bar{x}) \quad (l=l_k;\, k=1, \ldots, m).$$

Convergence is proved by the usual means. Let

$$x_1 + \cdots + x_m = X, \quad \chi_k \prec \frac{c_1 X^2}{1-c_1 X} \qquad (k=1, \ldots, m).$$

Since the real parts of $\lambda_1, \ldots, \lambda_p$ are all negative and (8) is satisfied, it follows that

$$g_1 + \cdots + g_p < c_2 \left| -\lambda_k + \sum_{l=1}^{p} g_l \lambda_l \right| \qquad (k=1, \ldots, m).$$

Consequently for the uniquely determined solution $\psi_1, \ldots, \psi_m$ of

$$\begin{cases} \sum_{l=1}^{p} \psi_{k x_l} x_l = c_2 \left(1 + \sum_{l=1}^{p} \psi_{k x_l}\right) \dfrac{c_1 X^2}{1-c_1 X} & (k=1, \ldots, m), \\ x_k = \psi_k(x_1, \ldots, x_p) & (k=p+1, \ldots, m) \end{cases} \tag{9}$$

we have the relation $\phi_k \prec \psi_k$ $(k=1, \ldots, m)$. By (9), however, $\psi_1 = \cdots = \psi_m$, and if in addition one sets $x_1 = \cdots = x_p = x$, it is evidently enough to prove convergence for the solution $\psi(x)$ of

$$x\psi_x = (1+\psi_x) \frac{c_3(x+\psi)^2}{1-c_4(x+\psi)}.$$

On the other hand, the recursion formulas for the coefficients of the power series ψ coming from the above equation show that $x^{-1}\psi$ is majorized by the convergent solution Ψ to the cubic equation

$$\Psi = \frac{c_3 x(1+\Psi)^3}{1-c_4 x(1+\Psi)}.$$

This completes the proof of convergence.

By (4), (5) one obtains for the given differential equation the particular solutions

$$u_k = \begin{cases} c_k e^{\lambda_k t} & (k=1,\dots,p) \\ 0 & (k=p+1,\dots,m)\,. \end{cases} \tag{10}$$

Since the real parts of $\lambda_1,\dots,\lambda_p$ are negative, the behavior as $t\to-\infty$ shows that if $p>0$ the equilibrium solution is not stable. On the other hand, an interchange of t and $-t$ replaces the eigenvalues λ_k by $-\lambda_k$, which shows that stability of the equilibrium solution can occur only when the real parts of all m eigenvalues are equal to 0. This is the second assertion of Liapunov's theorem.

Now suppose that the real parts $\varrho_1,\dots,\varrho_m$ are all different from 0, so that $\varrho_1\leqq\varrho_2\leqq\dots\leqq\varrho_p<0<\varrho_{p+1}\leqq\dots\leqq\varrho_m$. For ε a positive constant chosen sufficiently small, we will determine all real solutions to the given system which satisfy the condition

$$\sum_{k=1}^{m} |u_k|^2 < \varepsilon \tag{11}$$

for all $t\geqq 0$. By (4) the expression

$$\sum_{k=p+1}^{m} |u_k|^2 = w \tag{12}$$

satisfies the differential equation

$$\dot w = 2\sum_{k=p+1}^{m} \varrho_k|u_k|^2 + \sum_{k=p+1}^{m} \{u_k \bar{j}_k(\bar u) + \bar u_k\, j_k(u)\}\,,$$

where in view of (5) each term in the second summand on the right is divisible by a product of two of the variables $u_k, \bar u_k$ $(k=p+1,\dots,m)$. Since this summand begins with cubic terms, in view of (11), (12) its absolute value for ε sufficiently small is at most equal to $\varrho_{p+1}w$ and consequently

$$\dot w \geqq 2\sum_{k=p+1}^{m} \varrho_k|u_k|^2 - \varrho_{p+1}w \geqq \varrho_{p+1}w,$$

$$\frac{d(we^{-\varrho_{p+1}t})}{dt} \geqq 0\,.$$

Hence the expression $we^{-\varrho_{p+1}t}$ is monotonically increasing for all $t\geqq 0$; on the other hand, since $\varrho_{p+1}>0$ and $w<\varepsilon$, it tends to 0 as $t\to\infty$. Consequently, we have $w=0$, $u_k=0$ $(k=p+1,\dots,m)$ for the desired solutions, and (4), (5) imply (10). Conversely, from (10) one again obtains (11) provided one chooses

$$\sum_{k=1}^{p} |c_k|^2 < \varepsilon\,.$$

Thus we have found all solutions that remain near the equilibrium solution as $t \to \infty$. This result was used already in § 13. From the behavior of the solutions to (10) as $t \to -\infty$ it follows that only the equilibrium solution satisfies (11) for all t, and consequently we have instability. This also proves the first part of Liapunov's theorem. Furthermore we see that for stability with respect to future time it is necessary that no eigenvalue have a positive real part, and sufficient that all eigenvalues have negative real parts. In addition to the assumption about the simplicity of the eigenvalues made already at the beginning, in the course of the proof there arose also the inequality conditions (8). If these restrictive assumptions are dropped, the expression for the normal form given by (4), (5) has to be suitably extended. This, however, does not involve any new ideas, and will not be carried out here any further.

For the particular case when the eigenvalues all have negative real parts we have $p = m$ and (4) reduces to the linear system

(13) $$\dot{u}_k = \lambda_k u_k \quad (k = 1, \ldots, m)$$

provided that the conditions (8) are satisfied, while the case of all positive real parts is transformed into the above by an interchange of t with $-t$. On the other hand, the condition relating to the sign of the real parts of the eigenvalues played no role in the recursive determination of the power series $\phi_1, \ldots, \phi_m$, but only in the proof of convergence, and it is natural to ask whether one can not always obtain the linear normal form (13) by means of a convergent substitution, provided only that the eigenvalues are all distinct and satisfy conditions (8) with m in place of p. The investigation of this question leads to reasoning similar to that employed in the first two sections of this chapter for the function-theoretic center problem. In place of the divisors $\lambda^n - \lambda$ $(n = 2, 3, \ldots)$ we now have the expressions

$$-\lambda_k + \sum_{l=1}^{m} g_l \lambda_l = A_k(g_1, \ldots, g_m) = A_k \quad (k = 1, \ldots, m)$$

for nonnegative integers $g_1, \ldots, g_m$ with $g_1 + \cdots + g_m = h > 1$. As before, on the one hand an example can be constructed where in analogy to (25; 11) a subsequence of the $A_k(g_1, \ldots, g_m)$ converges very rapidly to 0, thus leading to divergence of the corresponding series $\phi_1, \ldots, \phi_m$; on the other hand, under the assumption $|A_k| > \varepsilon h^{-\mu}$ $(k = 1, \ldots, m)$ analogous to (25; 22) one can carry out a proof of convergence [2]. As in the case of Schröder's series, it then readily follows that for the transformation taking the given system (1) into the linear normal form (13) the case of divergence is exceptional.

§ 29. The Theorem of Dirichlet

The following sufficient condition for stability may be traced back to Lagrange. However, it was first proved in a somewhat more special form by Dirichlet [1] and subsequently generalized by Liapunov. We again consider a system

$$\dot{x}_k = f_k(x) \quad (k=1,\ldots,m), \tag{1}$$

where the $f_k(x)$ are power series in $x_1, \ldots, x_m$ without constant term, convergent in some neighborhood of the origin. The theorem on stability then says:

If the system (1) has an integral $g(x)$ that does not depend on t and has a relative extremum in the strong sense at $x=0$, then the equilibrium solution $x=0$ is stable.

Since $g(x)$ can be replaced by $-g(x)$, one can restrict oneself to the case of a minimum, so that for $\varrho>0$ sufficiently small we have $g(0)<g(x)$ whenever

$$0<\sum_{k=1}^{m} x_k^2 = r^2 \leqq \varrho^2 .$$

Let $x(t,\xi)$ again denote the solution to (1) with initial value $x(0,\xi)=\xi$, and let S_t be the mapping taking ξ into $x(t,\xi)$. Moreover, for $0<\varepsilon<\varrho$ let $\mu(\varepsilon)=\mu$ be the minimum of $g(x)$ on the spherical surface $r=\varepsilon$, so that $g(0)<\mu$. Let $\mathfrak{W}$ be the set of points in the interior of the sphere $r<\varepsilon$ at which $g(x)<\mu$, which then is an open set containing $x=0$, and therefore a neighborhood of $x=0$. If now ξ is in $\mathfrak{W}$ and $x=x(t,\xi)$, then since $g(x)$ is an integral we also have $g(x)<\mu$; and moreover, such a point x must also lie in the sphere $r<\varepsilon$, for otherwise by continuity we would have $r=\varepsilon$ for at least one value of t, and at this point we would have $g(x)\geqq\mu$. Thus $x(t,\xi)$ lies in $\mathfrak{W}$ whenever ξ does, and consequently $\mathfrak{W}$ is invariant under S_t for all t. This, on the other hand, implies stability.

We now wish to apply the above criterion to a Hamiltonian system

$$\dot{x}_k = H_{y_k}, \quad \dot{y}_k = -H_{x_k} \quad (k=1,\ldots,n), \tag{2}$$

and as before, we set $z_k=x_k$, $z_{k+n}=y_k$. Let z denote the column vector with components z_l $(l=1,\ldots,2n)$ and let the Hamiltonian function $H(x,y)=\frac{1}{2}z'\mathfrak{S}z+\cdots$ have a convergent power series in a neighborhood of $z=0$ with the coefficients real and the matrix $\mathfrak{S}$ symmetric. Then H is an integral of (2) and $z=0$ an equilibrium solution. If the matrix $\mathfrak{S}$ is positive then the function H has a strict relative minimum at $z=0$ and it follows that the solution $z=0$ is stable. On the other hand, it may very well happen that $z'\mathfrak{S}z$ is indefinite, yet we still have stability. This can be seen by, say, taking $2H=x_1^2+y_1^2-x_2^2-y_2^2$ with $n=2$. For the solutions of Lagrange treated in § 14, which appear as equilibrium solutions in a rotating coordinate system, the Hamiltonian function has a saddle point at the place of equilibrium, and consequently in that case one cannot make any assertion using Dirichlet's criterion.

To exhibit a connection between the theorems of Dirichlet and Liapunov for canonical systems of differential equations, let us consider the eigenvalues λ_k $(k=1, \ldots, 2n)$ of the system (2), which according to § 15 are the roots of the equation $|\lambda\mathfrak{J}+\mathfrak{S}|=0$. Let $z \neq 0$ be an eigenvector corresponding to $\lambda=\lambda_k$, so that $(\lambda\mathfrak{J}+\mathfrak{S})z=0$. Then

$$\bar{z}'\mathfrak{S}z = -\lambda\bar{z}'\mathfrak{J}z, \tag{3}$$

where $\bar{z}$ denotes the vector whose components are complex conjugate to those of z. Since the matrix $\mathfrak{J}' = -\mathfrak{J}$ is real and alternating, it follows that

$$\overline{\bar{z}'\mathfrak{J}z} = z'\mathfrak{J}\bar{z} = -\bar{z}'\mathfrak{J}z,$$

and therefore the number $\bar{z}'\mathfrak{J}z$ is purely imaginary. If $\mathfrak{S}$ is positive, then also $\bar{z}'\mathfrak{S}z>0$, so that by (3) the eigenvalue λ is purely imaginary. By the theorem of Liapunov this was already known as a necessary condition for stability. The previous simple example, however, shows that this condition of Liapunov can be fulfilled, and $z'\mathfrak{S}z$ nevertheless be indefinite.

§ 30. The Normal Form for Hamiltonian Systems

We again begin with a canonical system of differential equations

$$\dot{u}_k = H_{v_k}, \quad \dot{v}_k = -H_{u_k} \quad (k=1, \ldots, n), \tag{1}$$

where the Hamiltonian function H has a convergent power series expansion in a neighborhood of $u_k=0, v_k=0$ $(k=1, \ldots, n)$ that begins with quadratic terms and is independent of t. If w denotes the column vector with the $2n$ components $w_k=u_k$, $w_{k+n}=v_k$, the expansion of H takes the form $H=\frac{1}{2}w'\mathfrak{S}w+\cdots$ with $\mathfrak{S}$ a symmetric real matrix of $2n$ rows. The roots $\lambda_1, \ldots, \lambda_{2n}$ of the corresponding equation $|\lambda\mathfrak{J}+\mathfrak{S}|=0$ may be ordered so that $\lambda_{k+n}=-\lambda_k$ $(k=1, \ldots, n)$, and we will assume that they are all distinct.

Our aim in this section is to obtain a certain normal form for the given system (1) by means of a canonical substitution with a power series representation [1]. To this end, as in § 15, we first bring the linear terms on the right side of (1), or the quadratic terms of H, into their normal form. We denote the new variables by x_k, y_k and set $z_k=x_k$, $z_{k+n}=y_k$ $(k=1, \ldots, n)$, with z then being the column vector with components

z_l $(l=1,\dots,2n)$. After a suitable linear canonical substitution $w=\mathfrak{C}z$ the system (1) takes the form

$$\text{(2)}\qquad \dot{x}_k=H_{y_k},\quad \dot{y}_k=-H_{x_k}\qquad (k=1,\dots,n)$$

with

$$H=H_2+H_3+\cdots,\qquad H_2=\sum_{k=1}^{n}\lambda_k x_k y_k,$$

where H_l $(l=2,3,\dots)$ is a homogeneous polynomial of degree l in $z_1,\dots,z_{2n}$. We now subject the system (2) to another canonical substitution of the form

$$\text{(3)}\quad x_k=\phi_k(\xi,\eta)=\xi_k+\sum_{l=2}^{\infty}\phi_{kl},\ y_k=\psi_k(\xi,\eta)=\eta_k+\sum_{l=2}^{\infty}\psi_{kl}\ (k=1,\dots,n),$$

where the ϕ_{kl}, ψ_{kl} are homogeneous polynomials of degree l in the $2n$ new variables ξ,η. The system (2) then becomes a new Hamiltonian system

$$\text{(4)}\qquad \dot{\xi}_k=H_{\eta_k},\quad \dot{\eta}_k=-H_{\xi_k}\qquad (k=1,\dots,n)$$

with

$$\text{(5)}\qquad H=\sum_{l=2}^{\infty}H_l(\phi(\xi,\eta),\psi(\xi,\eta))=H_2(\xi,\eta)+\cdots.$$

Furthermore, we impose the additional restriction that a linear relation of the form

$$g_1\lambda_1+g_2\lambda_2+\cdots+g_n\lambda_n=0$$

with integral coefficients $g_1,g_2,\dots,g_n$ can hold only in the trivial case $g_1=g_2=\cdots=g_n=0$. We will show that the $2n$ formal power series ϕ_k,ψ_k can then be chosen so that the right side of (5) becomes a formal power series in the n products $\omega_k=\xi_k\eta_k$ alone.

To show this we represent the desired canonical transformation (3) in terms of a generating function $v(x,\eta)$ which we take to be a formal power series of the form

$$v(x,\eta)=v_2+v_3+\cdots,\qquad v_2=\sum_{k=1}^{n}x_k\eta_k,$$

where v_l $(l=3,4,\dots)$ is a homogeneous polynomial of degree l in the x_k,η_k $(k=1,\dots,n)$ with undetermined coefficients. In analogy to (3;4), the equations

$$\text{(6)}\quad \xi_k=v_{\eta_k}=x_k+\sum_{l=3}^{\infty}v_{l\eta_k},\quad y_k=v_{x_k}=\eta_k+\sum_{l=3}^{\infty}v_{lx_k}\quad (k=1,\dots,n)$$

then define a formal canonical substitution which, when solved for the x_k, takes the form (3), and following the reasoning of § 2 one readily shows that it takes (2) into (4), independently of possible convergence. If, in accordance with (3), one introduces the series ϕ_k, ψ_k in place of x_k, y_k it follows from (6) that for $l=2,3,\ldots$ each coefficient of the polynomials $\phi_{kl}+v_{l+1,\eta_k}(\xi,\eta)$, $\psi_{kl}-v_{l+1,x_k}(\xi,\eta)$ is a polynomial in the coefficients of $v_2,\ldots,v_l$ with integral coefficients. Now if

$$H=\sum_{l=2}^{\infty} K_l(\xi,\eta)$$

is the expansion of H in homogeneous polynomials in the ξ_k, η_k, then $K_2=H_2(\xi,\eta)$ while

$$K_l=\sum_{k=1}^{n} \lambda_k(\xi_k v_{lx_k}(\xi,\eta)-\eta_k v_{l\eta_k}(\xi,\eta))+\cdots \quad (l=3,4,\ldots),$$

where the coefficients of the additional terms on the right not written out explicitly are polynomials in the coefficients of $v_2,\ldots,v_{l-1}$ and linear functions in the coefficients of $H_3,\ldots,H_l$. If the product

$$P=\prod_{k=1}^{n} \xi_k^{\alpha_k}\eta_k^{\beta_k}$$

enters in $v_l(\xi,\eta)$ with the coefficient γ, then, because of the identity

$$\sum_{k=1}^{n} \lambda_k(\xi_k P_{\xi_k}-\eta_k P_{\eta_k})=P\sum_{k=1}^{n} \lambda_k(\alpha_k-\beta_k),$$

the product P in K_l has the coefficient

$$\varkappa=\gamma\lambda+\cdots,\quad \lambda=\sum_{k=1}^{n} \lambda_k(\alpha_k-\beta_k),$$

where the additional summands in $\varkappa$ are again polynomials in the coefficients of $v_2,\ldots,v_{l-1}$ and linear functions in the coefficients of $H_3,\ldots,H_l$. By the previous assumption about the linear independence of $\lambda_1,\ldots,\lambda_n$ we always have λ different from 0 unless $\alpha_k=\beta_k$ for $k=1,\ldots,n$, i.e. unless P is a product in powers of the $\omega_k=\xi_k\eta_k$ alone, so that when P is not such a product γ is uniquely determined by the requirement that $\varkappa=0$. In order to fix γ also for the remaining case $\alpha_k=\beta_k$ $(k=1,\ldots,n)$ we make the additional requirement that no product in powers of the ω_k alone appears in the expression

$$\Phi=\sum_{k=1}^{n} (\xi_k y_k-\eta_k x_k)$$

when expanded as a series in the ξ_k, η_k $(k=1, \ldots, n)$. Namely, the terms of degree l in Φ are determined from

$$\sum_{k=1}^{n} \{\xi_k v_{l x_k}(\xi, \eta) + \eta_k v_{l \eta_k}(\xi, \eta)\} + \cdots = l v_l(\xi, \eta) + \cdots \qquad (l=3, 4, \ldots),$$

so that indeed this fixes also the remaining coefficients γ. We have thus shown that there is exactly one power series v such that the formal canonical transformation defined by (6) takes the Hamiltonian function H into a power series in $\omega_1, \ldots, \omega_n$ alone while at the same time taking Φ into a series that contains no such products in the ω_k. The coefficients of v_l are uniquely determined in terms of the coefficients of $H_3, \ldots, H_l$, with the same then being true for the coefficients of $\phi_{k,l-1}, \psi_{k,l-1}$ $(k=1, \ldots, n;$ $l=3, 4, \ldots)$.

In order to discuss conditions for the solutions to be real, we note that $H(z) = H(\mathfrak{C}^{-1} w)$ is a real power series in $w_1, \ldots, w_{2n}$, while the matrices $\mathfrak{C}, \overline{\mathfrak{C}}$ and $\mathfrak{T} = \mathfrak{C}^{-1}\overline{\mathfrak{C}}$ are symplectic. Let the canonical transformation (3) be abbreviated as $z = \phi(\zeta)$, where ζ is the column vector with the $2n$ components ξ_k, η_k $(k=1, \ldots, n)$. Then $H(z) = H(\mathfrak{C}^{-1} w)$ $= \bar{H}(\overline{\mathfrak{C}}^{-1} w) = \bar{H}(\mathfrak{T}^{-1} z)$, while in addition $H(\phi(\zeta))$ is a power series in $\omega_1, \ldots, \omega_n$, and the series $\Phi(\zeta) = \zeta' \mathfrak{J} z = \zeta' \mathfrak{J}\, \phi(\zeta)$ contains no terms in the ω_k alone. By (16;5) the linear substitution $z = \mathfrak{T} z^*$ is given explicitly as $z_k^* = \varrho_k z_l$ $(l = l_k;\ k=1, \ldots, 2n)$ with $\varrho_k = -i$ when λ_k is purely imaginary and $\varrho_k = 1$ otherwise. It follows from this, or also from (15;22), (15;23) without prior normalization of the ϱ_k, that $\omega_k^* = \xi_k^* \eta_k^* = -\omega_k$ when λ_k is purely imaginary and $\omega_k^* = \omega_l$ otherwise. Consequently also $\bar{H}(\bar{\phi}(\mathfrak{T}^{-1}\zeta))$ $= H(\mathfrak{T}\bar{\phi}(\mathfrak{T}^{-1}\zeta))$ is a series in $\omega_1, \ldots, \omega_n$ alone while

$$\bar{\Phi}(\mathfrak{T}^{-1}\zeta) = (\mathfrak{T}^{-1}\zeta)' \mathfrak{J} \bar{\phi}(\mathfrak{T}^{-1}\zeta) = \zeta' \mathfrak{J}\mathfrak{T} \bar{\phi}(\mathfrak{T}^{-1}\zeta)$$

contains no terms in the ω_k. Since the substitution $z = \mathfrak{T}\bar{\phi}(\mathfrak{T}^{-1}\zeta)$ is likewise canonical and has the form (3), it follows from the previously established uniqueness that it agrees with $z = \phi(\zeta)$. Thus

$$\text{(7)} \qquad \phi(\zeta) = \mathfrak{T}\bar{\phi}(\mathfrak{T}^{-1}\zeta), \quad \bar{H}(\bar{\phi}(\mathfrak{T}^{-1}\zeta)) = H(\phi(\zeta)).$$

Suppose now that the substitution $z = \phi(\zeta)$ is convergent in a neighborhood of $\zeta = 0$. In order that w be real, we must have $\mathfrak{C} z = w = \bar{w} = \overline{\mathfrak{C}}\bar{z}$, so that $z = \mathfrak{T}\bar{z}$, which in view of the first equation in (7) is equivalent to the condition $\zeta = \mathfrak{T}\bar{\zeta}$. This means that $\eta_k = i\bar{\xi}_k$ when λ_k is purely imaginary and $\xi_l = \bar{\xi}_k$, $\eta_l = \bar{\eta}_k$ $(l = l_k;\ k=1, \ldots, n)$ otherwise, which in turn says that ω_k is purely imaginary when λ_k is, and $\omega_l = \bar{\omega}_k$ otherwise. Since H is a power series in $\omega_1, \ldots, \omega_n$ alone, the Hamiltonian system (4) becomes

$$\text{(8)} \qquad \dot{\xi}_k = H_{\omega_k} \xi_k, \quad \dot{\eta}_k = -H_{\omega_k} \eta_k \quad (k=1, \ldots, n),$$

from which it follows that

$$\dot{\omega}_k = \dot{\xi}_k \eta_k + \xi_k \dot{\eta}_k = 0\,.$$

Hence the ω_k are integrals of the system. The derivatives H_{ω_k} are then likewise independent of t, and one can immediately integrate (8) in the form

$$\xi_k = \alpha_k e^{H_{\omega_k} t}, \qquad \eta_k = \beta_k e^{-H_{\omega_k} t} \qquad (k = 1, \ldots, n)\,, \tag{9}$$

with α_k, β_k constant and $\omega_k = \alpha_k \beta_k$. Since α_k, β_k are the initial values of ξ_k, η_k at $t = 0$, the reality conditions require that $\beta_k = i\bar{\alpha}_k$ when λ_k is purely imaginary, and $\alpha_l = \bar{\alpha}_k$, $\beta_l = \bar{\beta}_k$ $(l = l_k)$ otherwise. Correspondingly, also ω_k is purely imaginary with λ_k, and $\omega_l = \bar{\omega}_k$ otherwise. Thus, in view of the second equation in (7) also H_{ω_k} is purely imaginary when λ_k is, and $H_{\omega_l} = \bar{H}_{\omega_k}$ otherwise, so that the solution (9) satisfies the reality conditions for all real t.

In the case when the transformation from (1) to the normal form (4) converges, the above procedure leads to a complete integration of the given system in a neighborhood of the equilibrium solution $w = 0$. In particular, since the series H_{ω_k} begins with λ_k, we obtain in this case once more the assertion of Liapunov's theorem. Moreover, we are in a position to assert that if, conversely, the eigenvalues λ_k are purely imaginary then the equilibrium solution is stable, and by inserting the exponential expressions (9) into $w = \mathfrak{C}\phi(\zeta)$ we can express the general solution u_k, v_k of (1) as trigonometric series [2–5].

One may conjecture that perhaps the unknown method of Dirichlet mentioned in § 5 is related to the process carried out here. Unfortunately, however, our method does not achieve what one might at first hope. Indeed, just as in the function-theoretic center problem in § 25, one can construct examples where the Hamiltonian function H, as a power series in u_k, v_k, is convergent, yet the series $v(x, \eta)$ fails to converge in any neighborhood of $x = 0$, $\eta = 0$. To do this one has only to take $n = 2$ and $\lambda_1 = i$, $\lambda_2 = i\varrho$ with ϱ a real irrational number that can be sufficiently well approximated by rational numbers, and then choose H appropriately. We will construct such an example in this section. It may nevertheless still appear plausible that divergence of the transformation taking a Hamiltonian system into its normal form constitutes an exceptional case, in the sense expressed by the result in § 26 for the Schröder series, or by the remark at the end of § 28 for the general system (27;1). However, it was shown [6] that even for $n = 2$ such a convergent transformation can exist only if the coefficients of H satisfy a certain denumerable set of analytic relations. Consequently, in general one has divergence, and in particular the above proof of stability fails. On the other hand, it is trivial that there do exist Hamiltonian systems for which the transformation into normal form is convergent. Indeed, one has only to take a convergent power series in $\omega_1, \ldots, \omega_n$ for H and transform the variables by an arbitrary convergent canonical substitution.

Although the transformation into the normal form diverges, we are nevertheless able to utilize it in studying the solutions of the Hamiltonian system (1) near the equilibrium solution. One sees from the first equation in (7) that in terms of $\sigma = \mathfrak{C}\zeta$ the canonical transformation $w = \mathfrak{C}\,\phi(\mathfrak{C}^{-1}\sigma) = \sigma + \cdots$ has only real coefficients, and in accordance with (3;4) we express this transformation in terms of a formal power series v as its generating function. If one truncates the power series v, keeping terms up to order $l > 1$, one obtains a convergent real transformation $w = g(\sigma) = \sigma + \cdots$ that agrees with the previous one up to terms of degree l, and consequently this transformation takes the given Hamiltonian function H into a real convergent power series whose terms agree with those of the formal series $H(\phi(\mathfrak{C}^{-1}\sigma))$ up to and including degree at least l. Now let us neglect the terms of degree higher than l in the series $H(\mathfrak{C}^{-1}g(\sigma))$ and apply the substitution inverse to $w = g(\sigma)$, whereby we are led to a real convergent series H^*. The Hamiltonian system

$$(10) \qquad \dot{u}_k = H^*_{v_k}, \quad \dot{v}_k = -H^*_{u_k} \quad (k = 1, \ldots, n)$$

then has the property that the right sides agree with those of (1) in terms of degree less than l. Moreover, it was constructed so that the convergent canonical transformation $w = g(\mathfrak{C}\zeta)$ takes it into normal form, and consequently it can be completely integrated in a neighborhood of the equilibrium solution $w = 0$ in accordance with (9). This fact, together with certain standard estimates from the theory of differential equations, can be used to approximate the solutions of the given system (1). It is not possible to determine from the remark we referred to, made by Dirichlet to Kronecker, whether there is a connection here with the method of Dirichlet, which allegedly did consist of stepwise approximation of the solutions to the differential equations of mechanics.

The convergent canonical transformation $w = g(\mathfrak{C}\zeta)$ takes the given Hamiltonian function H into a power series $H = F + G$ in $\zeta_1, \ldots, \zeta_{2n}$, where G begins with terms of degree $l+1$ while F is a polynomial of degree l and depends only on the products $\xi_k \eta_k = \omega_k$ $(k = 1, \ldots, n)$. Let the eigenvalues λ_k all be purely imaginary, so that for real solutions we have $i^{-1}\xi_k\eta_k = \xi_k\bar{\xi}_k \geqq 0$ $(k = 1, \ldots, n)$. Setting

$$\left(\sum_{k=1}^{n} |\xi_k|^2\right)^{\frac{1}{2}} = q \geqq 0,$$

in view of

$$(11) \quad \dot{\xi}_k = H_{\eta_k} = F_{\omega_k}\xi_k + G_{\eta_k}, \quad \dot{\eta}_k = -H_{\xi_k} = -F_{\omega_k}\eta_k - G_{\xi_k} \quad (k = 1, \ldots, n)$$

one obtains the differential equation

$$2iq\dot{q} = \sum_{k=1}^{n} (\eta_k G_{\eta_k} - \xi_k G_{\xi_k}).$$

If $\delta = \delta_l$ is now a sufficiently small positive number, which depends on l, it follows that

$$|\dot{q}| < A_l q^l \quad (0 < q < \delta),$$

where A_l and later B_l, C_l denote positive constants that also depend on l. Upon integration this gives

(12) $$|q^{1-l} - q_0^{1-l}| \leqq (l-1) A_l |t| \quad (-T < t < T),$$

provided that the function $q = q(t)$ remains smaller than δ throughout the interval $-T < t < T$, where $q_0 = q(0) > 0$ denotes the initial value. Suppose in addition that

(13) $$q_0 < \tfrac{1}{2}\delta, \quad (l-1) A_l q_0^{l-1} T < \tfrac{1}{2}.$$

Then using the continuity of $q(t)$ one obtains from (12) that

$$\tfrac{2}{3} q_0 < q < 2q_0 < \delta, \quad |q - q_0| \leqq (2l-2) A_l q_0^l |t| \quad (|t| < T),$$

and because

$$(\xi_k \eta_k)^{\cdot} = \eta_k G_{\eta_k} - \xi_k G_{\xi_k} \quad (k = 1, \ldots, n),$$

it further follows that

$$|\xi_k \eta_k - (\xi_k \eta_k)_0| \leqq B_l q_0^{l+1} |t| \quad (|t| < T).$$

Finally, upon integrating (11) we obtain in view of (13)

(14) $$|\xi_k - (\xi_k)_0 \, e^{(F_{\omega_k})_0 t}| \leqq C_l (q_0^l |t| + q_0^{l+2} t^2) \left(q_0 < \frac{\delta}{2}; |t| < \frac{q_0^{1-l}}{(2l-2) A_l} \right).$$

In (14) we have an estimate on how well the solutions of the given Hamiltonian system can be approximated by trigonometric series [7]. Because of the appearance of the constants C_l and A_l, which may possibly grow very rapidly with l, for q_0 fixed and $l \to \infty$ this approximation has in general only the character of semi-convergence, in the sense of, say, the Stirling series. On the other hand, we have in particular shown that

$$\frac{2}{3} q_0 < q < 2q_0 \quad \left(|t| < \frac{q_0^{1-l}}{(2l-2) A_l} \right),$$

and this offers a weak contribution to the unsolved problem of stability. Namely, returning to the original coordinates u_k, v_k and setting

$$\sum_{k=1}^{n} (u_k^2 + v_k^2) = \varrho^2, \quad \varrho \geqq 0,$$

we can assert that:

If at time $t=0$ the distance $\varrho = \varrho_0$ from the origin is smaller than ε_l, then $\varrho \leqq 2\varrho_0$ remains valid for at least the time interval of length $\delta_l \varrho_0^{1-l}$, where ε_l, δ_l $(l = 3, 4, \ldots)$ are positive constants that depend on l. To obtain the best possible estimate for a given ϱ_0 one has to determine the least upper bound for the constants $\delta_l \varrho_0^{1-l}$ over those values of l for which $\varepsilon_l > \varrho_0$.

In the case when the Hamiltonian system (2) is transformed into the normal form (4) by a convergent power series, the series

$$\omega_k = \xi_k \eta_k = x_k y_k + \cdots \quad (k = 1, \ldots, n) \tag{15}$$

represent n independent integrals of (2) that converge in a neighborhood of the origin. We will say that a formal power series $g(x, y)$ which formally satisfies the equation

$$\sum_{k=1}^{n} (g_{x_k} H_{y_k} - g_{y_k} H_{x_k}) = 0, \tag{16}$$

fulfilled by an integral, is likewise an integral of (2). Thus, in this sense, under the previously made assumption about the linear independence of $\lambda_1, \ldots, \lambda_n$, the Hamiltonian system (2) always has the n integrals ω_k $(k = 1, \ldots, n)$. We will now show that each integral $g(x, y)$ can be expressed as a formal power series in $\omega_1, \ldots, \omega_n$. Indeed, since the difference $\omega_k - x_k y_k$, as a power series in $x_1, \ldots, y_n$, begins with cubic terms, one can recursively construct a power series $P(\omega)$ in the ω_k so that the series $h(x, y) = g(x, y) - P(\omega)$ does not contain any terms of the form $c(x_1 y_1)^{\alpha_1} \ldots (x_n y_n)^{\alpha_n}$ in the variables $x_1, \ldots, y_n$. Since $h(x, y)$ is also an integral, it satisfies the formal relation

$$\sum_{k=1}^{n} (h_{x_k} H_{y_k} - h_{y_k} H_{x_k}) = 0. \tag{17}$$

Now if the series $h(x, y)$ were not identically 0, it would contain a term $c x_1^{\alpha_1} y_1^{\beta_1} \ldots x_n^{\alpha_n} y_n^{\beta_n}$ of smallest degree, with $c \neq 0$, and it would follow from (17) by comparison of coefficients that

$$c \sum_{k=1}^{n} (\alpha_k - \beta_k) \lambda_k = 0$$

and therefore $\alpha_k = \beta_k$ $(k = 1, \ldots, n)$. This, however, is impossible since by construction $h(x, y)$ no longer contains any terms of this form. Consequently $h(x, y) = 0$ and $g(x, y) = P(\omega)$, which proves our assertion.

We will now construct an example where H has a convergent power series but the integral $\omega_1 = x_1 y_1 + \cdots$ diverges. In particular, it will then follow that the corresponding Hamiltonian system cannot be transformed into normal form by a convergent canonical substitution. We specialize to $n = 2$ and $\lambda_1 = i$, $\lambda_2 = i\varrho$ with ϱ a real irrational number, so that the condition about linear independence of λ_1, λ_2 is satisfied. Furthermore, we set

$$H = i(x_1 y_1 + \varrho x_2 y_2) + \sum_{p,q} a_{pq}(x_1^p y_2^q + x_2^q y_1^p) \tag{18}$$

and allow for a_{pq} only the values $0, \pm 1$. In particular, we take $a_{pq} = 0$ unless both p and q are divisible by 4. The reality condition $y_k = i\bar{x}_k$ $(k = 1, 2)$ then makes H real valued. For ϱ we select an irrational number in the interval $0 < \varrho < 1$ that can be approximated sufficiently well by rational numbers – namely, the inequality

$$0 < |p - \varrho q| < \frac{1}{q!} \tag{19}$$

should have infinitely many solutions in natural numbers p, q divisible by 4. It is easily seen that the number α constructed in § 25 has this property. The integral

$$\omega_1 = g(x, y) = x_1 y_1 + \sum_{l=3}^{\infty} g_l(x, y),$$

with g_l a homogeneous polynomial of degree l, then satisfies (16) and comparison of coefficients leads to the relation

$$x_1 g_{l x_1} - y_1 g_{l y_1} + \varrho(x_2 g_{l x_2} - y_2 g_{l y_2}) + i \sum_{p+q=l} p a_{pq}(x_1^p y_2^q - x_2^q y_1^p) = \cdots,$$

where the right side is a homogeneous polynomial of degree l in x_1, y_1, x_2, y_2 whose coefficients are expressed only in terms of the coefficients of $g_3, \ldots, g_{l-1}$ and the a_{pq} with $p + q < l$. For the individual term $c_{pq} x_1^p y_2^q$ in g_l it then follows that

$$(p - \varrho q) c_{pq} + i p a_{pq} = \gamma_{pq}, \tag{20}$$

where γ_{pq} is expressed in terms of the coefficients of $g_3, \ldots, g_{l-1}$ and the a_{rs} with $r + s < l$. As was observed earlier, the coefficients in the terms of degree less than l in the canonical substitution (3) are uniquely determined by the coefficients of the terms of degree $\leq l$ in H, and consequently also $g_3, \ldots, g_{l-1}$ are determined by the a_{rs} with $r + s < l$, as are the γ_{pq} $(p+q=l)$. For p, q positive solutions to (19) and divisible by 4 we now choose $a_{pq}=\pm 1$ so that $|pa_{pq}+i\gamma_{pq}|\geqq p\geqq 1$, which by the triangle inequality is easily seen to be possible. By (19), (20) then

$$|c_{pq}|\geqq q!\,, \tag{21}$$

and indeed this is true for infinitely many q. For all other pairs p, q let $a_{pq}=0$. In view of (19), (21) the series $g(x, y)$ cannot converge in any full neighborhood of the origin.

Thus in this example the transformation into normal form does not converge. On the other hand the quadratic term

$$i(x_1 y_1 + \varrho x_2 y_2) = -(x_1 \bar{x}_1 + \varrho x_2 \bar{x}_2)$$

in H is negative definite, and therefore by the theorem of Dirichlet the equilibrium solution $x_1 = x_2 = y_1 = y_2 = 0$ is stable. This observation is noteworthy, since for the function-theoretic center problem stability implied convergence of the transformation into normal form, while this shows that there is no analogous theorem in the problem of stability for Hamiltonian systems.

By a construction similar to that in the above example one can show [8] that there exist canonical systems of differential equations with a convergent Hamiltonian function H which have no convergent integrals $g(x, y)$, other than H itself and the convergent power series in H. For the case $n=2$ one can construct such a function H by again starting with the expressions (18) and (19), although one must now replace $1/q!$ by a function of q that goes to 0 even more rapidly. To be more exact, it is even true that each Hamiltonian function with the fixed quadratic part $i(x_1 y_1 + \varrho x_2 y_2)$ can be converted by an arbitrarily small perturbation of the coefficients of the higher order terms into one which has the above mentioned property of having no other convergent integrals. In this connection one should mention a theorem of Poincaré [9], who considered Hamiltonian functions $H(z, \mu)$ which in addition to $z_1, \ldots, z_{2n}$ also depended analytically on a parameter μ near $\mu=0$. The theorem states that under certain assumptions about $H(z, 0)$ and the derivative $H_\mu(z, 0)$, which are in general satisfied, the Hamiltonian system corresponding to $H(z, \mu)$ can have no integrals represented as convergent series in the $2n+1$ variables $z_1, \ldots, z_{2n}$ and μ, other than the convergent series in H, μ. However this theorem of Poincaré does not assert anything about a fixed parameter value μ.

As was already observed, in the case of linear independence of the eigenvalues $\lambda_1, \ldots, \lambda_n$ a Hamiltonian system cannot be brought into normal form by a convergent canonical transformation if it does not possess n independent convergent integrals, and we gave an example of such a system. One might now suppose that perhaps, as in the case of the function-theoretic center problem, the set of purely imaginary λ_k $(k=1, \ldots, n)$ for which the transformation into normal form diverges has n-dimensional Lebesgue measure 0. This, however, is not so. Indeed, a deeper investigation shows that for any l each Hamiltonian function H that is not a power series in $H_2(\zeta)$ alone can, by an arbitrarily small perturbation of the coefficients in the terms of degree higher than l, be changed into one whose corresponding canonical system cannot be transformed convergently into normal form, and this statement clearly does not depend on the eigenvalues $\lambda_1, \ldots, \lambda_n$.

Let us summarize the main results on stability of Hamiltonian systems, where the n eigenvalues $\lambda_1, \ldots, \lambda_n$ are now assumed only to be distinct from one another and different from 0. If none of the eigenvalues is purely imaginary, then by Liapunov's theorem we have instability. If, on the other hand, at least one eigenvalue is purely imaginary, let λ_1 be the one of largest absolute value among these. Then none of the $n-1$ numbers $\dfrac{\lambda_k}{\lambda_1}$ $(k=2, \ldots, n)$ is an integer, and by the existence theorem in § 16 there is a one parameter family of periodic solutions in a neighborhood of the equilibrium solution. It follows from this that the equilibrium is not unstable. For stability, however, by Liapunov's theorem it is necessary that all the eigenvalues be purely imaginary, so that if there are eigenvalues that are purely imaginary and also others that are not, we have the mixed case. It remains to consider the case when all the eigenvalues are purely imaginary. Here, if there exists an integral that has a strict extremum at the equilibrium solution, then we have stability by Dirichlet's theorem. This, in particular, is so if the quadratic part of the Hamiltonian function is definite. If in addition the eigenvalues $\lambda_1, \ldots, \lambda_n$ are linearly independent, we always have stability when the transformation into normal form converges. Still, in this case there also exists an integral with a strict minimum at the origin – say, the integral

$$-i(\omega_1 + \cdots + \omega_n) = \xi_1 \bar{\xi}_1 + \cdots + \xi_n \bar{\xi}_n \qquad (\eta_k = i\bar{\xi}_k).$$

However, there is no known finite method for deciding whether the transformation to normal form converges or diverges. When the transformation diverges and the Hamiltonian function is indefinite, the existing methods do not allow us to decide whether we have the stable or mixed case when $n > 2$. The case of two degrees of freedom will be discussed in detail in § 35. To be sure, there is no known example with linearly independent purely imaginary eigenvalues $\lambda_1, \ldots, \lambda_n$ for which we actually have the mixed case, and it would seem that this cannot happen. However, it appears that the complete solution to the problem of stability for Hamiltonian systems still lies in the distant future.

We conclude here with an application of our sparse results to the planar three-body problem. As a starting point we choose the solutions of Lagrange which in accordance with § 18 are equilibrium solutions in the rotating coordinate system, while for the Hamiltonian system we take the six differential equations (18; 27) which arise from the equations of motion after the elimination of the center of mass integrals and the angular momentum integral. In the equilateral case, if

$$27(m_1 m_2 + m_2 m_3 + m_3 m_1) < (m_1 + m_2 + m_3)^2 , \tag{22}$$

then the eigenvalues are all purely imaginary but the Hamiltonian function is indefinite. In this case there is no known way for deciding about stability, although, at any rate, we do not have instability. If on the other hand

$$27(m_1 m_2 + m_2 m_3 + m_3 m_1) > (m_1 + m_2 + m_3)^2 ,$$

then not all the eigenvalues are purely imaginary, and consequently we do not have stability. For the collinear solutions there is always one real eigenvalue, so that again we do not have stability. In the solar system there actually are small planets that together with the sun and Jupiter form approximately an equilateral triangle and satisfy condition (22), while there are none that approximate the collinear solutions.

§ 31. Area-Preserving Transformations

We now wish to extend the definition of stability of an equilibrium solution to other solutions of a system of differential equations $\dot{x}_k = f_k(x)$ $(k = 1, \ldots, m)$. The m functions $f_k(x)$ are to satisfy a Lipschitz condition in a domain $\mathfrak{R}$ of the m-dimensional real x-space, and $x = x(t)$ shall be a solution of the system that remains in $\mathfrak{R}$ for all real time. By a neighborhood of such a solution we will mean any open subset $\mathfrak{U}$ of $\mathfrak{R}$ that contains the entire orbit $x = x(t)$ in its interior. It could happen that the orbit is dense in $\mathfrak{R}$, in which case $\mathfrak{R}$ itself would be the only neighborhood, and to avoid this and similar difficulties we will define stability only for periodic solutions $x(t)$. Making the obvious generalization, we say that such a periodic solution is stable if for each neighborhood $\mathfrak{U}$ of its orbit there exists another neighborhood $\mathfrak{B}$ such that any orbit passing through an arbitrary point of $\mathfrak{B}$ remains entirely in $\mathfrak{U}$. This, of course, means that then $\mathfrak{B} \subset \mathfrak{U}$. The definitions of instability and the mixed case given in § 25 are generalized similarly in the obvious way. In particular, if the periodic solution is an equilibrium solution the new definition agrees with the earlier one.

For a Hamiltonian system

$$\dot{x}_k = E_{y_k}, \quad \dot{y}_k = -E_{x_k} \quad (k=1, \dots, n) \tag{1}$$

we will also define a weaker form of stability of a periodic solution $x=x(t)$, $y=y(t)$. Let $\mathfrak{R}$ and $\mathfrak{U}$ have the same meaning as before, and let $E=\gamma$ along the solution in question. By neighborhoods we will now mean the $(2n-1)$-dimensional intersections $\mathfrak{U}_\gamma$ of $\mathfrak{U}$ with the surface $E=\gamma$, and we will speak of isoenergetic stability if for each neighborhood $\mathfrak{U}_\gamma$ of the given closed orbit there exists another neighborhood $\mathfrak{B}_\gamma$ such that any orbit through an arbitrary point of $\mathfrak{B}_\gamma$ remains entirely in $\mathfrak{U}_\gamma$. Clearly stability implies isoenergetic stability. Isoenergetic instability and the mixed case are defined analogously.

From here on we will consider the Hamiltonian system (1) only for $n=2$, and will assume that E is analytic in $\mathfrak{R}$. As described in § 22, one can associate with a periodic solution of such a system a two-dimensional area-preserving analytic mapping S that has the origin as a fixed-point, and the question of whether the original periodic solution is isoenergetically stable, unstable, or mixed then naturally reduces to that of whether the mapping S is stable, unstable, or mixed relative to the origin. Let the area-preserving transformation S be expressed in the form

$$x_1 = g(x,y) = ax+by+\cdots, \quad y_1 = h(x,y) = cx+dy+\cdots, \tag{2}$$

where the power series $g(x,y)$, $h(x,y)$ have real coefficients and converge in a neighborhood of the origin. Since $ad-bc=1$, the eigenvalues λ, μ of the matrix corresponding to the linear part satisfy $\lambda\mu=1$.

In the hyperbolic case λ, μ are real and distinct, and for this case instability of S was established already in § 23. Levi-Civita [1] has generalized this result to more than two variables, obtaining also an analogue to the first assertion of Liapunov's theorem.

In the parabolic case $\lambda=\mu=\pm 1$, and the case $\lambda=\mu=-1$ reduces to $\lambda=\mu=1$ if one considers S^2 in place of S. Here too we refer to the work of Levi-Civita, who showed that the coefficients of the quadratic terms must satisfy a certain relation if the transformation S is to be stable.

In the elliptic case $|\lambda|=1$ and $\lambda^2 \neq 1$. First, we consider the special case where λ is a root of unity. Let $\lambda^q=1$ and $\lambda^k \neq 1$ $(k=1, \dots, q-1)$ with $q>2$, so that λ is a primitive q-th root of unity. By again considering S^q in place of S, one is led back to the parabolic case $\lambda=\mu=1$. However, a simple computation shows that for the transformation S^q all terms from degree 2 up to degree $q-2$ vanish, and consequently for $q>3$ the above mentioned result of Levi-Civita applies only trivially. For $q=3$ the situation is different, and Levi-Civita applied this case also to the restricted three-body problem. The area-preserving mapping in this connection was introduced already at the end of § 24, where for $\tau = 2\pi|\omega|^{-1}$ denoting the period of the initial solution, we had $\lambda = e^{i\tau}$, so that $q = 3$ when $\omega = 3$. Upon computing the quadratic terms of the mapping S^3 for $\omega = 3$, Levi-Civita showed that the required relation was not satisfied, and consequently we do not have stability.

For the case $q = 3$ one can even show that S is unstable if a certain simple relation does not hold between the coefficients of the quadratic terms. Indeed, expressing the mapping S in the complex form

$$(3) \quad z_1 = \lambda z + az^2 + bz\bar{z} + c\bar{z}^2 + \cdots, \quad z = x + iy, \quad z_1 = x_1 + iy_1,$$

where the power series in $z, \bar{z}$ converges when $r^2 = x^2 + y^2 = z\bar{z}$ is sufficiently small, we assert that if $c \neq 0$ then S is unstable – and here it is not even necessary to assume that S is area-preserving. To show this, iterating the mapping S and using $\lambda^2 + \lambda + 1 = 0$, we obtain

$$z_2 = \lambda^2 z + (\lambda^2 + \lambda)az^2 + (\lambda + 1)bz\bar{z} + 2\lambda c\bar{z}^2 + \cdots,$$
$$z_3 = z + 3\lambda^2 c\bar{z}^2 + \cdots, \quad z_3^3 = z^3 + 9\lambda^2 c(z\bar{z})^2 + \cdots,$$

whereupon by replacing z, z_1 by $\varrho z, \varrho z_1$ with $\varrho\bar{\varrho}^{-2} = 9\lambda^2 c$ we get

$$(4) \quad z_3^3 = z^3 + (z\bar{z})^2 + \cdots = z^3 + (z\bar{z})^2(1 + \eta r)$$

with $|\eta| r < \frac{1}{2}$ when $r < r_0$ is sufficiently small. Let

$$S^n z = z_n, \quad z_{3n}^3 = Z_n = X_n + iY_n, \quad |z_{3n}|^3 = |Z_n| = R_n \quad (n = 0, \pm 1, \ldots),$$

and suppose that for a certain $z = z_0$ we have $|z_n| < r_0$ for all n. It then follows from (4) that

$$(5) \quad X_{n+1} \geqq X_n + \tfrac{1}{2} R_n^{\frac{4}{3}} \geqq X_n,$$

so that the sequence X_n is monotonic, while in view of $|X_n| \leqq |Z_n| < r_0^3$ it is also bounded. Consequently the difference $X_{n+1} - X_n$ goes to 0 as $n \to \infty$ or as $n \to -\infty$, and therefore by (5) R_n and with it X_n tend to 0, implying that $X_n = 0$ and with it $R_n = 0$ for all n. Thus z must necessarily be 0, which proves instability. Moreover, it is easily seen that even for area-preserving mappings S in general $c \neq 0$.

Similarly, we will exhibit now for arbitrary $q > 0$ an example of an unstable area-preserving mapping for which λ is a primitive q-th root of unity [2]. As we saw in § 23, in two dimensions one generates area-preserving mappings through the expression

$$(6) \quad y = w_x, \quad \xi = w_\eta,$$

with the generating function $w = w(x, \eta)$ satisfying $w_{x\eta} \neq 0$. First we assume that $q \neq 4$, so that $\lambda^2 \neq -1$, and set

$$\mu = \bar{\lambda}, \quad 2\sigma = \lambda + \mu \neq 0, \quad \sigma u = x + i\lambda\eta, \quad \sigma v = x - i\mu\eta,$$

$$2iw = \frac{\sigma}{2}(\mu u^2 - \lambda v^2) + f(u, v), \tag{7}$$

where $f(u, v)$ is a polynomial in u, v that begins with cubic terms and satisfies the relation $f(v, u) = -\bar{f}(u, v)$. Then w is a polynomial in x, η with real coefficients, and $w_{x\eta} = 1 + \cdots$ so that $w_{x\eta} \neq 0$ for $x = 0, \eta = 0$. With this generating function w one obtains from (6) the formulas

$$2iy = \mu u - \lambda v + \sigma^{-1}(f_u + f_v), \quad 2\xi = u + v + \sigma^{-1}(\lambda f_u - \mu f_v),$$

while also $2i\eta = u - v$, $2x = \mu u + \lambda v$. Introducing $z = x + iy$, $\zeta = \xi + i\eta$, we can express the transformation in the complex form

$$\begin{cases} z = \mu u + \dfrac{1}{2\sigma}(f_u + f_v), & \bar{z} = \lambda v - \dfrac{1}{2\sigma}(f_u + f_v), \\ \zeta = u + \dfrac{1}{2\sigma}(\lambda f_u - \mu f_v), & \bar{\zeta} = v + \dfrac{1}{2\sigma}(\lambda f_u - \mu f_v), \end{cases} \tag{8}$$

$$\zeta = \lambda z - f_v, \tag{9}$$

which in particular shows that λ and μ are the eigenvalues of the mapping. Specializing to

$$f(u, v) = q^{-1} uv(u^q - v^q)$$

we obtain now a function with the required properties. Indeed, with A_l, B_l henceforth denoting convergent power series in $z, \bar{z}$ that begin with terms of degree l, from (8) one obtains by inversion

$$u = \lambda z + A_{q+1}, \quad v = \mu \bar{z} + B_{q+1},$$

and since $\lambda^q = 1$, we have

$$f_v = q^{-1} u(u^q - (q+1) v^q) = q^{-1} \lambda z\{z^q - (q+1)\bar{z}^q\} + A_{2q+1},$$

so that (9) now becomes the explicit transformation

$$\zeta = \lambda z(1 + q^{-1}\{(q+1)\bar{z}^q - z^q\}) - A_{2q+1}. \tag{10}$$

Next we show that with this as our mapping S, we have instability at $z = 0$.

From (10) it follows that

$$\zeta^q = z^q\{1 + (q+1)\bar{z}^q - z^q\} + A_{3q}. \tag{11}$$

Abbreviating

$$z_n = S^n z\,, \quad z_n^q = Z_n = X_n + iY_n\,, \quad R_n = |Z_n| = |z_n|^q \quad (n = 0, \pm 1, \ldots)$$

and using (11) we obtain the estimate

$$X_{n+1} \geqq X_n + (q+1)\,R_n^2 - R_n^2 - |A_{3q}(z_n, \bar{z}_n)| \geqq X_n + \tfrac{1}{2} R_n^2$$

provided that $R_n < \delta$ for a sufficiently small $\delta > 0$. Finally, the reasoning following (5) now shows that the mapping (10) is unstable.

In expression (7) it was necessary to have $q \neq 4$, for otherwise we would have $\sigma = 0$. However, if one constructs the above transformation S for $q = 8$, then the eigenvalues of S^2 are primitive fourth roots of unity, while both mappings clearly have the same behavior with regard to stability. Thus we have shown that for each root of unity λ there is an area-preserving mapping which has eigenvalues $\lambda, \bar{\lambda}$ and is unstable. The mappings constructed have the additional property of being algebraic.

It remains to discuss the case where $|\lambda| = 1$ and λ is not a root of unity. As shown in § 23, the mapping (2) can then be transformed by an area-preserving substitution C, represented as a formal power series, into the normal form

$$U = C^{-1} S C\,, \quad \xi_1 = u\xi\,, \quad \eta_1 = v\eta\,, \tag{12}$$

where $u = \lambda + \cdots$, $v = \mu + \cdots$ are formal power series in $\omega = \xi\eta$ satisfying $uv = 1$ and the reality condition $v = \bar{u}$. In the case of convergence of C, real values of the original variables x, y correspond to complex conjugate values for $\xi, \eta = \bar{\xi}$, and (12) shows that $|\xi_1| = |\xi|$. Consequently, the mapping leaves all concentric circles in the ξ-plane with center at $\xi = 0$ invariant. From this it is evident that when the transformation into the normal form U converges, the mapping S is always stable at $x = 0$, $y = 0$. The convergence behavior of C, however, is similar to that of the transformation to the normal form for Hamiltonian systems discussed in § 30. Namely, one can give examples of elliptic area-preserving mappings for which C diverges, and one can even set up infinitely many independent analytic relations that must be satisfied by the coefficients of $g(x, y)$ and $h(x, y)$ in (2), showing that divergence is the rule and convergence the exception. In the function-theoretic center problem we were able to show that, conversely, stability also implies the convergence of the Schröder series and with it convergence of the transformation taking the conformal mapping into normal form. The method of proof in § 25, however, cannot be carried over to the present case, since for area-preserving mappings there is no analogue to the Riemann mapping theorem for conformal mappings. There is not nearly as far reaching a theory on the differential equation $\phi_x \psi_y - \phi_y \psi_x = 1$ as on the system $\phi_x = \psi_y, \phi_y = -\psi_x$.

Let us assume now that the formal power series u appearing in the normal form (12) is not constant, so that $u=\lambda$ does not hold identically. Then according to the Birkhoff fixed-point theorem of § 24, in each neighborhood $\mathfrak{U}$ of the origin and for all sufficiently large natural numbers $n>n_0(\mathfrak{U})$ there are fixed-points of the mapping S^n distinct from the origin whose images under S^k $(k=1,\ldots,n)$ all lie in $\mathfrak{U}$. In particular it follows from this that S is not unstable. Consequently in general, namely whenever the power series u is not identically constant, one does not have instability. We will show in the following sections that we actually have stability when $u \neq \lambda$. As we already noted, there are no known examples of the mixed case, and it is not known whether in the case $u=\lambda$ one can actually have instability. If that were the case, one would have an example of convergent series for u and a divergent substitution C, also a situation not yet known to be possible.

If the product $\omega=\xi\eta$ of the variables appearing in the normal form (12) is expressed in terms of the old variables x, y, one obtains a formal power series $\omega=\phi(x,y)$ which, in view of the identity $\xi_1\eta_1=\xi\eta$, has the property that it remains invariant under the mapping S. Thus $\phi(x,y)$ is an analogue of an integral for differential equations, and corresponding to the theorem of Dirichlet one can easily show that if there exists a convergent power series in x, y that is invariant under S and has a strict extremum at the origin, then S is always stable with respect to the origin. However, one can again give examples of elliptic area-preserving transformations with $\lambda^n \neq 1$ $(n=3,4,\ldots)$ for which no such invariant series exists.

§ 32. Existence of Invariant Curves

In the previous section we saw that an elliptic fixed-point of an area-preserving mapping need not be stable; indeed, we constructed counterexamples for any root of unity as the eigenvalue λ. However, we will now show that these examples are actually the exceptions. As was noted before, one cannot express a possible stability criterion in terms of the eigenvalues λ, μ of the linearized mapping alone, but has to consider the nonlinear terms as well. Following the discussion of formal normal forms in § 23, we may assume that our mapping has the form

$$(1)\qquad \begin{cases} r_1 = r\cos w - s\sin w + O_{2l+2} \\ s_1 = r\sin w + s\cos w + O_{2l+2} \\ w = \sum_{k=0}^{l} \gamma_k (r^2+s^2)^k \end{cases}$$

where O_{2l+2} denotes a convergent power series in r, s with terms of order $\geqq 2l+2$ only. More precisely, if $\lambda, \lambda^2, \ldots, \lambda^{2l+2} \neq 1$ then we can always transform our mapping by a convergent area-preserving substitution into the above form.

Rather than have restrictions on $\lambda = e^{i\gamma_0}$ alone, we will express our stability criterion in terms of the coefficients $\gamma_0, \gamma_1, \ldots$ appearing in the normal form (1). The aim of this and the following section is to prove: If at least one of the coefficients $\gamma_1, \gamma_2, \ldots, \gamma_l$ is not zero, then the mapping (1) is stable at the origin. This, of course, implies that an area-preserving mapping with $\lambda, \lambda^2, \ldots, \lambda^{2l+2} \neq 1$, and $\gamma_l \neq 0$ the first nonvanishing coefficient in the normal form, has the origin as a stable fixed-point, stability being a property preserved under convergent coordinate transformations. Incidentally, this is precisely the assumption under which we established the Birkhoff fixed-point theorem in § 24.

If λ is not a root of unity this result asserts stability whenever the normal form is not linear, i.e. if not all of the coefficients $\gamma_1, \gamma_2, \ldots$ vanish. In this case, following Birkhoff, we speak of a fixed-point of the general elliptic type, the other case being clearly exceptional. We can therefore say that a fixed-point of the general elliptic type is always stable.

For later applications it is important to observe that for a fixed l in a particular example only a finite number of roots of unity have to be excluded, and hence only finitely many conditions verified. For example, if $\lambda^3 \neq 1$, $\lambda^4 \neq 1$, and $\gamma_1 \neq 0$ then the mapping is stable.

The proof of the stability criterion will be based on a theorem on the existence of closed invariant curves, which will be the topic of this section. In each neighborhood of the fixed-point we will construct closed invariant curves surrounding the fixed-point, and their interiors will then form invariant neighborhoods of the fixed-point. This will show that each neighborhood of the fixed-point contains an invariant neighborhood of this point, a fact that, as we saw in § 25, assures stability.

If the error terms O_{2l+2} in (1) are neglected, clearly the concentric circles, given by $r^2 + s^2 = \text{constant}$, provide such invariant curves. Our aim is to show that some of these circles for which $w/2\pi$ is irrational can be continued into invariant curves of the actual mapping (1). For the success of this approach it is crucial that the angle of rotation w vary with the radius $(r^2 + s^2)^{\frac{1}{2}}$ of the circle, and this is assured by the assumption $\gamma_l \neq 0$.

After these preliminary remarks we will now formulate and prove the existence theorem for such invariant curves in a somewhat simpler setting, and return to the stability proof of (1) in § 34. Let r, θ denote polar coordinates in the plane, and consider the mapping

$$\theta_1 = \theta + \alpha(r), \qquad r_1 = r$$

of the annulus $A: 0 \leqq a_0 \leqq r \leqq b_0$ into itself. This mapping leaves each circle about the origin invariant, rotating it through an angle $\alpha(r)$ which we assume increases with r, so that $\alpha(r)$ is a monotone increasing function. We will refer to such a mapping as a twist mapping, and we will study a mapping M:

$$(2) \qquad \begin{cases} \theta_1 = \theta + \alpha(r) + f(\theta, r) \\ r_1 = r + g(\theta, r), \end{cases}$$

which is close to this twist mapping. Here we assume that α, f, g are real analytic and have period 2π in θ. We wish to construct an invariant curve of the form $r = \psi(\theta) = \psi(\theta + 2\pi)$ for the perturbed mapping (2). It is clear that a smallness condition alone on f, g will not suffice for this, as can be seen by taking g to be a small positive constant. In that case r always increases under application of the mapping, and no closed invariant curve can exist. Rather than require the mapping to be area-preserving, we will assume that M has the property that any curve $\Gamma: r = \phi(\theta) = \phi(\theta + 2\pi)$ always intersects its image curve $M\Gamma$. Under this assumption, together with the hypothesis that $\alpha'(r) \neq 0$ in $a_0 \leqq r \leqq b_0$ and a smallness restriction on f, g, we will establish the existence of an invariant curve M in the annulus $a_0 \leqq r \leqq b_0$. Incidentally, M is not required to map the annulus into itself.

To simplify the situation so that $\alpha(r)$ is replaced by a linear function, we set

$$y = \frac{\alpha(r)}{\gamma}, \qquad \gamma = |\alpha(b_0) - \alpha(a_0)| > 0$$

and let $x = \theta$. In the new variables our transformation takes the form

$$(3) \qquad \begin{cases} x_1 = x + \gamma y + f(x, y) \\ y_1 = y + g(x, y) \end{cases}$$

with f, g in general different from the corresponding functions in (2), but still real analytic and of period 2π in x. The dependence of f, g on γ is not indicated. The variable y ranges over an interval $a \leqq y \leqq b$ of length $b - a = 1$, while the parameter γ measures the length of the interval over which the angle $\alpha(r)$ ranges. We may assume that $\gamma \leqq 1$, since this can always be achieved by restricting our original annulus so that $\alpha(r)$ varies over an interval of length $\leqq 1$.

Because f, g are real analytic functions, they can be extended to a complex domain $\mathfrak{D}$ which we may take to be of the form

$$(4) \qquad \mathfrak{D}: |\mathrm{Im}\, x| < r_0; \quad y \in \mathfrak{D}',$$

with $\mathfrak{D}'$ a complex neighborhood of the interval $a \leqq y \leqq b$. Moreover, we will continue to assume that each curve $y = \phi(x) = \phi(x + 2\pi)$ intersects its image under the mapping (3), and we may also take $0 < r_0 \leqq 1$.

Under these hypotheses we assert the following theorem: For each positive ε there exists a positive δ, depending on ε and $\mathfrak{D}$ but not on γ, such that for

$$|f|+|g|<\gamma\delta$$

in $\mathfrak{D}$ the mapping M in (3) admits an invariant curve of the form

$$x=\xi+u(\xi), \quad y=v(\xi) \tag{5}$$

with u, v real analytic functions of period 2π in the complex domain $|\mathrm{Im}\,\xi|<r_0/2$. Moreover, the parametrization is chosen so that the induced mapping on the curve (5) is given by

$$\xi_1=\xi+\omega$$

with ω a constant incommensurable with 2π, and the functions u, v satisfy

$$|u|+|v-\gamma^{-1}\omega|<\varepsilon\,.$$

Before turning to the proof of this theorem, which will be given in this and the next section, it is instructive to construct the invariant curve by a power series expansion under the assumption that the functions f, g in (3) depend analytically on a parameter λ. Expressing this dependence in the form

$$f=\sum_{\nu=1}^{\infty}\lambda^{\nu}f_{\nu}(x,y), \quad g=\sum_{\nu=1}^{\infty}\lambda^{\nu}g_{\nu}(x,y)\,,$$

we require that the intersection property hold for all real λ sufficiently small in absolute value, and try to determine the functions $u=u(\xi,\lambda)$, $v=v(\xi,\lambda)$ in (5) by a power series expansion. The requirement that the induced mapping on the curve (5) be given by

$$x_1=\xi+\omega+u(\xi+\omega,\lambda), \quad y_1=v(\xi+\omega,\lambda)$$

leads to the functional equations

$$\begin{aligned} u(\xi+\omega,\lambda)&=u+\gamma v-\omega+f(\xi+u,v,\lambda)\\ v(\xi+\omega,\lambda)&=v \qquad\qquad\quad\;\, +g(\xi+u,v,\lambda) \end{aligned}$$

for u, v. Setting

$$u=\sum_{n=1}^{\infty}\lambda^{n}u_{n}(\xi), \quad v=\omega\gamma^{-1}+\sum_{n=1}^{\infty}\lambda^{n}v_{n}(\xi)$$

we obtain for u_n, v_n the equations

$$\begin{aligned} u_n(\xi+\omega)-u_n(\xi)-\gamma v_n(\xi)&=F_n(\xi)\\ v_n(\xi+\omega)-v_n(\xi)\qquad\qquad\;\;&=G_n(\xi) \end{aligned} \qquad (n=1,2,\ldots)$$

with $F_n(\xi) - f_n(\xi, \omega\gamma^{-1})$, $G_n(\xi) - g_n(\xi, \omega\gamma^{-1})$ depending only on f_l, g_l, u_l, v_l for $l < n$, which we may consider as already known. We are looking for solutions u_n, v_n of period 2π, assuming that F_n, G_n have the same period and are real analytic. Clearly such a solution can exist only if the mean value of G_n vanishes, and as we shall show presently, the intersection property implies that it does. Assuming for the time being this to be so, we proceed to solve the above equations by Fourier series expansions. Suppressing the subscript n, we denote the Fourier coefficients of u_n, v_n, F_n, G_n by $\hat{u}_k, \hat{v}_k, \hat{F}_k, \hat{G}_k$ $(k = 0, \pm 1, \pm 2, \ldots)$ and obtain the equations

$$\begin{aligned} e^{ik\omega}\hat{u}_k - \hat{u}_k - \gamma\hat{v}_k &= \hat{F}_k \\ e^{ik\omega}\hat{v}_k - \hat{v}_k \qquad &= \hat{G}_k . \end{aligned}$$

The last equation gives

$$\hat{v}_k = \frac{\hat{G}_k}{e^{ik\omega} - 1} \qquad (k \neq 0)$$

provided that the denominator does not vanish, which will be so if $\omega/2\pi$ is irrational. For $k = 0$ the right hand side $\hat{G}_0$ is the mean value of G_n, which was assumed to be 0, so that no restriction is placed on $\hat{v}_0$ by the second equation, while the first equation requires that we take

$$\hat{v}_0 = -\hat{F}_0 \gamma^{-1} .$$

Finally, solving the first equation for the coefficients of u_n, we get

$$\hat{u}_k = \frac{\gamma\hat{v}_k + \hat{F}_k}{e^{ik\omega} - 1} = \frac{\hat{F}_k}{e^{ik\omega} - 1} + \frac{\gamma\hat{G}_k}{(e^{ik\omega} - 1)^2} \qquad (k \neq 0),$$

with $\hat{u}_0$ remaining arbitrary. The Fourier series u_n, v_n so obtained are real for real values of ξ.

These formal Fourier series will actually converge only if the denominators $e^{ik\omega} - 1$ do not approach zero too rapidly. This will be so if $\omega/2\pi$ is required to satisfy a condition like (25; 22) for α, while F_n, G_n are assumed to be analytic. Indeed, then $|e^{ik\omega} - 1|^{-2}$ grows at most like a power of k, while the Fourier coefficients of analytic functions decay exponentially, so that the Fourier series for u_n, v_n will converge and even be real analytic. This consideration will be made more precise when we come to the actual convergence proof. The above reasoning does, however, suggest that we choose $\omega/2\pi$ irrational, and even so as to satisfy (25; 22).

We now show that the mean value $\frac{1}{2\pi}\int_0^{2\pi} G_n d\xi$ does indeed vanish for all $n > 0$. If not, let n be the smallest index for which $\frac{1}{2\pi}\int_0^{2\pi} G_n d\xi = m$

is different from 0. Replacing g by $g-m\lambda^n$, so that G_n is replaced by G_n-m with mean value 0, we can then determine the truncated series

$$\tilde{u}=\sum_{\nu=1}^{n}\lambda^\nu u_\nu,\quad \tilde{v}=\omega\gamma^{-1}+\sum_{\nu=1}^{n}\lambda^\nu v_\nu$$

so as to satisfy our previous equations for terms of order $\leqq n$. Geometrically this means that the curve

$$x=\xi+\tilde{u}(\xi,\lambda),\quad y=\tilde{v}(\xi,\lambda)$$

is mapped into the curve

$$\begin{aligned} x_1&=\xi+\tilde{u}+\gamma\tilde{v}+f(\xi+\tilde{u},\tilde{v},\lambda)=\xi+\omega+\tilde{u}(\xi+\omega,\lambda)+O(\lambda^{n+1})\\ y_1&=\tilde{v}(\xi,\lambda)+g(\xi+\tilde{u},\tilde{v},\lambda)\qquad=\tilde{v}(\xi+\omega,\lambda)+m\lambda^n+O(\lambda^{n+1}),\end{aligned}$$

and if this and the original curve are to intersect for all small real λ, there must exist parameter values ξ,ξ' such that

$$\begin{aligned} \xi+\tilde{u}(\xi,\lambda)&=\xi'+\tilde{u}(\xi',\lambda)+O(\lambda^{n+1})\\ \tilde{v}(\xi,\lambda)\qquad&=\tilde{v}(\xi',\lambda)+m\lambda^n+O(\lambda^{n+1}).\end{aligned}$$

The first of these equations, however, implies that $\xi'=\xi+O(\lambda^{n+1})$, whereupon the second implies that $m=0$, which contradicts our assumption.

To sum up, if we choose a number $\omega/2\pi$ satisfying a condition (25; 22) and require the intersection property for our mapping, then we obtain a formal power series for the invariant curve. Unfortunately we cannot verify in general the convergence of the series u, v so obtained, and have to proceed differently.

We now turn to the actual proof of the theorem, for which we first take $\gamma=1$. The basic tool here will be the rapidly convergent iteration scheme proposed by Kolmogorov in a related setting, and already used for the simpler problem in § 26. The scheme consists of a succession of coordinate transformations that simplify the given mapping, and in the present case we will try to get closer and closer to the twist mapping $x\to x+y$, $y\to y$ by diminishing the error terms f, g in (3). This will be achieved by an infinite sequence of coordinate changes, with the range of these coordinates at the same time restricted to an annulus that shrinks down to the desired curve.

First, for $0<s_0<\frac{1}{4}$, we choose ω in the interval

$$a+s_0<\omega<b-s_0$$

so as to satisfy the infinitely many inequalities

$$\left|\frac{\omega}{2\pi}q-p\right|\geqq\frac{c_0}{q^\mu}\qquad(p,q=1,2,\ldots).\tag{6}$$

Here c_0 is a positive constant. For $\mu \geqq 2$ the existence of such ω and c_0 was proved at the end of §25. We will fix ω and c_0 from now on.

The basic step in the iteration scheme is contained in the lemma to follow. It will be important, however, to estimate the various quantities in the complex plane, and therefore we consider the mapping (3) in the complex domain

$$\mathfrak{A}: |\mathrm{Im}\, x| < r; \quad |y-\omega| < s, \tag{7}$$

where it will be assumed that

$$|f| + |g| < d.$$

Here r, s, d are positive constants to be restricted presently. Within $\mathfrak{A}$ we will consider a smaller domain

$$\mathfrak{B}: |\mathrm{Im}\, \xi| < \varrho; \quad |\eta - \omega| < \sigma \tag{8}$$

with $0 < \varrho < r, 0 < \sigma < s$, and three intermediate domains

$$\mathfrak{A}^{(\nu)}: \begin{cases} |\mathrm{Im}\, x| < r^{(\nu)} = r - \dfrac{r-\varrho}{4}\nu; \\ |y-\omega| < s^{(\nu)} = s - \dfrac{s-\sigma}{4}\nu \end{cases} \quad (\nu = 1, 2, 3), \tag{9}$$

so that

$$\mathfrak{A} \supset \mathfrak{A}^{(1)} \supset \mathfrak{A}^{(2)} \supset \mathfrak{A}^{(3)} \supset \mathfrak{B}.$$

We will assume that

$$0 < r \leqq 1, \quad 0 < 3\sigma < s < \frac{r-\varrho}{4}, \quad d < \frac{s}{6} \tag{10}$$

and require that

$$\vartheta = c_3(r-\varrho)^{-2(\mu+1)}\frac{d}{s} < \frac{1}{7}, \tag{11}$$

where c_3 is a positive constant, to be determined later, that depends only on c_0 and the exponent μ in (6). In what follows we will denote by $c_1, c_2, \ldots, c_6$ certain positive constants that we do not choose to determine, insisting, however, that they depend on c_0, μ only.

We are now in a position to state our lemma:

Under the above assumptions for the mapping M in (3), with $\gamma = 1$, there exists a coordinate transformation U of the form

$$x = \xi + u(\xi, \eta), \quad y = \eta + v(\xi, \eta) \tag{12}$$

with u, v real analytic functions in $\mathfrak{A}^{(1)}$ of period 2π in ξ, such that the transformed mapping $U^{-1}MU$ takes the form

$$\xi_1 = \xi + \eta + \phi(\xi, \eta), \quad \eta_1 = \eta + \psi(\xi, \eta) \tag{13}$$

with ϕ, ψ real analytic functions defined in $\mathfrak{B}$ where they satisfy the estimate

$$|\phi| + |\psi| < c_6 \left\{ (r-\varrho)^{-\varkappa} \left(\frac{d^2}{s} + sd \right) + \left(\frac{\sigma}{s} \right)^2 d \right\} \tag{14}$$

for $\varkappa = 2\mu + 3$. To be more exact: U maps $\mathfrak{B}$ into $\mathfrak{A}^{(3)}$, M takes $\mathfrak{A}^{(3)}$ into $\mathfrak{A}^{(2)}$, and U^{-1} takes $\mathfrak{A}^{(2)}$ into $\mathfrak{A}^{(1)}$, so that $U^{-1}MU$ is well defined in $\mathfrak{B}$. Moreover, in $\mathfrak{A}^{(1)}$ the functions u, v satisfy the inequality

$$|u| + |v| < \vartheta s \tag{15}$$

for ϑ the quantity defined in (11).

Postponing the proof of the lemma to the next section, we now use it to prove the theorem on the existence of an invariant curve. To this end we make a sequence of successive applications of the lemma, starting with the given mapping (3), now denoted by $M_0 = M$ and restricted to the domain

$$\mathfrak{A}_0 : |\mathrm{Im}\, x| < r_0; \quad |y - \omega| < s_0,$$

which for s_0 sufficiently small is contained in the domain $\mathfrak{D}$ given by (4). By assumption we have there

$$|f| + |g| < \delta,$$

where δ can be chosen sufficiently small, and to conform with the notation of the lemma we set $\delta = d_0$. Transforming the mapping M_0 by the coordinate transformation $U = U_0$ provided by the lemma, we obtain a mapping $M_1 = U_0^{-1} M_0 U_0$ defined in the domain

$$\mathfrak{A}_1 : |\mathrm{Im}\, x| < r_1; \quad |y - \omega| < s_1,$$

where r_1, s_1 correspond to the parameter ϱ, σ of the lemma. Applying the lemma to the new mapping M_1, we obtain another coordinate transformation U_1 and a transformed mapping $M_2 = U_1^{-1} M_1 U_1$, and proceeding in this way we are led to a sequence of mappings

$$M_{n+1} = U_n^{-1} M_n U_n \quad (n = 0, 1, \ldots) \tag{16}$$

whose domains $\mathfrak{A}_{n+1}$ are defined like $\mathfrak{A}$ by (7) with r_{n+1}, s_{n+1} replacing r, s. We have to verify, of course, that this sequence of transformations is well defined, and that M_n approximates the twist mapping with increasing precision. For this we fix the parameters r_n, s_n, d_n $(n = 0, 1, \ldots)$

by setting

$$(17) \qquad r_n = \frac{r_0}{2}\left(1 + \frac{1}{2^n}\right), \quad s_n = d_n^{\frac{2}{3}}, \quad d_{n+1} = r_0^{-\varkappa} c_7^{n+1} d_n^{\frac{4}{3}}$$

with $c_7 \geqq 2$ a suitably chosen constant. Thus r_n is a decreasing sequence converging to the positive value $r_0/2$, and all functions to be considered will be analytic in ξ for $|\mathrm{Im}\,\xi| < r_0/2$. The sequence d_n converges to 0 provided d_0 is chosen sufficiently small. Indeed, the sequence $e_n = r_0^{-3\varkappa} c_7^{3(n+4)} d_n$ satisfies

$$e_{n+1} = e_n^{\frac{4}{3}}$$

and therefore converges to zero if we take $0 \leqq e_0 < 1$, or $0 \leqq d_0 < r_0^{3\varkappa} c_7^{-12}$.

To show that the mapping M_n is well defined in $\mathfrak{A}_n$ and satisfies there the appropriate estimate, we proceed by induction. Assuming M_n to be defined in $\mathfrak{A}_n$ and satisfying there the estimate

$$|f| + |g| < d_n,$$

we will verify the corresponding statement for M_{n+1}. For this we apply the lemma with $r = r_n$, $s = s_n$, $d = d_n$, $\varrho = r_{n+1}$, $\sigma = s_{n+1}$, and therefore have to verify first the inequalities (10), (11). The inequality $3\sigma < s$ follows from

$$(18) \qquad \left(\frac{s_{n+1}}{s_n}\right)^{\frac{3}{2}} = \frac{d_{n+1}}{d_n} = c_7^{-3} e_n^{\frac{1}{3}} \leqq c_7^{-3} \leqq \frac{1}{8},$$

and since d_n goes to 0 faster than exponentially while

$$r - \varrho = r_n - r_{n+1} = r_0 2^{-n-2}$$

decays only exponentially, the whole second inequality in (10) will hold for d_0 sufficiently small. Clearly also the last inequality of (10) can be met by a suitable restriction on d_0, while for $\vartheta = \vartheta_n$ in (11) we have

$$\vartheta_n = c_3 (r_n - r_{n+1})^{-2\mu-2} d_n^{\frac{1}{3}},$$

and this too can be made less than $\frac{1}{7}$ by choosing d_0 small. Thus there exists a positive constant $d^* = d^*(r_0, c_0, \mu)$ such that for $d_0 < d^*$ the inequalities (10), (11) hold and the lemma is applicable. From the lemma we now obtain the transformation U_n taking $\mathfrak{B} = \mathfrak{A}_{n+1}$ into $\mathfrak{A} = \mathfrak{A}_n$ and the transformed mapping $M_{n+1} = U_n^{-1} M_n U_n$ defined in $\mathfrak{A}_{n+1}$. Moreover, representing M_{n+1} in the form (13), by (14) we have the estimate

$$|\phi| + |\psi| < c_6 \left\{ (r_n - r_{n+1})^{-\varkappa} \left(d_n^{\frac{4}{3}} + d_n^{\frac{5}{3}}\right) + \left(\frac{d_{n+1}}{d_n}\right)^{\frac{4}{3}} d_n \right\}$$

$$= c_6 \left\{ (r_n - r_{n+1})^{-\varkappa} d_n^{\frac{4}{3}} \left(1 + d_n^{\frac{1}{3}}\right) + \left(\frac{d_{n+1}}{d_n}\right)^{\frac{1}{3}} d_{n+1} \right\}$$

which in view of (17), (18) gives

$$|\phi| + |\psi| \leqq c_6 \left\{ 2^{\varkappa(n+2)} c_7^{-n-1} (1 + d_n^{\frac{1}{3}}) + \left(\frac{d_{n+1}}{d_n} \right)^{\frac{1}{3}} \right\} d_{n+1}$$

$$\leqq c_6 \{ 2^{\varkappa} (2^{\varkappa} c_7^{-1})^{n+1} (1 + d_n^{\frac{1}{3}}) + c_7^{-1} \} d_{n+1} ,$$

and since the sequence d_n is bounded, the coefficient of d_{n+1} in the last inequality can be made less than 1 by taking c_7 large. With such a choice we finally have

$$|\phi| + |\psi| < d_{n+1} ,$$

and the induction is complete.

Since U_k maps the domain $\mathfrak{A}_{k+1}$ into $\mathfrak{A}_k$ ($k = 0, 1, 2, \ldots$), the transformation $V_n = U_0 U_1 \ldots U_n$ is well defined in $\mathfrak{A}_{n+1}$ and is seen to take M_0 into

$$M_{n+1} = V_n^{-1} M_0 V_n .$$

Moreover, if we express V_n in the form

$$x = \xi + p_n(\xi, \eta) , \quad y = \eta + q_n(\xi, \eta) ,$$

then p_n, q_n are analytic functions in the domain $\mathfrak{A}_{n+1}$ which, in view of $s_n \to 0$, $r_n \to r_0/2$, shrinks down to the domain

$$|\mathrm{Im}\, \xi| < \frac{r_0}{2} , \quad |\eta - \omega| = 0 .$$

We will show that as $n \to \infty$ the sequences $p_n(\xi, \omega)$, $q_n(\xi, \omega)$ converge to analytic functions of ξ for $|\mathrm{Im}\, \xi| < r_0/2$. Indeed, from $V_n = V_{n-1} U_n$ it follows that

$$p_n = u_n + p_{n-1}(\xi + u_n, \eta + v_n)$$
$$q_n = v_n + q_{n-1}(\xi + u_n, \eta + v_n) ,$$

where u_n, v_n correspond to U_n as u, v to U in (12), and consequently

$$p_n = u_n + u_{n-1} + \cdots + u_0$$
$$q_n = v_n + v_{n-1} + \cdots + v_0 .$$

In the last two sums we have suppressed the different arguments, but this is irrelevant for the proof of convergence since we can estimate $|u_n|$, $|v_n|$ by their supremum over $\mathfrak{A}_{n+1}$. Namely, from (15) it follows that

$$|u_n| + |v_n| < \vartheta_n s_n < \tfrac{1}{7} s_n ,$$

and this implies uniform convergence of p_n, q_n in $|\mathrm{Im}\, \xi| < r_0/2$ for $\eta = \omega$. We denote the limit of p_n by $u(\xi)$ and of q_n by $v(\xi) - \omega$, which then are real analytic functions in $|\mathrm{Im}\,\xi|<r_0/2$ and of period 2π. Also, from $s_{n+1}<\frac{1}{3}s_n$ we have

$$|p_n|+|q_n|<\frac{1}{7}\sum_{\nu=0}^{n}s_\nu<\frac{1}{7}\sum_{\nu=0}^{\infty}\frac{1}{3^\nu}s_0<s_0,$$

and we choose $d_0=s_0^{\frac{3}{2}}$ so small that $s_0<\varepsilon$. Thus there exists a positive $\delta=\delta(\varepsilon,r_0,c_0,\omega)<d^*$ such that for $d_0<\delta$ we have

$$|u|+|v-\omega|<\varepsilon$$

in $|\mathrm{Im}\,\xi|<\dfrac{r_0}{2}$, as asserted in the theorem, and this δ can be used as the constant restricting the size of $|f|+|g|$.

To verify that $u(\xi)$, $v(\xi)$ are the desired functions describing the invariant curve, we consider the relation

$$V_nM_{n+1}=M_0V_n$$

in the limit as $n\to\infty$. Since $d_{n+1}\to 0$ we conclude that M_{n+1} tends to the twist mapping $\xi_1=\xi+\eta$, $\eta_1=\eta=\omega$, while the convergence of V_n we just discussed. Consequently in the limit we have the relations

$$\begin{aligned}\xi+\omega+u(\xi+\omega)&=\xi+u(\xi)+v(\xi)+f(\xi+u,v)\\ v(\xi+\omega)&=v(\xi)+g(\xi+u,v),\end{aligned}$$

expressing precisely that the curve (5) is invariant under M_0, and that the induced mapping on it is given by $\xi_1=\xi+\eta=\xi+\omega$. This concludes the proof of our theorem, for $\gamma=1$, on the existence of an invariant curve-apart from the lemma, which will be proved in the next section.

§ 33. Proof of the Lemma

To prove the lemma used in the previous section we have to construct the transformation U in (32;12) and carry out the required estimates. The choice of U is motivated by the desire to transform the mapping M in (32;3) into the twist mapping $\xi_1=\xi+\eta$, $\eta_1=\eta$, and as one readily verifies, this would require that the functions u,v in (32;12) satisfy the functional equations

$$\begin{aligned}u(\xi+\eta,\eta)&=u+v+f(\xi+u,\eta+v)\\ v(\xi+\eta,\eta)&=v+g(\xi+u,\eta+v).\end{aligned}$$

These equations, however, are nonlinear and not directly solvable, and we replace them by the linear equations

$$(1) \qquad \begin{cases} u(\xi+\omega,\eta)-u(\xi,\eta)-v(\xi,\eta)=f(\xi,\eta) \\ \qquad\qquad v(\xi+\omega,\eta)-v(\xi,\eta)=g(\xi,\eta)-g^*(\eta) \end{cases}$$

which we will solve presently and use for the definition of U. Here g^* denotes the mean value $(2\pi)^{-1}\int_0^{2\pi} g(\xi,\eta)\,d\xi$, which we subtracted from the right side of the second equation in order to be able to solve it. Of course, now $U^{-1}MU$ will no longer be the twist mapping, but the approximation will be good enough to yield the estimate (32;14).

The study of (1) immediately reduces to that of the difference equation

$$(2) \qquad w(x+\omega)-w(x)=h(x),$$

where w, h are assumed to be analytic in $|\operatorname{Im} x|<r$ and of period 2π in x. Obviously, this equation admits a solution w only if the mean value h^* of h vanishes. Under these assumptions, we solve (2) by means of a Fourier expansion. Setting

$$h=\sum_{k\neq 0} h_k e^{ikx}, \qquad w=\sum_{k\neq 0} w_k e^{ikx},$$

we obtain from (2) the relation

$$w_k=\frac{h_k}{e^{ik\omega}-1},$$

and have only to verify convergence of the series for w. Since ω satisfies (32;6), the denominators $e^{ik\omega}-1$ do not vanish for $k\neq 0$ and, indeed, can be estimated by

$$(3) \qquad |e^{ik\omega}-1|=2\left|\sin\frac{k\omega}{2}\right|\geqq 4c_0|k|^{-\mu}.$$

On the other hand, since h is real analytic, the Fourier coefficients h_k of h decay exponentially. Indeed, we have

$$h_k=\frac{1}{2\pi}\int h(x)\,e^{-ikx}\,dx,$$

where the integral is taken over a path $\operatorname{Im} x=r'$, $0\leqq \operatorname{Re} x\leqq 2\pi$, which in view of Cauchy's theorem is independent of r' for $|r'|<r$. If $|h|<K$ in $|\operatorname{Im} x|<r$, this gives

$$|h_k|\leqq K e^{kr'},$$

and letting r' approach $\pm r$ we obtain

(4) $$|h_k| \leqq K e^{-|k|r}.$$

From (3), (4) convergence of the Fourier series for w in $|\mathrm{Im}\, x| < r$ becomes evident. Moreover, in a narrower strip $|\mathrm{Im}\, x| < \varrho$ $(0 < \varrho < r)$ we have

$$|w| \leqq \sum_{k \neq 0} \left| \frac{h_k}{e^{ik\omega} - 1} e^{ikx} \right| \leqq \frac{K}{c_0} \sum_{k=1}^{\infty} k^{\mu} e^{-k(r-\varrho)},$$

which leads to the estimate

$$|w| \leqq c_1 K (r-\varrho)^{-\mu-1} \qquad (|\mathrm{Im}\, x| < \varrho).$$

The solution we constructed was normalized to have mean value zero, and it is apparent that (2) actually determines w up to an additive constant. This is true even in the class of continuous functions. Indeed, $w(x+\omega) - w(x) = 0$ implies $w(k\omega) = w(0)$ for all integers k, and since the numbers $k\omega$ are dense modulo 2π, we conclude that a continuous solution to the homogeneous system is constant. We denote the normalized solution of (2) with mean value zero by $w = Lh$, and express the last estimate in the form

(5) $$|Lh| \leqq c_1 K (r-\varrho)^{-\mu-1} \qquad (|\mathrm{Im}\, x| < \varrho,\ 0 < \varrho < r).$$

These considerations can be extended to the difference equation

$$w(x+\omega, y) - w(x, y) = h(x, y) - h^*(y)$$

in which y enters as a parameter. Again we have a unique solution w, denoted by Lh, with mean value zero, and the estimate carries over in an obvious way.

To solve the equations (1) we begin with the second one, whose solution is given by

$$v(\xi, \eta) = v^*(\eta) + Lg$$

with the mean value $v^*(\eta)$ arbitrary. Adjusting v^* to meet the compatibility requirement for the first equation, we set $-v^* = f^*$, so that $v = -f^* + Lg$, and obtain

$$u = L(v+f) = L^2 g + Lf$$

as a solution to the first equation. Thus a solution of (1) is given by

(6) $$u = Lf + L^2 g, \qquad v = -f^* + Lg,$$

and we use these functions to define the transformation U in (32;12). Note that u, v are real analytic and of period 2π, and it remains to verify the required estimate (32;15) for u, v as well as (32;14) for the corresponding ϕ, ψ.

Retaining the assumptions and notation of the previous section, we have $|f|+|g|<d$ in $\mathfrak{A}$ so that by successively applying (5) with $r-\frac{r-\varrho}{16}$ and $r-\frac{r-\varrho}{8}$ in place of ϱ we obtain from (6) the estimate

$$|u|+|v|<c_2(r-\varrho)^{-2\mu-2}d \qquad \left(|\operatorname{Im}\xi|<r-\frac{r-\varrho}{8},\ |\eta-\omega|<s\right)$$

with c_2 depending only on c_0, μ. Using Cauchy's estimate on the derivatives of u, v in the domain $\mathfrak{A}^{(1)}$ defined in (32;9), we find that

$$|u_\xi|+|v_\xi|<c_3(r-\varrho)^{-2\mu-3}d$$
$$|u_\eta|+|v_\eta|<c_3(r-\varrho)^{-2\mu-2}\frac{d}{s},$$

where $c_3>c_2$; we also used that $3\sigma<s$. With this constant c_3 we define

$$\vartheta=c_3(r-\varrho)^{-2\mu-2}\frac{d}{s}$$

as in (32;11) and rewrite these inequalities in the form

$$(7)\qquad \begin{cases} |u|+|v|<\vartheta s \\ |u_\xi|+|v_\xi|<\vartheta\dfrac{s}{r-\varrho}<\vartheta, \quad |u_\eta|+|v_\eta|<\vartheta \end{cases}$$

valid in $\mathfrak{A}^{(1)}$, whereby we have made use of (32;10).

The first inequality in (7) verifies (32;15), from which it also follows that U maps $\mathfrak{B}$ into $\mathfrak{A}^{(3)}$. Indeed, for $|\operatorname{Im}\xi|<\varrho$, $|\eta-\omega|<\sigma$ we have for the image (x, y) of the point (ξ, η) the estimate

$$|\operatorname{Im}x|<\varrho+\vartheta s, \quad |y-\omega|<\sigma+\vartheta s,$$

and to show that (x, y) lies in $\mathfrak{A}^{(3)}$ one only has to verify that

$$\vartheta s<\frac{r-\varrho}{4}, \quad \vartheta s<\frac{s-\sigma}{4}.$$

This, however, is an immediate consequence of (32;10) provided that $\vartheta<\frac{1}{6}$. Similarly, to check that M maps $\mathfrak{A}^{(3)}$ into $\mathfrak{A}^{(2)}$ we use (32;3), (32;9) and find that

$$|\operatorname{Im}x_1|<r^{(3)}+|\operatorname{Im}y|+d<r^{(3)}+s^{(3)}+d$$
$$|y_1-\omega|<s^{(3)}+d,$$

so that we only have to verify the inequalities

$$s^{(3)}+d<\frac{r-\varrho}{4}, \quad d<\frac{s-\sigma}{4}.$$

The second one is an immediate consequence of (32;10), while from this and again (32;10) we get

$$s^{(3)}+d=s-\frac{3(s-\sigma)}{4}+d<s<\frac{r-\varrho}{4},$$

as required.

Finally, we show that U^{-1} is defined in the domain $\mathfrak{A}^{(2)}$ and maps it into $\mathfrak{A}^{(1)}$. In other words, we assume that $(x, y) \in \mathfrak{A}^{(2)}$ and have to construct a solution (ξ, η) of the equation (32;12) in $\mathfrak{A}^{(1)}$. For this purpose we use the usual iteration scheme and define ξ_k, η_k inductively by $\xi_0 = x$, $\eta_0 = y$ and

$$\xi_{k+1}=x-u(\xi_k,\eta_k), \quad \eta_{k+1}=y-v(\xi_k,\eta_k) \quad (k=0,1,\ldots).$$

We must show that the (ξ_k, η_k) remain in $\mathfrak{A}^{(1)}$. For $k=0$ this is obviously the case, and assuming this to be true for (ξ_ν, η_ν) with $\nu \leqq k$, we find from (7) that

$$|\xi_{k+1}-\xi_k|+|\eta_{k+1}-\eta_k|<\vartheta(|\xi_k-\xi_{k-1}|+|\eta_k-\eta_{k-1}|) \quad (k=1,2,\ldots)$$

and hence

$$|\xi_{k+1}-x|+|\eta_{k+1}-y| \leqq \sum_{n=0}^{k}(|\xi_{n+1}-\xi_n|+|\eta_{n+1}-\eta_n|)$$
$$\leqq \frac{1}{1-\vartheta}(|u|+|v|)<\frac{\vartheta}{1-\vartheta}s.$$

Since we assumed $\vartheta<\frac{1}{7}$, the last quantity is $<\frac{s}{6}$ and, as before, we verify that

$$\frac{s}{6}<\frac{r-\varrho}{4}, \quad \frac{s-\sigma}{4},$$

which guarantees that $(\xi_{k+1}, \eta_{k+1}) \in \mathfrak{A}^{(1)}$. Thus all the iterates (ξ_k, η_k) remain in $\mathfrak{A}^{(1)}$ and, since $\vartheta<1$, converge to a solution (ξ, η) in $\mathfrak{A}^{(1)}$. This solution is clearly unique, and U^{-1} maps $\mathfrak{A}^{(2)}$ into $\mathfrak{A}^{(1)}$ as asserted.

Now we come to the main part of the lemma – namely, the verification of the estimate (32;14). It is here that the intersection property of M comes into play, being used to show that the error term g^* introduced in (1) is small enough to cause no harm.

We recall that the mapping (32;13), which we denote symbolically by N, is given by $N = U^{-1}MU$, and the equation $UN = MU$ takes the form

$$\xi + \eta + \phi + u_1 = \xi + u + \eta + v + f(\xi + u, \eta + v)$$
$$\eta + \psi + v_1 = \eta + v + g(\xi + u, \eta + v)$$

with $u_1 = u(\xi + \eta + \phi, \eta + \psi)$, $v_1 = v(\xi + \eta + \phi, \eta + \psi)$. Simplifying this to the equations

$$\phi = u - u_1 + v + f(\xi + u, \eta + v)$$
$$\psi = v - v_1 + g(\xi + u, \eta + v),$$

which serve to define ϕ, ψ implicitly in $\mathfrak{B}$, and taking into account (1), which defines u, v, we obtain

$$(8) \qquad \begin{cases} \phi = u(\xi + \omega, \eta) - u_1 + f(\xi + u, \eta + v) - f(\xi, \eta) \\ \psi = v(\xi + \omega, \eta) - v_1 + g(\xi + u, \eta + v) - g(\xi, \eta) + g^*(\eta), \end{cases}$$

on which we base our estimates. The variables (ξ, η) are, of course, assumed to lie in $\mathfrak{B}$ and therefore satisfy (32;8).

The contribution from the functions u, v on the right hand side of (8) can be estimated using the mean value theorem, yielding

$$|u(\xi + \omega, \eta) - u_1| \leqq \sup |u_\xi| \, (|\eta - \omega| + |\phi|) + \sup |u_\eta| \, |\psi|$$
$$< \vartheta \frac{s}{r - \varrho} |\eta - \omega| + \vartheta(|\phi| + |\psi|),$$

where the suprema of the partial derivatives of u, v are taken over $\mathfrak{A}^{(1)}$, and the final inequality follows from (7). The same final estimate is obtained also for the corresponding contribution from v. Recalling that $|f| + |g| < d$ in $\mathfrak{A}$, we can use Cauchy's estimate to bound the derivatives of f, g by $2\frac{d}{s}$ in $\mathfrak{A}^{(3)}$, so that, again applying the mean value theorem followed by (7), we obtain

$$|f(\xi + u, \eta + v) - f(\xi, \eta)| < 2\frac{d}{s}(|u| + |v|) < 2\vartheta d,$$

with the same final estimate for the corresponding contribution from g. The troublesome mean value $g^*(\eta)$ will be approximated by the linear function

$$h(\eta) = g^*(\omega) + g^*_\eta(\omega)(\eta - \omega),$$

which we will estimate later using the intersection property. From (8) and the previous estimates we now have

$$|\phi|+|\psi-h|<2\vartheta(|\phi|+|\psi-h|)+2\vartheta|h|$$
$$+2\vartheta\frac{s}{r-\varrho}|\eta-\omega|+4\vartheta d+|g^*-h|\,.$$

Since $2\vartheta<\frac{1}{3}<1$ we can eliminate $|\phi|+|\psi-h|$ from the right hand side, and recalling that $|\eta-\omega|<\sigma<s$, we can express this in the form

$$|\phi|+|\psi-h|<c_4\left\{\vartheta\frac{s}{r-\varrho}s+\vartheta|h|+\vartheta d+|g^*-h|\right\}.$$

A preliminary estimate of $h(\eta)$ and $|g^*-h|$ is obtained by observing that for $|\eta-\omega|<s$ we have $|g^*(\eta)|<d$ and therefore by Cauchy's estimate $|g^*_\eta(\omega)|<d/s$, while for $|\eta-\omega|<\sigma$ also $|g^*_{\eta\eta}|<2d/(s-\sigma)^2$. Consequently, for $|\eta-\omega|<\sigma$, we have

$$|h(\eta)|<d+\frac{d}{s}\sigma<2d$$

and

$$|g^*-h|\leqq\frac{\sigma^2}{2}\sup|g^*_{\eta\eta}|<\left(\frac{\sigma}{s-\sigma}\right)^2 d<3\left(\frac{\sigma}{s}\right)^2 d\,,$$

where we have used that $3\sigma<s$. Combining these with the previous estimates, we now have

$$\begin{aligned}|\phi|+|\psi-h|&<c_4\left\{\vartheta\frac{s}{r-\varrho}s+2\vartheta d+\vartheta d+3\left(\frac{\sigma}{s}\right)^2 d\right\}\\&=c_4\left\{\frac{\vartheta}{r-\varrho}(s^2+3(r-\varrho)\,d)+3\left(\frac{\sigma}{s}\right)^2 d\right\}\\&\leqq c_5\left\{\frac{\vartheta}{r-\varrho}(s^2+d)+\left(\frac{\sigma}{s}\right)^2 d\right\},\end{aligned}$$

and with ϑ given by (32;11), this becomes

$$|\phi|+|\psi-h|<c_5\left\{c_3(r-\varrho)^{-2\mu-3}\left(sd+\frac{d^2}{s}\right)+\left(\frac{\sigma}{s}\right)^2 d\right\}=Q\,, \tag{9}$$

where the right hand side of this inequality also serves to define Q.

The preliminary estimate of $2d$ for $|h|$, however, is insufficient for decreasing the error term, and to obtain a better estimate we use the intersection property of M, or that of $N=U^{-1}MU$. Accordingly, each curve $\eta=$ constant, in particular, has to intersect its image curve under N, and at such a point of intersection we have $\eta_1=\eta$ or $\psi=0$, so that for each real η in $|\eta-\omega|<\sigma$ there exists a real $\xi=\xi_0(\eta)$ such that $\psi(\xi_0(\eta),\eta)=0$. Applying (9) at such points $(\xi_0(\eta),\eta)$, we find that

$$|h(\eta)|<Q \qquad (\omega-\sigma<\eta<\omega+\sigma).$$

Consequently, setting $\eta=\omega$ in the definition of h, we get $|g^*(\omega)|<Q$, and letting η approach $\omega+\sigma$ in the same definition, we obtain

$$|g^*(\omega)+g^*_\eta(\omega)\,\sigma|\leqq Q,$$

so that

$$|g^*_\eta(\omega)\,\sigma|<2Q.$$

From this we conclude that for complex η in the disk $|\eta-\omega|<\sigma$ we have

$$|h(\eta)|\leqq|g^*(\omega)|+|g^*_\eta(\omega)|\,|\eta-\omega|<3Q,$$

which in view of (9) gives

$$|\phi|+|\psi|<4Q.$$

This yields the desired estimate (32;14) with $c_6=4c_5(c_3+1)$ and completes the proof of the lemma, and therefore also of the theorem on the existence of an invariant curve for the case $\gamma=1$.

The proof of the theorem for the general case $0<\gamma\leqq 1$ follows exactly the same lines, except that we have to see how γ enters into the various estimates. First, to find an ω in the interval

$$\Delta: a\gamma<\omega<b\gamma$$

of length γ one has to modify inequality (32;6). If, for example, this interval contains an integer p, the inequality excludes all ω at a distance less than $2\pi c_0$ from $2\pi p$; consequently, if in addition $\gamma<2\pi c_0$, the whole interval Δ is excluded. We therefore replace (32;6) by

$$\left|\frac{\omega}{2\pi}q-p\right|\geqq\gamma\frac{c_0}{q^\mu}, \tag{10}$$

where c_0 is a positive constant independent of γ.

To show that for $\mu>1$ and c_0 sufficiently small there exists in any interval of length γ a number ω satisfying (10), consider the complementary set Σ of those ω in Δ that violate (10) for at least one pair of integers p,q with $q\geqq 1$. To estimate the Lebesgue measure of Σ, we fix q and consider all p for which the interval

$$\left|\frac{\omega}{2\pi}-\frac{p}{q}\right|<\gamma\frac{c_0}{q^{\mu+1}}$$

intersects Δ. Since the length of Δ is $\gamma\leqq 1$, it is clear that for c_0 sufficiently small there will be at most $q+1$ such integers p, so that the measure of Σ

can be estimated by

$$m(\Sigma) < \sum_{q=1}^{\infty} (q+1)\, 2\gamma \frac{c_0}{q^{\mu+1}} < 4\gamma c_0 \sum_{q=1}^{\infty} \frac{1}{q^{\mu}},$$

which can be made less than $\gamma = m(\Delta)$ by taking c_0 sufficiently small. Hence the set $\Delta - \Sigma$ is not empty, and there exists an ω in Δ satisfying (10).

Having replaced (32;6) by (10), one now has to replace (5) by

$$|Lh| \leqq \gamma^{-1} c_1 K (r-\varrho)^{-\mu-1}$$

and (6) by

$$u = Lf + \gamma L^2 g$$
$$v = -\gamma^{-1} f^* + Lg\,.$$

However, if we introduce $f_0 = \gamma^{-1} f$, $g_0 = \gamma^{-1} g$, and $L_0 = \gamma L$, this relation can be expressed in the form

$$u = L_0 f_0 + L_0^2 g_0$$
$$v = -f_0^* + L_0 g_0$$

which does not contain γ. In fact, the entire proof, with all the estimates, carries over if we replace f, g by $\gamma^{-1} f, \gamma^{-1} g$ and similarly ϕ, ψ by $\gamma^{-1}\phi$, $\gamma^{-1}\psi$. The condition $|f| + |g| < \delta$ for $\gamma = 1$ then, of course, has to be replaced by $|f| + |g| < \gamma\delta$ for $0 < \gamma \leqq 1$. With this the proof readily extends to cover the general case.

In conclusion, we observe that if the mapping (32;3) depends continuously on a real parameter λ, i.e. if $f = f(x, y, \lambda)$, $g = g(x, y, \lambda)$, $\gamma = \gamma(\lambda) \neq 0$ are continuous functions of λ for, say, $|\lambda| \leqq 1$, then also the invariant curve depends continuously on λ. This follows simply from the fact that our approximations will then converge uniformly with respect to λ. This observation will prove useful when we apply the existence theorem in the ensuing sections to questions relating to stability. The proof given in this and the preceding section is an adaptation to the analytic case of that in [1], where the analogous theorem is proved for mappings with only finitely many derivatives.

§ 34. Application to the Stability Problem

As we indicated already in § 32, the theorem on the existence of invariant curves can be applied to the problem of stability of an elliptic fixed-point, to which we now turn. We consider an area-preserving mapping near a fixed-point of the general elliptic type, which in suitable coordinates can be expressed in the form

$$\text{(1)} \qquad \begin{cases} u_1 = u\cos w - v\sin w + O_{2l+2} \\ v_1 = u\sin w + v\cos w + O_{2l+2}\,, \end{cases}$$

where

$$w = \gamma_0 + \gamma_l (u^2 + v^2)^l, \quad \gamma_l > 0 \quad (l > 0),$$

and O_{2l+2} stands for a power series in u, v containing terms of order $\geqq 2l+2$ only. We will show that for each sufficiently small $\varepsilon > 0$ the punctured disk

$$0 < u^2 + v^2 < \varepsilon^2$$

contains an invariant curve surrounding the fixed-point $u = v = 0$. For this purpose we introduce polar coordinates x, y by

$$u = \varepsilon y^{\frac{1}{2l}} \cos x, \quad v = \varepsilon y^{\frac{1}{2l}} \sin x$$

so that the mapping takes the form

$$\begin{aligned} x_1 &= x + \gamma_0 + \gamma_l \varepsilon^{2l} y + O(\varepsilon^{2l+1}), \\ y_1 &= y + O(\varepsilon^{2l+1}), \end{aligned}$$

as one easily verifies. Here the error terms are real analytic functions in x, y for $0 < y < 1$ and have period 2π in x. Restricting y to a closed subinterval, we can now apply the theorem of § 32 with $\gamma = \gamma_l \varepsilon^{2l}$ and y replaced by $y + \gamma_0 \gamma^{-1}$. The corresponding error terms are estimated by

$$\frac{|f| + |g|}{\gamma} = \frac{O(\varepsilon^{2l+1})}{\gamma_l \varepsilon^{2l}} = O(\varepsilon)$$

and can be made arbitrarily small by choosing ε small, even with the variables extended to a suitable complex domain containing the real annulus in question. Finally, the intersection property follows from the area-preserving character of the mapping, since if a closed curve $y = \psi(x)$ for x, y real did not intersect its image curve, the two regions enclosed by these curves would not have the same area. From the existence theorem we conclude that for each sufficiently small $\varepsilon > 0$ the punctured disk $0 < u^2 + v^2 < \varepsilon^2$ contains an invariant curve Γ surrounding the fixed-point $u = v = 0$, and this in turn implies stability of the mapping (1) at this point.

The above argument shows the existence of a sequence of invariant curves converging to the fixed-point. Without much more effort one can show that there are uncountably many such curves in the vicinity of the fixed-point. Indeed, the construction in the previous two sections led to an invariant curve for each ω satisfying the inequalities (33;10), and we observe that to each such curve the number ω can be associated in a unique way as

$$\lim_{k \to \infty} \frac{x_k}{k} = \lim_{k \to \infty} \frac{\xi + k\omega + u(\xi + k\omega)}{k} = \omega.$$

Here x_k is the angular coordinate of the k-th iterate M^kP of any point P on the curve, while $u(\xi)$ is the function appearing in (32;5). Moreover no two of these curves can intersect, for otherwise the iterates M^kP of a point P in the intersection would lie densely on both curves, and the curves would be one and the same. Thus, curves corresponding to different values of ω are disjoint, and since the set of admissible values for ω is a Cantor set of positive measure, it is certainly uncountable. Actually one can show that, as point sets in the plane, these curves form a set of positive measure whose complement in $u^2+v^2<r^2$ has measure $o(\pi r^2)$. Consequently, we can say that the majority of points near a fixed-point belong to the set of invariant curves.

To visualize the situation geometrically, let us consider the mapping (1) without the error term O_{2l+2}. This leaves invariant each of the concentric circles $u^2+v^2=\text{constant}$, and if we admit only those circles for which the angle $w=\gamma_0+\gamma_l(u^2+v^2)^l$ satisfies the conditions

$$\left|\frac{w}{2\pi}q-p\right| \geqq \gamma_l(u^2+v^2)^l\, c_0 q^{-\mu}$$

for all integers p, q with $q \geqq 1$, we obtain a Cantor set $\mathfrak{B}$ of circles which is in one-to-one correspondence with our set of invariant curves of the full mapping (1). In other words, if we multiply the error terms in (1) by a small parameter τ then the circles belonging to the Cantor set $\mathfrak{B}$ can be continued as invariant curves belonging to the same rotation number for small values of τ. In the regions outside our Cantor set of invariant curves, however, the behavior is quite different, and invariance properties there disintegrate under small perturbations of the system corresponding to $\tau=0$. Indeed, as we shall see, for the perturbed system these regions cannot in general be covered by a one-parameter family of invariant curves.

To see what happens in this excluded set, let us look at those circles for which $w/2\pi$ is rational, say p/q. For $\tau=0$ these are circles consisting of fixed-points of the q-th iterate of our mapping. Under small perturbation, however, such a curve of fixed-points will, in general, break up into a finite set of fixed-points, and we may say that these circles disintegrate under perturbation. We shall illustrate this situation with an example below. Some of these fixed-points, whose existence follows from Birkhoff's theorem of § 24, may be of general elliptic type, in which case they in turn possess a neighborhood covered largely by invariant curves belonging to different fixed-points. This leads to a hierarchy of fixed-points and invariant curves, and the geometrical distribution of the invariant curves near an elliptic fixed-point clearly is quite intricate.

The regions outside the Cantor set will, in general, contain also hyperbolic fixed-points, which give rise to a very complicated motion in the large. These so-called regions of instability may conceivably contain open sets in which the images of the iterates of a single point are dense. However, not much is known about the intricate behavior in these regions.

It may seem surprising that the invariant curves can be obtained by a convergent iteration scheme, while the transformation into normal form generally diverges – especially since the construction of the invariant curves is also based on a transformation technique. The answer to this apparent paradox is simply that in the study of normal forms we expanded our power series about the fixed-point and asked for convergence in a whole neighborhood of it, while our construction of an invariant curve can be viewed as an expansion near the corresponding unperturbed curve.

We will now give a simple example which illustrates some of the previous points, and for which the transformation into normal form actually diverges while the invariant curves nevertheless exist. For M we take the simple polynomial mapping

$$x_1 = (x + y^3)\cos\alpha - y\sin\alpha$$
$$y_1 = (x + y^3)\sin\alpha + y\cos\alpha\,,$$

which for $\sin 2\alpha \neq 0$ is readily seen to transform into (1) with $l = 1$, $\gamma_0 = \alpha$, $\gamma_1 = -\frac{3}{8}$. Thus by our earlier result there are infinitely many invariant curves surrounding the origin, which is therefore a stable fixed-point of the mapping. On the other hand, for α incommensurable with 2π the transformation into normal form is divergent, as shown by the following argument. Assuming that the transformation converges, there exists a neighborhood of the origin covered by a one-parameter family of invariant curves, namely the images under the coordinate transformation of the concentric circles $\xi^2 + \eta^2 = \text{constant}$. Moreover, if P_k is a fixed-point of the iterate M^k for some $k \geqq 1$ then the entire curve through P_k consists of fixed-points of M^k, as is clear from the nature of the normal form, and since the rotation angle $\gamma_0 + \gamma_1(\xi^2 + \eta^2) + \cdots$ is not constant, such fixed-points do exist in any neighborhood of the origin. Thus, if the transformation into normal form were to converge, there would exist infinitely many k for which M^k has a continuum of fixed-points, and we may take k to be even. We will show, however, that the number of fixed-points of M^k is finite, in fact at most 3^k, and this contradiction will prove our contention.

For this we observe that the inverse M^{-1} of our mapping is again a polynomial mapping, a fact that is true for any area-preserving polynomial mapping. For any integer k, we denote the image of the point (x, y) under the mapping M^k by (x_k, y_k), so that a fixed-point of M^k satisfies $x_k = x$, $y_k = y$. If $k = 2q \geqq 2$ is even we can replace these two equations by the equivalent ones $x_q - x_{-q} = 0$, $y_q - y_{-q} = 0$. We now apply a theorem of Bézout stating that two polynomials in x, y have only finitely many roots in common, at most equal to the product of their degrees, unless they have a common factor. One readily verifies that for our example the polynomials $x_q - x_{-q}$, $y_q - y_{-q}$ have degree 3^q, with their respective leading terms being

$$a_q\{\cos\alpha\, y^{3^q} - (x\sin\alpha - y\cos\alpha)^{3^q}\}, \quad a_q \sin\alpha\, y^{3^q}$$

where

$$a_q = (\sin\alpha)^\beta \neq 0, \quad \beta = 3 + 3^2 + \cdots + 3^{q-1}.$$

Thus they do not have a common factor, and we conclude that the mapping M^{2q} has at most $3^{2q} = 3^k$ fixed-points.

This example shows clearly that the question of convergence does not depend on the number-theoretic properties of the eigenvalues, but rather on the nature of the nonlinear terms. This is in contrast to the situation for the conformal mapping discussed in § 25, where stability as well as convergence of the transformation into normal form was completely determined by the linear part of the mapping.

The theorem developed in the previous two sections has a number of applications to Hamiltonian systems of two degrees of freedom – in particular, to the question of stability of periodic solutions. As we saw before, the problem of isoenergetic stability of such a periodic orbit can be reduced to the question of whether or not a fixed-point of a certain two dimensional area-preserving mapping is stable. As an application, we return once more to the much discussed restricted three-body problem.

In § 21 we constructed for small values of the mass parameter μ a family of solutions to the restricted three-body problem which for $\mu = 0$ represented the circular orbits

$$x_1 = r\cos(\omega t), \quad x_2 = r\sin(\omega t), \quad r^3(\omega + 1)^2 = 1.$$

This family of closed orbits is parametrized by the frequency ω, which is related to the energy. One readily verifies that on any such energy surface there exist exactly two orbits of this kind, one with $\omega + 1 > 0$ and the other with $\omega + 1 < 0$. We first wish to investigate the isoenergetic stability of these solutions for small values of μ, and for this we have to look at the associated area-preserving mapping which we already studied at the end of § 24 when applying the Birkhoff fixed-point theorem to the problem. It is crucial here that the stability conditions are expressed in terms of finitely many inequalities

$$\lambda^k \neq 1 \ (k = 1, \ldots, 2l + 2), \quad \gamma_l \neq 0$$

for some $l \geqq 1$. Since $\lambda, \gamma_1, \gamma_2, \ldots$ are continuous and actually analytic functions of μ if $\lambda^2 \neq 1$, it is enough to verify the conditions for $\mu = 0$. For that case we found under assumption (24; 31) that

$$l = 1, \quad \gamma_0 = \frac{2\pi}{\omega}, \quad \gamma_1 = -\frac{3\pi}{r^2 \omega^3},$$

and therefore if we assume

$$\omega \neq \frac{3}{g}, \frac{4}{g}, 0 \quad (g = \pm 1, \pm 2, \ldots) \tag{2}$$

we have isoenergetic stability of these solutions for $|\mu| \leqq \mu_0(\omega)$. Actually, under the same conditions we even have unrestricted stability, as we will now show.

If we follow the construction of the area-preserving mapping near a periodic orbit but retain the energy as an independent variable which we call w, we are led to a mapping of the form

$$u_1 = F(u, v, w, \mu), \quad v_1 = G(u, v, w, \mu), \quad w_1 = w,$$

where the last equation simply expresses the conservation of energy. The points on the line $u = v = 0$ are fixed-points whose stability under this mapping implies orbital stability of the periodic solutions they represent. To show stability of a fixed-point $p_0 = (0, 0, w_0)$ we construct in any preassigned neighborhood $\mathfrak{U}$ of p_0 an invariant neighborhood $\mathfrak{B}$ of p_0 as follows. In $\mathfrak{U}$ we can find an invariant curve which we write in the form

$$u^2 + v^2 = R(\theta), \quad w = w_0,$$

where $\theta = \arctan v/u$. Since our mapping depends analytically, hence continuously, on the parameter w, we can actually find a family of such curves

$$u^2 + v^2 = R(\theta, w)$$

that depend continuously on w and therefore remain in $\mathfrak{U}$ if $|w - w_0| < \delta$ for a sufficiently small positive δ. But then

$$u^2 + v^2 < R(\theta, w), \quad |w - w_0| < \delta$$

defines an invariant neighborhood $\mathfrak{B}$ of p_0 contained in $\mathfrak{U}$, and thus p_0 is a stable fixed-point. This shows that under the conditions (2) our periodic solutions to the restricted three-body problem are stable for sufficiently small values of the mass parameter μ.

As we mentioned in § 31, for $\omega = 3/g$, with g an integer not divisible by 3, Levi-Civita has shown that the corresponding orbits are indeed not stable. This leaves the cases $\omega = 4/g$ undecided. However, if $\omega = 4/g$ and g is odd, a case to which our result does not apply directly, one can actually show stability for sufficiently small μ.

To interpret these stability conditions, we drop the normalization of the frequency of the particles P_1, P_2 and call it ν_1. For $\mu = 0$ the point P_3 of zero mass will describe a circular orbit whose frequency in the stationary coordinate system we denote by ν_3. The frequency ω appearing above is then given by

$$\omega = \frac{\nu_3}{\nu_1} - 1 = \frac{\nu_3 - \nu_1}{\nu_1}$$

and our condition (2) is equivalent to

$$\text{(3)} \qquad \frac{\nu_1}{\nu_3} \neq \frac{p}{q}, \quad |p-q| \leqq 4$$

for p, q relatively prime integers. Picturing the one-parameter family of periodic orbits as a family of closed curves that cover the plane in a manner similar to that of the concentric circular orbits for $\mu = 0$, we thus obtain the stable orbits by removing from this family those solutions that correspond to (3).

We conclude this discussion with an interesting application to the motion of asteroids [1]. The asteroids are small planets that move in large numbers primarily between Mars and Jupiter and form an approximate ring about the sun. If we neglect the influence of the planets other than Jupiter, this configuration can be considered as a model for the restricted three-body problem with P_1 being Jupiter, P_2 the sun, and P_3 an asteroid whose mass is neglected entirely. Assuming that the majority of the asteroids move in nearly circular periodic orbits in the same plane as the sun and Jupiter, we can try to apply the above stability criterion. For most of the observed asteroids the ratios of their frequencies ν_3 to that of Jupiter satisfy

$$\frac{1}{4} \leqq \frac{\nu_1}{\nu_3} \leqq \frac{1}{2}.$$

In this interval the values of ν_1/ν_3 that violate our criterion are $\frac{1}{4}, \frac{2}{5}, \frac{1}{3}, \frac{1}{2}$, and these are values for which the asteroids are particularly sparse. Actually for $\frac{\nu_1}{\nu_3} = \frac{3}{7}$ one finds a less pronounced gap which, however, corresponds to a stable periodic orbit. These gaps in the distribution of asteroids, which were observed as early as 1866 by Kirkwood, seem therefore to be due to an instability caused by Jupiter, if we ignore the exceptional value $\frac{3}{7}$. Of course this interpretation is only of a qualitative nature and does not allow any prediction about the width of the gaps, nor have we checked whether the mass parameter μ is small enough. However, the criterion does provide the correct frequency ratios for the most pronounced gaps.

§ 35. Stability of Equilibrium Solutions

As another application of the result of § 32 we consider the problem of stability of an equilibrium for a Hamiltonian system with two degrees of freedom. We assume that the corresponding linearized system is stable and, even more, that the eigenvalues $\pm\lambda_1, \pm\lambda_2$ are purely imaginary and distinct. After a suitable linear transformation the Hamiltonian function can then be expressed in the form $H = H_2 + H_3 + \cdots$ with

$$H_2 = \lambda_1 x_1 y_1 + \lambda_2 x_2 y_2 ,$$

and the reality conditions developed in § 15 take the form $y_k = i\bar{x}_k$ $(k=1,2)$. There are essentially two different cases to be distinguished. In the first, H_2 is either positive or negative definite, and stability follows directly from the theorem of Dirichlet in § 29. In the second, H_2 is indefinite, meaning that 0 lies between λ_1 and λ_2 on the imaginary axis, and one cannot determine the question of stability from the linear terms of the differential equations alone. It is this case to which we now turn.

Assuming that $\lambda_1/\lambda_2 \neq -p/q$ $(p, q = 1, 2, 3, 4)$, we can find coordinates, which we again denote by x_k, y_k, so that the Hamiltonian takes the form

$$H = \lambda_1 w_1 + \lambda_2 w_2 + \tfrac{1}{2}\{\mu_{11} w_1^2 + 2\mu_{12} w_1 w_2 + \mu_{22} w_2^2\} + H_5 + \cdots \tag{1}$$

with $w_k = x_k y_k$, the μ_{kl} real, and the reality conditions still given by $y_k = i\bar{x}_k$. We will show that under the additional assumption

$$D = \mu_{11}\lambda_2^2 - 2\mu_{12}\lambda_1\lambda_2 + \mu_{22}\lambda_1^2 \neq 0 , \tag{2}$$

which is equivalent to requiring that the polynomial

$$H_4 = \tfrac{1}{2}\{\mu_{11} w_1^2 + 2\mu_{12} w_1 w_2 + \mu_{22} w_2^2\}$$

not be divisible by $H_2 = \lambda_1 w_1 + \lambda_2 w_2$, the origin is a stable solution of the corresponding Hamiltonian system. This result is due to Arnold [1].

For the proof we may assume that $|\lambda_1| > |\lambda_2|$ and $\mathrm{Im}\,\lambda_2 > 0 > \mathrm{Im}\,\lambda_1$. To establish stability it is enough to prove that any solution of the above system with initial values in $|x_1|^2 + |x_2|^2 < \varepsilon^2$ will remain for all real time in $|x_1|^2 + |x_2|^2 < c^2\varepsilon^2$, where $c = 3\sqrt{-\lambda_1/\lambda_2}$ and ε is a sufficiently small positive number. Here we admit only solutions that satisfy the reality conditions $y_k = i\bar{x}_k$ $(k=1,2)$, so that $w_k = i|x_k|^2$ and $H_2 = |\lambda_1||x_1|^2 - |\lambda_2||x_2|^2$.

The proof will be indirect. Namely, assuming the above to be false, we can find a solution with $r^2(t)=|x_1(t)|^2+|x_2(t)|^2$ satisfying $r(0)<\varepsilon$, while $r(\tau)=c\varepsilon$ for some real τ. It is convenient to replace x_k, y_k by the magnified variables $\varepsilon^{-1}x_k$, $\varepsilon^{-1}y_k$ $(k=1,2)$, in which case the system of differential equations remains canonical and the new Hamiltonian takes the form

$$H(x,y,\varepsilon)=H_2(x,y)+\varepsilon^2 H_4(x,y)+\varepsilon^3 H_5(x,y)+\cdots,$$

where the $H_v(x,y)$ are the same homogeneous polynomials of degree v that appear in the previous Hamiltonian. Our assumption then asserts the existence of a solution to the new system satisfying $r(0)<1$, while $r(\tau)=c$ for some real τ, an assertion that for sufficiently small values of ε will lead to a contradiction.

With this in mind, we reduce the differential equation in question to a mapping that admits application of the theorem in § 32. First we observe that the function $H(x,y,\varepsilon)$ is constant, say equal to h, along the solution we are considering. To estimate its value we set $t=0$ and, since $|\lambda_1|>|\lambda_2|$ and $r(0)<1$, we find that

$$|h|<\big||\lambda_1||x_1|^2-|\lambda_2||x_2|^2\big|+c_1\varepsilon^2<|\lambda_1|+c_1\varepsilon^2<\tfrac{3}{2}|\lambda_1|$$

for sufficiently small ε. Here c_1 and later c_2, c_3, c_4, c_5 denote positive constants that are independent of ε and the particular solution. On the other hand, setting $t=\tau$, we have

$$|\lambda_1||x_1|^2-|\lambda_2||x_2|^2\geqq-|h|-c_2\varepsilon^2>-2|\lambda_1|$$

for ε sufficiently small. Recalling that for $t=\tau$ we have

$$|x_1|^2+|x_2|^2=c^2=\frac{9|\lambda_1|}{|\lambda_2|},$$

we eliminate $|x_2|^2$ from the last two relations to obtain

$$(|\lambda_1|+|\lambda_2|)|x_1|^2\geqq|\lambda_2|c^2-2|\lambda_1|=7|\lambda_1| \quad (t=\tau),$$

so that from $|\lambda_1|+|\lambda_2|\leqq 2|\lambda_1|$ we finally have

$$|x_1|^2\geqq\tfrac{7}{2}>3 \quad (t=\tau).$$

From this we conclude that for the solution in question the function $|x_1(t)|^2$ assumes all values ϱ in the interval $1\leqq\varrho\leqq 3$, and therefore certainly in $2\leqq\varrho\leqq 3$.

Next, on the energy surface $H=h$ with $|h|<\frac{3}{2}|\lambda_1|$ we consider the set Ω defined by $2\leqq|x_1|^2\leqq 3$ and the reality conditions $y_k=i\bar{x}_k$ $(k=1,2)$. From $|H|<\frac{3}{2}|\lambda_1|$ we have on this three-dimensional set the estimate

$$\begin{aligned}|\lambda_2||x_2|^2&\geqq|\lambda_1||x_1|^2-\tfrac{3}{2}|\lambda_1|-c_3\varepsilon^2\\&\geqq\tfrac{1}{2}|\lambda_1|-c_3\varepsilon^2\geqq\tfrac{1}{4}|\lambda_1|,\end{aligned}$$

so that $|x_2| \geqq \frac{1}{2}$. Similarly one sees that $|x_2|$ is also bounded from above in Ω. Consequently, in Ω one can determine an argument $\theta = \operatorname{Im} \log x_2$ up to an integral multiple of 2π, and we denote by Σ the two-dimensional surface in Ω defined by $\theta \equiv 0 (\operatorname{mod} 2\pi)$. The points in Σ are parametrized by the real and imaginary parts of x_1, with $x_2 = |x_2|$ being determined implicitly from the relation $H = h$. From the estimate

$$\dot{\theta} = \operatorname{Im} \frac{\dot{x}_2}{x_2} = \operatorname{Im} \frac{H_{y_2}}{x_2} \geqq \operatorname{Im} \lambda_2 - c_4 \varepsilon^2 \geqq \frac{2}{3} |\lambda_2| ,$$

valid for $\varepsilon > 0$ sufficiently small, we see that any solution of our system which stays in Ω for a time interval of length $\geqq 3\pi/|\lambda_2|$ intersects Σ at least once, and nontangentially.

We now define the mapping S by following a real solution from a point in Σ with increasing t to the next intersection with Σ, whenever it exists. If $t_0 < t_1$ are the consecutive times of intersection of such a solution, we have $0 < t_1 - t_0 \leqq 3\pi/|\lambda_2|$ and

$$\frac{d}{dt} |x_1|^2 = 2 \operatorname{Re}(\bar{x}_1 H_{y_1}) \leqq c_5 \varepsilon^2 .$$

From this inequality we conclude that for any δ in $0 < \delta < \frac{1}{3}$ there exists an $\varepsilon_\delta > 0$ such that for $0 < \varepsilon < \varepsilon_\delta$ the mapping S is actually defined for $2 + \delta \leqq |x_1|^2 \leqq 3 - \delta$ and that

$$-\delta < |x_1(t_1)|^2 - |x_1(t_0)|^2 < \delta .$$

We recall that for the solution which was presumed to exist the values of $|x_1(t)|^2$ cover the interval $2 \leqq \varrho \leqq 3$, and therefore every annulus $a < |x_1|^2 < b$ in Σ with $0 < b - a < \delta$ contains at least one intersection of this solution with Σ. This implies that such an intersection of the solution with Σ in $2 \leqq |x_1|^2 \leqq 2 + \delta$ is mapped by some iterate of S or S^{-1} into a point in $3 - \delta \leqq |x_1|^2 \leqq 3$, which clearly is impossible if there exists a closed curve invariant under S in the annulus $2 + \delta < |x_1|^2 < 3 - \delta$. Thus, our proof will be complete once we show the existence of such an invariant curve.

Having reduced the assertion about stability to the existence of an invariant curve for the above mapping S in the annulus $2 + \delta \leqq |x_1|^2 \leqq 3 - \delta$, we will now verify the hypothesis necessary for an application of the existence theorem in § 32. First, from the discussion in § 22 we see that S preserves the area integral

$$\frac{1}{2i} \oint \bar{x}_1 \, dx_1$$

taken about any closed curve, and therefore satisfies the required intersection property. Thus it remains only to verify that S can be closely approximated by a twist mapping and, since the estimates have to be carried out in a complex neighborhood of the annulus, we begin by extending the real manifold Σ to a complex domain. The manifold was defined in Ω by the conditions $H=h$ and $\theta\equiv 0\ (\mathrm{mod}\, 2\pi)$, and we can replace the last condition by $y_2=ix_2$ which, in conjunction with the reality condition $y_2=i\bar{x}_2$, implies that x_2 is real. From the energy relation

$$H=\lambda_1 x_1 y_1+\lambda_2 x_2 y_2+\varepsilon^2 H_4+\cdots=h$$

together with $y_2=ix_2$ we obtain

(3) $$i\lambda_2 x_2^2=-\lambda_1 x_1 y_1+h+O(\varepsilon^2),$$

and restricting x_1, y_1 to the complex domain

(4) $$2<|x_1|^2<3, \quad 2<|y_1|^2<3, \quad |y_1-i\bar{x}_1|<|x_1|,$$

we can solve (3) for $x_2=\phi(x_1, y_1, h, \varepsilon)$ as an analytic function of the other variables. Indeed, since

$$|\lambda_1 x_1 y_1-h|\geqq 2|\lambda_1|-|h|>|\lambda_1|/2,$$

we see that for ε small the right hand side of (3) stays away from 0 and, since the last relation in (4) implies that $\mathrm{Im}\, x_1 y_1>0$, there is a single valued branch of the square root of $-\lambda_1 x_1 y_1+h$ defined in all of (4). Thus, for small values of ε, the function ϕ is defined in all of (4) and, in agreement with the definition of Σ, we choose that branch which is positive when $y_1=i\bar{x}_1$, $h=\bar{h}$. We now define the complex extension Σ_c of Σ by the analytic equations $y_2=ix_2$, $x_2=\phi(x_1, y_1, h, \varepsilon)$, with the complex variables x_1, y_1 ranging over the domain (4).

The complex extension of the mapping S is obtained by following the complex solution of our system from a point on Σ_c to its next intersection with Σ_c. To approximate this mapping we replace $H(x, y, \varepsilon)$ by $H^*=H_2+\varepsilon^2 H_4$, which depends on $w_k=x_k y_k$ $(k=1,2)$ only, and denote the surface corresponding to Σ by Σ^* and the corresponding mapping on Σ^* by S^*. The mapping S^* is determined by solving the system of differential equations

$$\dot{x}_k=H^*_{y_k}=H^*_{w_k}x_k, \quad \dot{y}_k=-H^*_{x_k}=-H^*_{w_k}y_k \quad (k=1,2)$$

in the form

$$x_k(t)=x_k(0)\, e^{tH^*_{w_k}}, \quad y_k(t)=y_k(0)\, e^{-tH^*_{w_k}},$$

where in the arguments of $H^*_{w_k}$ one has to insert $w_1(0)$, $w_2(0)$. Choosing the initial values on $y_2 = ix_2$, we determine T as a function of these initial values such that $y_2 = ix_2$ when $t = T$ and such that T is near $2\pi/|\lambda_2|$. Clearly this requires that

$$H^*_{w_2} T = 2\pi i .$$

The mapping thus takes the form

$$x_1(T) = x_1(0)\, e^{iQ}, \quad y_1(T) = y_1(0)\, e^{-iQ}$$

with

$$Q = 2\pi \frac{H^*_{w_1}}{H^*_{w_2}} = 2\pi \frac{\lambda_1 + \varepsilon^2(\mu_{11} w_1 + \mu_{12} w_2)}{\lambda_2 + \varepsilon^2(\mu_{12} w_1 + \mu_{22} w_2)},$$

and upon elimination of w_2 by use of the energy relation $H^* = h$ the last expression becomes

$$Q = 2\pi \left\{ \frac{\lambda_1}{\lambda_2} + \frac{\varepsilon^2}{\lambda_2^3} (D w_1 + (\mu_{12}\lambda_2 - \mu_{22}\lambda_1)\, h) \right\} + O(\varepsilon^3)$$

where D is the quantity defined in (2). The complex extension of S^* is then given by

$$\text{(5)} \qquad \begin{cases} x_1(T) = x_1(0)\, e^{i\alpha + \beta x_1(0) y_1(0)} + O(\varepsilon^3) \\ y_1(T) = y_1(0)\, e^{-i\alpha - \beta x_1(0) y_1(0)} + O(\varepsilon^3) \end{cases}$$

with

$$\alpha = 2\pi \left(\frac{\lambda_1}{\lambda_2} + \frac{\varepsilon^2}{\lambda_2^3} (\mu_{12}\lambda_2 - \mu_{22}\lambda_1)\, h \right), \quad \beta = \frac{2\pi i D}{\lambda_2^3} \varepsilon^2 \neq 0 .$$

Finally we are in a position to apply the existence theorem of § 32. Since $H(x, y, \varepsilon)$ and H^* differ only in terms of third order, one easily verifies that S, too, has the form (5), and in $y_1(0) = i\bar{x}_1(0)$, $2 \leqq |x_1(0)|^2 \leqq 3$ this mapping is seen to satisfy the necessary hypothesis of the theorem. Thus we conclude that the annulus $2 + \delta \leqq |x_1|^2 \leqq 3 - \delta$ contains a closed curve invariant under S. This completes the proof of stability of an equilibrium satisfying (2).

One might expect an analogous result on stability if the quadratic form $\mu_{11} w_1^2 + 2\mu_{12} w_1 w_2 + \mu_{22} w_2^2$ is nondegenerate, i.e. if $\mu_{11}\mu_{22} - \mu_{12}^2 \neq 0$. This, however, is not the case, as the following example will show. Turning immediately to the general case of n degrees of freedom, we let $g = (g_1, g_2, \dots, g_n)$ be a vector whose components are nonnegative integers satisfying $|g| = \sum_{k=1}^{n} g_k \geqq 3$ and set

$$H^*(w) = \sum_{k=1}^{n} \lambda_k w_k + \sum_{k,l=1}^{n} \mu_{kl} w_k w_l ,$$

with the λ_k purely imaginary, the μ_{kl} real, and

$$\sum_{k=1}^{n} \lambda_k g_k = 0, \qquad \sum_{k,l=1}^{n} \mu_{kl} g_k g_l = 0. \tag{6}$$

Denoting the product

$$\prod_{k=1}^{n} x_k^{g_k}$$

by x^g, we will show that for the Hamiltonian system with

$$H = H^*(w) + x^g + c y^g, \quad w_k = x_k y_k, \quad c = i^{-|g|}$$

subject to the reality conditions $y_k = i\bar{x}_k$ $(k = 1, \ldots, n)$ the origin is not stable.

To do this, we note that

$$\begin{cases} \dfrac{dw_k}{dt} = -g_k(x^g - cy^g) \\ \dfrac{d}{dt}(x^g + cy^g) = (x^g - cy^g) \displaystyle\sum_{k=1}^{n} H^*_{w_k} g_k \end{cases} \tag{7}$$

and seek solutions that satisfy the relations

$$w_k = i g_k \varrho^2 \quad (k = 1, \ldots, n), \quad x^g = -cy^g$$

for some positive scalar function $\varrho = \varrho(t)$. In view of (6), (7), if these relations are satisfied initially they persist for all t for which the solution exists, provided that ϱ satisfies the differential equation

$$\dot{\varrho} = a\varrho^{|g|-1}, \quad a = \pm\sqrt{g^g}.$$

Here the sign of a depends on the initial values and can be reversed by replacing x_k, y_k by $\tau x_k, \bar{\tau} y_k$ with $\tau^{|g|} = -1$. In any event these solutions approach the origin as $t \to +\infty$ or $t \to -\infty$, and the origin is not stable with respect to both future and past.

In this example the lack of stability is clearly due to the rational dependence of the eigenvalues $\lambda_1, \ldots, \lambda_n$. It is interesting to note here that the fourth order terms $\sum_{k,l} \mu_{kl} w_k w_l$ do not really affect the situation, as was the case for the flow of two degrees of freedom near a periodic solution. It is easy to see that the conditions (6) are not incompatible with the assumption that the matrix (μ_{kl}) be nondegenerate. For example, if $n = 2$ and

$$H^* = (\lambda_1 w_1 + \lambda_2 w_2)(1 + k_1 w_1 + k_2 w_2),$$

the second condition in (6) is satisfied as soon as the first one is, while $\mu_{11}\mu_{22} - \mu_{12}^2 \neq 0$ if $k_1\lambda_2 \neq k_2\lambda_1$. Here, of course, condition (2) is violated.

The stability criterion (2) can easily be generalized to the case where the fourth order terms violate it but the terms of sixth or higher order satisfy a similar nondegeneracy condition. Clearly what is of importance is that the function $H^*_{w_1}/H^*_{w_2}$ not be constant along the surface $H^*=0$, where H^* is the function of w_1, w_2 obtained by truncating the Hamiltonian after it has been brought into normal form to a sufficiently high order. When the ratio $\lambda_2\lambda_1^{-1}$ is irrational this nondegeneracy condition can be given a very simple form. Indeed, by the results of § 30 we can then introduce by a formal canonical transformation new variables $\xi_1, \eta_1, \xi_2, \eta_2$ so that the Hamiltonian $H(x, y)$ becomes a power series $K=K(\omega_1, \omega_2)$ in the products $\omega_1=\xi_1\eta_1$, $\omega_2=\xi_2\eta_2$, and in terms of this normal form the nondegeneracy condition can be expressed by the requirement that $K(\omega_1, \omega_2)$ not be divisible by $\lambda_1\omega_1+\lambda_2\omega_2$, or in other words, that $K(-\lambda_2\omega_1, \lambda_1\omega_1)$ not be identically 0. Under this condition one verifies that $K_{\omega_1}/K_{\omega_2}$ is not constant along $\lambda_1\omega_1+\lambda_2\omega_2=0$ and therefore, by the previous arguments, one has a stable equilibrium.

We conclude this discussion with an application of our result to Lagrange's equilateral solutions for the restricted three-body problem. These solutions, which we discussed for the general three-body problem, retain their meaning also for the restricted case. We assume that P_1, P_2 are particles of mass $\mu, 1-\mu$ respectively and consider the motion of the point P_3 of zero mass in a rotating coordinate system in which P_1, P_2 are at rest. As we saw in (24; 29), the equations for the coordinates (x_1, x_2) of P_3 are given by the Hamiltonian system

$$\dot{x}_k = H_{y_k}, \quad \dot{y}_k = -H_{x_k} \qquad (k=1,2)$$

with

$$H=\tfrac{1}{2}(y_1^2+y_2^2)+x_2y_1-x_1y_2-F(x_1,x_2)$$

$$F=\frac{1-\mu}{\sqrt{(x_1+\mu)^2+x_2^2}}+\frac{\mu}{\sqrt{(x_1+\mu-1)^2+x_2^2}}\,.$$

This system has an equilibrium at the points

$$x_1=\frac{1}{2}-\mu, \quad x_2=\pm\frac{\sqrt{3}}{2}$$

which form an equilateral triangle with the points $(-\mu, 0)$, $(1-\mu, 0)$, the respective positions of P_2, P_1. The eigenvalues λ of the linearized system are seen to be roots of the fourth order equation

$$\lambda^4+\lambda^2+\tfrac{27}{4}\mu(1-\mu)=0$$

which corresponds to (18; 4) with $m_1 = \mu$, $m_2 = 1-\mu$, $m_3 = 0$ and has two pairs of purely imaginary roots if

$$\mu(1-\mu) < \tfrac{1}{27}.$$

However, the quadratic part H_2 of the Hamiltonian turns out to be indefinite, as one readily verifies by computing, and therefore to decide about stability one has to take into account the higher order terms.

If those values of μ for which $\lambda_1 q - \lambda_2 p = 0$ with $|p| + |q| \leqq 4$ are excluded, the Hamiltonian can be brought into the form (1) with the quantity D in (2) being

$$-\frac{1}{8}\,\frac{36 - 541\lambda_1^2\lambda_2^2 + 644\lambda_1^4\lambda_2^4}{(1-4\lambda_1^2\lambda_2^2)(4-25\lambda_1^2\lambda_2^2)},$$

and this vanishes only for one exceptional value μ_0 in the interval of interest $0 < \mu < \mu_1$, where $\mu_1 = \frac{1}{2}(1 - \frac{1}{9}\sqrt{69})$ is the smaller of the roots of $27\mu(1-\mu) = 1$. The values of μ for which $\lambda_1 q - \lambda_2 p = 0$ with $|p| + |q| \leqq 4$ are easily determined. If we choose $|\lambda_1| > |\lambda_2|, q > 0$, then $1 \leqq q < p$, and the restriction $|p| + |q| \leqq 4$ leads to the two cases $(p, q) = (2, 1), (3, 1)$. On account of $\lambda_1^2 + \lambda_2^2 = -1$ this corresponds to $\lambda_2^2 = -\frac{1}{5}, -\frac{1}{10}$ which occurs for $\mu = \mu_2 = \frac{1}{2}(1 - \frac{1}{45}\sqrt{1833})$, $\mu_3 = \frac{1}{2}(1 - \frac{1}{15}\sqrt{213})$. Thus if we exclude these three values μ_0, μ_2, μ_3 from the interval $0 < \mu < \mu_1$ we have stability. This application of Arnold's theorem is due to Leontovitch [2], who merely verified that D does not vanish identically in μ. The actual determination of D and of the exceptional values of μ was carried out by Deprit and Deprit [3].

§ 36. Quasi-Periodic Motion and Systems of Several Degrees of Freedom

So far we have used the invariant curves found in the previous sections only for investigating stability. However, it is actually possible to give a very precise description of the qualitative nature of the corresponding orbits. Just as the fixed-points of the mapping associated with the flow gives rise to periodic motions, we will show that the points on our invariant curves correspond to a special class of almost periodic solutions, namely the so-called quasi-periodic motions. The significance of these solutions was recognized already by Bohl. We turn to the discussion of such solutions and to the generalization of the results of the previous sections to several degrees of freedom.

A complex function $f(t)$ will be called quasi-periodic if it can be represented as a Fourier series of the type

(1) $$f(t) = \sum_{k_1,\ldots,k_s} a_{k_1\ldots k_s} e^{i(k_1\omega_1 + \ldots + k_s\omega_s)t},$$

where $k_1, \ldots, k_s$ range over all integers and the coefficients decay exponentially with $|k| = |k_1| + \cdots + |k_s|$. To simplify the notation, we introduce the vectors $k = (k_1, \ldots, k_s)$, $\omega = (\omega_1, \ldots, \omega_s)$ together with their inner product $(k, \omega) = \sum_{\nu=1}^{s} k_\nu \omega_\nu$, and accordingly denote the coefficients by a_k. The real numbers $\omega_1, \ldots, \omega_s$ may be assumed to be independent over the rationals, since otherwise their number can be reduced. Real quasi-periodic functions $f(t)$ are characterized by the relations $a_{-k} = \bar{a}_k$.

To each quasi-periodic function $f(t)$ of the above type we associate a corresponding function

$$F(\theta_1, \ldots, \theta_s) = \sum_k a_k e^{i(k, \theta)}$$

of the s variables $\theta_1, \ldots, \theta_s$, where $\theta = (\theta_1, \ldots, \theta_s)$. The function $F(\theta)$ has period 2π in each variable, and is real if $a_{-k} = \bar{a}_k$. Moreover, by our assumption on the exponential decay of the coefficients, $F(\theta)$ is real analytic. The function $f(t)$ is obtained from $F(\theta)$ by replacing θ by ωt. Conservely, every real analytic function $F(\theta)$ of period 2π in its variables gives rise to a real quasi-periodic function $f(t)$ if θ is again replaced by ωt. To see this, one only has to show that the Fourier coefficients

$$a_k = (2\pi)^{-s} \int_0^{2\pi} \cdots \int_0^{2\pi} F(\theta) e^{-i(k, \theta)} d\theta_1 \ldots d\theta_s$$

decay exponentially with $|k|$. Since $F(\theta)$ is real analytic and periodic, it is defined for complex values of θ_ν in the region $|\operatorname{Im} \theta_\nu| \leqq \varrho$, for $\varrho > 0$ sufficiently small, and is bounded there in absolute value by a constant M. By Cauchy's theorem, the path of integration in the above integral can be shifted to $\theta_\nu = x_\nu \pm i\varrho$, $0 \leqq x_\nu \leqq 2\pi$ $(\nu = 1, \ldots, s)$, and choosing the sign equal to the sign of $-k_\nu$ if $k_\nu \neq 0$ and arbitrarily if $k_\nu = 0$, we find that

$$|a_k| \leqq M e^{-|k|\varrho}. \tag{2}$$

This proves the exponential decay of the a_k, while the real character of F implies that

$$\bar{a}_k = (2\pi)^{-s} \int_0^{2\pi} \cdots \int_0^{2\pi} F(\theta) e^{i(k, \theta)} d\theta_1 \ldots d\theta_s = a_{-k}.$$

The class of quasi-periodic functions so defined will be denoted by $\mathfrak{Q}(\omega)$. It is a subclass of the set of almost periodic functions introduced by Bohr [1], and is more restrictive in two respects: first, the frequencies are linear combinations with integer coefficients of only finitely many frequencies $\omega_1, \ldots, \omega_s$, whereas in Bohr's theory any denumerable set of real numbers is admitted for the frequencies; second, the coefficients a_k

are required to decay exponentially with $|k|$, making $f(t)$ as well as $F(\theta)$ real analytic, whereas Bohr's almost periodic functions are merely bounded and continuous. This narrower class $\mathfrak{Q}(\omega)$, however, is particularly well adapted for the nonlinear problems of celestial mechanics.

The class of functions $\mathfrak{Q}(\omega)$ obviously depends on the choice of the numbers $\omega_1, \ldots, \omega_s$. More precisely, it depends on the lattice generated by the $\omega_1, \ldots, \omega_s$, since $\mathfrak{Q}(\omega)$ is unaltered if the vector ω is replaced by $\omega' = U\omega$, where U is a matrix with integer entries and determinant ± 1. The requirement of rational independence of the $\omega_1, \ldots, \omega_s$ is clearly preserved under such transformations. Also it is clear that every quasi-periodic function in $\mathfrak{Q}(\omega)$ can be uniformly approximated by a finite trigonometric sum in $\mathfrak{Q}(\omega)$, for one merely has to truncate the Fourier series of a given function $f(t)$ by dropping all terms with $|k| \geqq N$ for a sufficiently large natural number N. On account of the exponential decay of the Fourier coefficients, the resulting trigonometric sums will then converge uniformly to $f(t)$.

To every quasi-periodic function $f(t)$ we also associate the mean value

$$f^* = \lim_{T \to \infty} \frac{1}{T} \int_0^T f(t)dt .$$

To prove the existence of this limit, we observe that for a trigonometric polynomial in $\mathfrak{Q}(\omega)$ the above limit clearly exists and is equal to

$$a_0 = (2\pi)^{-s} \int_0^{2\pi} \cdots \int_0^{2\pi} F(\theta) d\theta_1 \ldots d\theta_s .$$

For an arbitrary quasi-periodic function $f(t)$ in $\mathfrak{Q}(\omega)$, existence of the above limit follows by approximation of $f(t)$ by trigonometric polynomials $f_N(t)$ in $\mathfrak{Q}(\omega)$, where $\sup_t |f(t) - f_N(t)| = \varepsilon_N \to 0$ as $N \to \infty$. Indeed, by subtracting a constant from f, we may assume that the coefficient a_0 corresponding to the constant term is zero, and the same then holds for the trigonometric polynomials f_N obtained from $f(t)$ by truncation. Given $\delta > 0$, let N be so large that $\varepsilon_N < \delta/2$, whereby

$$\left| \frac{1}{T} \int_0^T f dt - \frac{1}{T} \int_0^T f_N dt \right| < \frac{\delta}{2} .$$

Since $f_N^* = 0$, we can find a T_δ such that the second term in the above inequality is less than $\delta/2$ when $T > T_\delta$, and therefore

$$\left| \frac{1}{T} \int_0^T f dt \right| < \frac{\delta}{2} + \frac{\delta}{2} = \delta \quad (T > T_\delta) .$$

This shows that the limit of the left side as $T \to \infty$ exists and is equal to 0.

Before investigating the significance of quasi-periodic functions for nonlinear Hamiltonian systems, we consider the simplest linear equation

$$\dot{y} = f(t)$$

containing a quasi-periodic function $f(t)$ in $\mathfrak{Q}(\omega)$. If $f(t)$ is actually periodic, any solution of this equation is the sum of the linear function $f^* t$ and a periodic function of the same period as $f(t)$, so that the solutions are periodic if and only if $f^* = 0$. Assuming now that f is quasi-periodic in $\mathfrak{Q}(\omega)$ and has mean value $f^* = 0$, we investigate whether the solutions of the above differential equation likewise belong to $\mathfrak{Q}(\omega)$. In general the answer is no, but if for some positive constants c, μ the frequencies (k, ω) are required to satisfy the inequalities

$$|(k, \omega)| \geqq c^{-1} |k|^{-\mu} \tag{3}$$

for all integer vectors $k \neq 0$, then the answer is affirmative.

To prove the last statement, let the function $f(t)$ again be represented as $F(\omega t)$ and the unknown solution $y(t)$ as $Y(\omega t)$, so that $Y(\theta)$ has to satisfy the partial differential equation

$$\sum_{\nu=1}^{s} \omega_\nu \frac{\partial Y}{\partial \theta_\nu} = F(\theta) . \tag{4}$$

Both functions $Y(\theta)$, $F(\theta)$ are assumed to be real analytic and of period 2π in $\theta_1, \ldots, \theta_s$ and, moreover, the mean value of $F(\theta)$ is assumed to be 0. This partial differential equation is readily solved in terms of Fourier series. Indeed, if we set

$$F(\theta) = \sum_k a_k e^{i(k, \theta)} ,$$

since $a_0 = 0$ the solution $Y(\theta)$ of (4) with mean value zero has the form

$$Y(\theta) = \sum_{k \neq 0} \frac{a_k}{i(k, \omega)} e^{i(k, \theta)} .$$

In view of (2), (3) the coefficients $b_k = -i(k, \omega)^{-1} a_k$ decay exponentially, while the real character of $F(\theta)$ implies that $\bar{b}_k = b_{-k}$, so that the above series represents a real analytic function $Y(\theta)$. Thus $y(t) = Y(\omega t)$ is a solution of $\dot{y} = f(t)$ in $\mathfrak{Q}(\omega)$. Clearly, the most general solution differs from the one just constructed only by an additive constant.

The solution to (4) will actually belong to $\mathfrak{Q}(\omega)$ under weaker restrictions on the small divisors (k, ω) than the inequalities (3). However, if no restrictions at all are placed on these frequencies, the corresponding solution may even be unbounded. To exhibit such an example, we choose a number $b > 1$ and construct a positive irrational number α satisfying the inequalities

$$|k_\nu \alpha - l_\nu| < b^{-k_\nu}$$

for infinitely many positive integers k_ν, l_ν ($\nu = 1, 2, \ldots$). Such a number α can easily be constructed following the reasoning at the end of § 25. Choosing a number a with $1 < a^{\alpha+1} < b$, set

$$f(t) = \sum_{k,l=1}^{\infty} a^{-k-l} \eta_{kl} \sin(k\alpha - l)\,t\,,$$

where η_{kl} is 1 if $k\alpha - l > 0$ and -1 if $k\alpha - l < 0$. The function $f(t)$ clearly belongs to $\mathfrak{Q}(\omega_1, \omega_2)$ with $\omega_1 = \alpha$, $\omega_2 = 1$ and has mean value 0. However the integral

$$g(t) = \int_0^t f(s)\,ds = \sum_{k,l=1}^{\infty} a^{-k-l} |k\alpha - l|^{-1} (1 - \cos(k\alpha - l)\,t)$$

is not even bounded, and therefore certainly not quasi-periodic. Indeed, setting $\delta_\nu = |k_\nu \alpha - l_\nu|$ and $t_\nu = \dfrac{\pi}{2\delta_\nu}$, we have

$$g(t_\nu) \geqq a^{-k_\nu - l_\nu} \delta_\nu^{-1}\,,$$

and since $k_\nu + l_\nu = k_\nu(1+\alpha) - (k_\nu\alpha - l_\nu) \leqq k_\nu(1+\alpha) + 1$ while $\delta_\nu < b^{-k_\nu}$, it follows that

$$g(t_\nu) > a^{-1} \left(\frac{b}{a^{1+\alpha}} \right)^{k_\nu} \to \infty \quad (k_\nu \to \infty)\,.$$

From now on we impose the restrictions (3) on the small divisors (k, ω). Measure-theoretical considerations similar to those of § 25 imply that almost all real vectors ω satisfy such inequalities for some positive constant c if $\mu \geqq s$.

It is convenient for later use to list certain operations that preserve quasi-periodicity, independently of the inequalities (3). First, $\mathfrak{Q}(\omega)$ clearly forms a vector space over the real numbers and is closed under differentiation as well as under forming products and quotients, in the latter case provided that the denominator is bounded away from 0. Second, if $\phi(y_1, \ldots, y_r)$ is a real analytic function defined for all real $y_1, \ldots, y_r$, and $f_1, \ldots, f_r$ are in $\mathfrak{Q}(\omega)$, then the composite function $\phi(f_1, \ldots, f_r)$ is likewise in $\mathfrak{Q}(\omega)$. Also, if f, g belong to $\mathfrak{Q}(\omega)$ then so does the function $f(t + g(t))$, and indeed, if $f(t), g(t)$ are represented by the multiply periodic functions $F(\theta), G(\theta)$ then $f(t + g(t))$ is represented by $F(\theta_1 + \omega_1 G, \ldots, \theta_s + \omega_s G)$.

Finally, we assert that if f is in $\mathfrak{Q}(\omega)$ then the inverse function of $\tau = \alpha t + f(t)$ can be written in the form $t = \alpha^{-1}(\tau + g(\tau))$ with g in $\mathfrak{Q}(\omega/\alpha)$, provided that $\alpha + \frac{df}{dt}$ is bounded away from 0 and (3) holds. This provision, of course, implies that $\alpha \neq 0$, since the mean value of $\frac{df}{dt}$ is zero.

To prove the last assertion, we may assume that $\alpha = 1$, or else replace τ by $\alpha\tau$. If $f(t)$ is represented by $F(\theta)$ and the unknown function $g(\tau)$ by $G(\theta)$, where $g(\tau) = G(\omega\tau)$, the condition for G becomes

(5) $$G(\theta) + F(\theta + \omega G) = 0\,.$$

We replace this equation by

$$G + \sigma F(\theta + \omega G) = 0$$

and seek a solution $G = G(\theta; \sigma)$ for $0 \leqq \sigma \leqq 1$ with period 2π in each of the variables $\theta_1, \ldots, \theta_s$. Differentiating this last equation with respect to σ, we are led to the differential equation

(6) $$\frac{\partial G}{\partial \sigma} = \phi(\theta + \omega G; \sigma)\,, \qquad G(\theta; 0) = 0\,,$$

where

$$\phi(\theta; \sigma) = -F(\theta)\left(1 + \sigma \sum_{\nu=1}^{s} \omega_\nu F_{\theta_\nu}\right)^{-1}.$$

By the assumption on $\alpha + \frac{df}{dt}$, the denominator on the right of the last expression is bounded away from 0 when $\sigma = 1$ and $\theta_\nu = \omega_\nu t$ $(\nu = 1, \ldots, s)$, and is indeed positive, since its mean value is 1. On the other hand, the vectors with components $\omega_\nu t + k_\nu 2\pi$ $(\nu = 1, \ldots, s)$ for integer k_ν and real t are dense in s-dimensional Euclidean space, and therefore the denominator is actually positive and bounded away from 0 for all real θ and $0 \leqq \sigma \leqq 1$. It follows that $\phi(\theta; \sigma)$ is real analytic and of period 2π in the θ_ν.

The solution $G(\theta; \sigma)$ of (6) is now constructed by means of the standard existence theorem for ordinary differential equations. To show that $G(\theta; \sigma)$ is analytic for $\sum_{\nu=1}^{s} |\operatorname{Im} \theta_\nu|$ sufficiently small and for $0 \leqq \sigma \leqq 1$, it suffices to verify that as we continue the solution we do not leave the region of analyticity of the differential equation. Assuming that ϕ is analytic in $|\operatorname{Im} \theta_\nu| < \delta$ $(\nu = 1, \ldots, s)$, $0 \leqq \sigma \leqq 1$ where it satisfies $\sum_{\nu=1}^{s} |\phi_{\theta_\nu}| < M$, we set

$$\varrho = \delta e^{-M\|\omega\|}\,,$$

where $\|\omega\| = \max_\nu |\omega_\nu|$, and claim that in the region $|\operatorname{Im} \theta_\nu| < \varrho$, $0 \leqq \sigma \leqq 1$ the function $G(\theta; \sigma)$ must satisfy the estimate $\varrho + \|\omega\|\,|\operatorname{Im} G| < \delta$. This, of course, will imply that the solution exists and is analytic in this region. We now verify the claim. Since ϕ is real analytic, we have $\overline{\phi(\theta;\sigma)}=\phi(\bar{\theta};\sigma)$, and therefore

$$\begin{aligned}\left|\frac{\partial}{\partial\sigma}\operatorname{Im}G\right| &= \frac{1}{2}\left|\frac{\partial}{\partial\sigma}(G-\bar{G})\right| \\ &= \frac{1}{2}|\phi(\theta+\omega G;\sigma)-\phi(\bar{\theta}+\omega\bar{G};\sigma)| \\ &\leqq M\max_{\nu}|(\operatorname{Im}\theta_\nu+\omega_\nu\operatorname{Im}G)| \\ &< M(\varrho+\|\omega\|\,|\operatorname{Im}G|)\end{aligned}$$

for as long as the estimate $\varrho+\|\omega\|\,|\operatorname{Im}G|<\delta$ remains valid. Suppose the estimate did not hold over the entire interval $0\leqq\sigma\leqq1$. Since it certainly holds initially, there would exist a smallest number $0<\sigma^*\leqq1$ for which

$$\max_{\theta}|\operatorname{Im}G|=\frac{\delta-\varrho}{\|\omega\|}\qquad(\sigma=\sigma^*),$$

while the above differential inequality remains valid for $0\leqq\sigma<\sigma^*$. By a comparison argument we would then conclude that

$$|\operatorname{Im}G|<h(\sigma)\qquad(0<\sigma\leqq\sigma^*),$$

where $h(\sigma)$ is the solution of the equation

$$\frac{dh}{d\sigma}=M(\varrho+\|\omega\|h),\quad h(0)=0,$$

and since

$$h(\sigma^*)=\frac{\varrho}{\|\omega\|}(e^{M\|\omega\|\sigma^*}-1)\leqq\frac{\varrho}{\|\omega\|}(e^{M\|\omega\|}-1)=\frac{\delta-\varrho}{\|\omega\|},$$

we would have

$$|\operatorname{Im}G|<\frac{\delta-\varrho}{\|\omega\|}\qquad(\sigma=\sigma^*),$$

contrary to the choice of σ^*. This verifies our claim, and we conclude that the solution $G(\theta;\sigma)$ of (6) exists for $0\leqq\sigma\leqq1$, is real analytic, and, by the uniqueness theorem, has period 2π in the θ_ν. It is easily verified that $G=G(\theta;1)$ is the solution of our original equation (5). Thus $g(\tau)=G(\omega\tau;1)$ belongs to $\mathfrak{Q}(\omega)$, and the assertion about the form of the inverse function is proved.

We conclude these preliminaries by applying the last result to the nonlinear scalar differential equation

$$\frac{d\tau}{dt} = f(\tau),$$

where f belongs to $\mathfrak{Q}(\omega)$ and satisfies $|f(\tau)| \geqq \delta > 0$ for all real τ. We will show that every solution of this equation has the form

$$\tau = \alpha t + g(t),$$

where α^{-1} is the mean value of f^{-1} and g belongs to $\mathfrak{Q}(\alpha\omega)$. Indeed, the inverse function $t = t(\tau)$ satisfies

$$\frac{dt}{d\tau} = f^{-1}(\tau)$$

and therefore, by earlier remarks, has the form

$$t = \alpha^{-1}\tau + h(\tau),$$

where $h(\tau)$ is in $\mathfrak{Q}(\omega)$ if condition (3) is satisfied. Consequently, by our assertion about the inverse function, $\tau(t)$ has the stated form.

After these preliminary remarks we turn to the construction of quasi-periodic solutions for Hamiltonian systems, beginning with non-autonomous systems of one degree of freedom, then treating systems of two and more degrees of freedom, and finally discussing some applications to the three-body problem. First, we consider a system

$$\dot{x} = \frac{\partial H}{\partial y}, \quad \dot{y} = -\frac{\partial H}{\partial x}, \tag{7}$$

where the Hamiltonian $H = H\,(t, x, y; \varepsilon)$ is periodic of period 2π in the independent variable t and in the scalar x, and is real analytic in all four of its arguments, with y restricted to a real interval $a < y < b$ and the parameter ε to a neighborhood of $\varepsilon = 0$. The critical additional assumption is that for $\varepsilon = 0$ the Hamiltonian

$$H(t, x, y; 0) = H^0(y)$$

is independent of t, x and satisfies

$$\frac{\partial^2 H^0}{\partial y^2} \neq 0 \quad (a < y < b). \tag{8}$$

We will show that under these conditions the above system possesses, for sufficiently small $|\varepsilon|$, solutions of the form

$$\begin{cases} x = \theta_2 + F(\theta; \varepsilon), \quad y = G(\theta; \varepsilon) \\ \theta = (\theta_1, \theta_2), \quad \theta_1 = t, \quad \theta_2 = \alpha t + \theta_2(0), \end{cases} \tag{9}$$

where F, G are real analytic in both θ and ε and have period 2π in θ_1, θ_2, while the number α is irrational. In other words, $x-\alpha t, y$ are quasi-periodic functions belonging to $\mathfrak{Q}(1,\alpha)$. Even though $x(t)$ itself is not quasi-periodic, we will call these solutions quasi-periodic of class $\mathfrak{Q}(1,\alpha)$. Indeed, since x is an angular variable, it is natural in this context to call a solution $x(t), y(t)$ quasi-periodic of class $\mathfrak{Q}(\omega_1,\omega_2)$ if for any real analytic function $f(t,x,y)$ of period 2π in t and x the function $f(t,x(t),y(t))$ belongs to $\mathfrak{Q}(\omega_1,\omega_2)$.

The existence of such quasi-periodic solutions is readily derived from the existence theorem in § 32. For this we construct the area-preserving mapping taking the initial values $x(0), y(0)$ of the solution at $t=0$ into the values $x(2\pi), y(2\pi)$ at $t=2\pi$. This mapping is defined in a closed subinterval $a+\delta \leqq y \leqq b-\delta$ for $\delta>0$ sufficiently small, is real analytic there, and has the form

$$x(2\pi)=x(0)+2\pi H_y^0(y(0))+O(\varepsilon)$$
$$y(2\pi)=y(0)+O(\varepsilon)\,.$$

By (8), the derivative of H_y^0 is different from 0, so that the theorem of § 32 applies, and we conclude that the mapping possesses an invariant closed curve

$$x(0)=\xi+u(\xi;\varepsilon)\,, \quad y(0)=v(\xi;\varepsilon)$$

in $a+\delta \leqq y(0) \leqq b-\delta$, with the image of the mapping on this curve given by

$$x(2\pi)=\xi+2\pi\alpha+u(\xi+2\pi\alpha;\varepsilon)\,,\; y(2\pi)=v(\xi+2\pi\alpha;\varepsilon)\,. \tag{10}$$

The irrational number α here plays the role of $\dfrac{\omega}{2\pi}$ in § 32, and satisfies the condition (32;6).

We now show that the solutions whose initial values lie on this invariant curve are quasi-periodic of class $\mathfrak{Q}(1,\alpha)$. If $|\varepsilon|$ is sufficiently small, these solutions will remain in the domain $a<y<b$ for $0\leqq t\leqq 2\pi$, and consequently for all real t. Denoting these solutions by

$$x=\xi+u(t,\xi;\varepsilon)\,, \quad y=v(t,\xi;\varepsilon)\,, \tag{11}$$

we observe that u, v are real analytic in t, ξ, ε and have period 2π in ξ, while in view of (10), we have

$$u(2\pi,\xi;\varepsilon)=\omega+u(0,\xi+\omega;\varepsilon),\;\; v(2\pi,\xi;\varepsilon)=v(0,\xi+\omega;\varepsilon)\,, \tag{12}$$

where $\omega=2\pi\alpha$. More generally, we can conclude that

$$\begin{cases} u(t+2\pi,\xi-\omega;\varepsilon)-\omega=u(t,\xi;\varepsilon) \\ \quad v(t+2\pi,\xi-\omega;\varepsilon)=v(t,\xi;\varepsilon)\,. \end{cases} \tag{13}$$

Indeed, if $u(t, \xi; \varepsilon)$, $v(t, \xi; \varepsilon)$ in (11) are replaced by the respective left hand sides of (13), the resulting functions x, y are again solutions of our system whose initial values, in view of (12), agree with those of the original solutions. Consequently, by the uniqueness theorem for differential equations, the solutions are the same, and (13) follows. Thus, the functions

$$F(\theta; \varepsilon) = u(\theta_1, \theta_2 - \alpha\theta_1; \varepsilon) - \alpha\theta_1$$

$$G(\theta; \varepsilon) = v(\theta_1, \theta_2 - \alpha\theta_1; \varepsilon)$$

have period 2π in θ_1, θ_2, and using these functions F, G we can express the solutions (11) in the form (9) with $\theta_1 = t$, $\theta_2 = \xi + \alpha t$. This completes the proof.

We extend the previous result to the autonomous Hamiltonian system

$$\dot{x}_k = H_{y_k}, \quad \dot{y}_k = -H_{x_k} \quad (k = 1, 2) \tag{14}$$

of two degrees of freedom, where $H = H(x, y; \varepsilon)$ is a real analytic function in the variables $x_1, x_2, y_1, y_2, \varepsilon$, of period 2π in x_1, x_2, and is defined for $y = (y_1, y_2)$ restricted to an open domain $\mathfrak{D}$ and the parameter ε to a neighborhood of the origin. The critical assumption here is that

$$H^0(y) = H(x, y; 0)$$

is independent of x and satisfies

$$H^0_{y_2 y_2}(H^0_{y_1})^2 - 2H^0_{y_1 y_2} H^0_{y_1} H^0_{y_2} + H^0_{y_1 y_1}(H^0_{y_2})^2 \neq 0 \tag{15}$$

in $\mathfrak{D}$. We consider now an energy surface $H(x, y; \varepsilon) = h$, where h is a constant such that $H(x, y; 0) = H^0(y) = h$ intersects $\mathfrak{D}$, and assert that on such an energy surface there are solutions $x(t), y(t)$ which are quasi-periodic in the sense that $f(x(t), y(t))$ is quasi-periodic for any real analytic function $f(x, y)$ having period 2π in x_1, x_2.

This result is derived from the previous one. In view of (15), we may assume that $H^0_{y_1}$ is bounded away from 0 in an open subdomain $\mathfrak{D}'$ of $\mathfrak{D}$, and choosing the domain $\mathfrak{D}'$ as well as $|\varepsilon|$ sufficiently small, we can solve the equation $H(x, y; \varepsilon) = h$ for $y_1 = -K(x_1, x_2, y_2; \varepsilon)$ with K again of period 2π in x_1, x_2. Moreover, since

$$\dot{x}_1 = H_{y_1} \tag{16}$$

is bounded away from 0, we can eliminate the variable t from the differential equation and use x_1 as the independent variable. The resulting differential equations

$$\frac{dx_2}{dx_1} = \frac{H_{y_2}}{H_{y_1}}, \quad \frac{dy_2}{dx_1} = -\frac{H_{x_2}}{H_{y_1}}$$

can be expressed in the Hamiltonian form

$$\frac{dx_2}{dx_1} = \frac{\partial K}{\partial y_2}, \quad \frac{dy_2}{dx_1} = -\frac{\partial K}{\partial x_2}, \tag{17}$$

as one verifies by differentiating the identity $H(x_1, x_2, -K, y_2; \varepsilon) = h$. One also checks that $K(x_1, x_2, y_2; 0) = K^0(y_2)$ is independent of x_1, x_2, while in view of (15), we have

$$\frac{\partial^2 K^0}{\partial y_2^2} \neq 0.$$

Consequently, by the previous result, there exist solutions to (17) of the type

$$\begin{cases} x_2 = \theta_2 + F(\theta; \varepsilon), & y_2 = G(\theta; \varepsilon) \\ \theta_1 = x_1, \quad \theta_2 = \alpha x_1 + \theta_2(0), \end{cases} \tag{18}$$

where F, G are again real analytic in $\theta_1, \theta_2, \varepsilon$ and of period 2π in θ_1, θ_2. Thus these solutions are quasi-periodic in their dependence on x_1. In order to investigate their dependence on t we solve (16), with the right side viewed as a known quasi-periodic function in x_1 of class $\mathfrak{Q}(1, \alpha)$. Since H_{y_1} is bounded away from 0 and α satisfies (32; 6), we conclude from the preliminary considerations that $x_1(t)$ has the form

$$x_1 = \beta(t - t_0) + f(t - t_0),$$

where f is in $\mathfrak{Q}(\beta, \beta\alpha)$ and β^{-1} is the mean value of $H_{y_1}^{-1}$ as a function of x_1. Combining this with (18) and using the preliminary results on composition of quasi-periodic functions, we finally obtain

$$\begin{cases} x_k(t) = \theta_k + \tilde{F}_k(\theta; \varepsilon), & y_k(t) = \tilde{G}_k(\theta; \varepsilon) \quad (k = 1, 2) \\ \theta_k = \omega_k t + \theta_k(0), & \omega_1 = \beta, \quad \omega_2 = \alpha\beta, \end{cases} \tag{19}$$

where $\tilde{F}, \tilde{G}$ are real analytic in $\theta_1, \theta_2, \varepsilon$ and of period 2π in θ_1, θ_2, while the ratio of the frequencies ω_1, ω_2 is irrational. This completes the proof.

To interpret the above result geometrically, we can view the first line of (19) as an imbedding of a two-dimensional torus into the four-dimensional phase space. This torus lies on a fixed energy surface $H(x, y; \varepsilon) = h$, and is invariant under the flow (14) in the sense that the vector $(\dot{x}_1, \dot{x}_2, \dot{y}_1, \dot{y}_2)$ is tangential to the torus. Consequently every solution of (14) with initial values on this torus remains on it for all real t, and the system (14) actually gives rise to a differential equation on the torus. By the second line of (19), this equation is given by $\dot{\theta}_k = \omega_k$, and since ω_1/ω_2 is irrational, every orbit is dense on the torus.

These results can also be generalized to several degrees of freedom. For this, let $x = (x_1, \ldots, x_n)$, $y = (y_1, \ldots, y_n)$, and let $H(x, y; \varepsilon)$ be real analytic in its $2n+1$ arguments and have period 2π in each of the variables $x_1, \ldots, x_n$. Thus H is defined for all real x, while we restrict y to an n-dimensional domain $\mathfrak{D}$ and the parameter ε to a neighborhood of the origin. Moreover, for $\varepsilon=0$ the function

$$H(x, y; 0) = H^0(y)$$

is assumed to be independent of x, so that for this value of the parameter the corresponding Hamiltonian system is simply

$$\dot{x}_k = H^0_{y_k}(y), \quad \dot{y}_k = -H^0_{x_k}(y) = 0 \qquad (k=1, \ldots, n).$$

The solutions to this system are given by

$$x_k(t) = H^0_{y_k}(v)t + x_k(0), \quad y_k = v_k$$

for any n-vector $v=(v_1, \ldots, v_n)$ in $\mathfrak{D}$, and the n equations $y_k = v_k (k=1, \ldots, n)$ can be viewed as defining an n-dimensional torus on which the differential equations are given by $\dot{x}_k = H^0_{y_k}(v)$. Thus, if v is chosen so that the $H^0_{y_k}(v)$ are rationally independent, the solutions are dense on this torus.

Our aim is to find solutions of this type for small values of ε, and for this we make the additional assumption

$$|H^0_{y_k y_l}| \neq 0 \tag{20}$$

about the Hessian determinant. This guarantees that the gradient $H^0_y(y)$ varies with y over an open n-dimensional set. In particular, the constants $v_1, \ldots, v_n$ can be chosen so that the numbers $\omega_k = H^0_{y_k}(v)$ are rationally independent and, moreover, satisfy the infinitely many inequalities

$$\left|\sum_{k=1}^{n} j_k \omega_k\right| \geqq c|j|^{-\mu} \tag{21}$$

for all integers j_k with $|j| = \sum_{k=1}^{n} |j_k| > 0$, where c, μ are certain fixed positive constants. Under these hypotheses we can assert that for small values of $|\varepsilon|$ there exist solutions to the corresponding Hamiltonian system which have the form

$$\begin{cases} x_k = \theta_k + F_k(\theta; \varepsilon), & y_k = G_k(\theta_k; \varepsilon) \\ \theta_k = \omega_k t + \theta_k(0) \end{cases} \qquad (k=1, \ldots, n), \tag{22}$$

where F_k, G_k again are real analytic functions of period 2π in $\theta_1, \ldots, \theta_n$. These solutions are quasi-periodic in the sense that for any real analytic function $f(x, y)$ of period 2π in $x_1, \ldots, x_n$ the function $f(x(t), y(t))$ is quasi-periodic of class $\mathfrak{Q}(\omega)$. Moreover, it can be shown that these quasi-periodic solutions form a set of positive measure in the phase space. More precisely, if Ω denotes the $2n$-dimensional open set of points (x, y) for which $y \in \mathfrak{D}$, and if δ is a given positive constant, then there exists a positive $\varepsilon_0 = \varepsilon_0(\delta)$ such that for $|\varepsilon| < \varepsilon_0$ the above quasi-periodic solutions in Ω fill out a closed set S with the measure $m(\Omega - S) < \delta m(\Omega)$. In this sense we may say that the majority of solutions in Ω are quasi-periodic.

The above theorem is due to Kolmogorov and Arnold [2–5]. It is worth noting that the irrationality condition (21) is not imposed on the given Hamiltonian system, but on the vector $\omega = (\omega_1, \ldots, \omega_n)$ chosen in the range of H_y^0. The theorem asserts the existence of quasi-periodic solutions in $\mathfrak{Q}(\omega)$, and in this connection condition (20) is crucial in that it asserts that the frequencies $\omega_k = H_{y_k}^0(v)$ can be controlled by the choice of v. There is another version of this theorem in which (20) is replaced by a different assumption. Instead of requiring the frequencies $H_{y_k}^0$ $(k = 1, \ldots, n)$ to vary over an n-dimensional open set, we ask that on a fixed energy surface $H^0(y) = h$ the ratios $H_{y_k}^0 / H_{y_1}^0$ $(k = 2, \ldots, n)$ with $H_{y_1}^0 \neq 0$ vary over an $(n-1)$-dimensional set. This leads to the assumption

$$\text{(23)} \qquad \begin{vmatrix} H_{y_k y_l}^0 & H_{y_k}^0 \\ H_{y_l}^0 & 0 \end{vmatrix} \neq 0$$

which is, in fact, identical with (15) for $n = 2$. Under this assumption, on any preassigned energy surface there exist quasi-periodic solutions of the form (22) with prescribed frequency ratios. We observe that neither of the conditions (20), (23) implies the other. For example, for $H^{(0)}(y) = y_1 + P(y_2, \ldots, y_n)$ condition (20) is violated but (23) holds if the Hessian of P is different from 0; on the other hand, for $n = 2$, the function $H^0(y) = \log(y_1 y_2^{-1})$ with y_1, y_2 positive satisfies (20) but not (23).

The proof of these theorems is based on the same rapidly convergent iteration scheme used in § 26 and § 32, and will not be carried out here. Instead we will only derive a formal expansion in powers of ε for the functions $F_k(\theta; \varepsilon)$, $G_k(\theta; \varepsilon)$ in (22) under assumption (20), omitting the convergence proof.

We begin by choosing a vector $\omega = (\omega_1, \ldots, \omega_n)$ that lies in the range of $H_y^0(y)$ for y in $\mathfrak{D}$ and satisfies (21), the existence of which is guaranteed by measure-theoretical considerations similar to those at the end of § 25, provided $\mu \geqq n$. By a suitable translation of the y-coordinates, we may assume that $H_{y_k}^0(0) = \omega_k$ $(k = 1, \ldots, n)$. To find the desired invariant torus, we propose to construct a canonical transformation that takes the variables (x, y) into (ξ, η) and the desired torus (22) into $\eta = 0$. To be more precise, following the developments in § 3, we represent the canonical transformation by means of a generating function $S(x, \eta; \varepsilon)$ in the form

(24) $$y_k = S_{x_k}, \quad \xi_k = S_{\eta_k} \quad (k=1,\dots,n),$$

where $S(x,\eta;0) = \sum_{k=1}^{n} x_k\eta_k$ and, to preserve the desired periodicity, we require that $S_{x_k}, S_{\eta_k} - x_k$ have period 2π in $x_1,\dots,x_n$. Denoting the transformed Hamiltonian by $\Phi(\xi,\eta;\varepsilon)$, we have

(25) $$H(x, S_x; \varepsilon) = \Phi(S_\eta, \eta; \varepsilon),$$

and we will try to determine S so that

(26) $$\Phi(\xi,0;\varepsilon) = \gamma(\varepsilon), \quad \Phi_{\eta_k}(\xi,0;\varepsilon) = \omega_k \quad (k=1,\dots,n),$$

where γ is independent of ξ. If this is possible, the transformed equations

$$\dot{\xi}_k = \Phi_{\eta_k}, \quad \dot{\eta}_k = -\Phi_{\xi_k} \quad (k=1,\dots,n)$$

clearly possess the solutions

$$\xi_k = \omega_k t + \xi_k(0), \quad \eta_k = 0 \quad (k=1,\dots,n).$$

By expressing these solutions back in the original coordinates x, y and setting $\theta_k = \xi_k$, we then obtain the desired quasi-periodic solutions (22).

Thus our problem is reduced to finding a generating function $S(x,\eta;\varepsilon)$ for which the transformed Hamiltonian $\Phi(\xi,\eta;\varepsilon)$ satisfies (26). For this purpose it is enough to admit only linear functions in η of the form

(27) $$S(x,\eta;\varepsilon) = a_0(x;\varepsilon) + \sum_{k=1}^{n} (x_k + a_k(x;\varepsilon))\eta_k,$$

where the $a_k(x;\varepsilon)$ $(k=1,\dots,n)$ have period 2π in $x_1,\dots,x_n$ while $a_0(x;\varepsilon)$ is of the form

$$a_0 = \sum_{k=1}^{n} c_k(\varepsilon)x_k + b_0(x;\varepsilon)$$

with $b_0(x;\varepsilon)$ of period 2π in $x_1,\dots,x_n$ and the c_k independent of x, η. The canonical transformations generated by such functions $S(x,\eta;\varepsilon)$ can be readily characterized. Indeed, if $S(x,\eta;0) = \sum_{k=1}^{n} x_k\eta_k$, such a transformation takes the explicit form

$$x = u(\xi;\varepsilon), \quad y = v(\xi,\varepsilon) + V(\xi;\varepsilon)\eta,$$

where $u(\xi;\varepsilon), v(\xi;\varepsilon)$ are n-vectors while $V(\xi;\varepsilon)$ is an $n\times n$ matrix, with $u(\xi;\varepsilon) - \xi, v(\xi;\varepsilon), V(\xi;\varepsilon)$ having period 2π in $\xi_1,\dots,\xi_n$. This transformation is canonical, which implies that V^{-1} is equal to the transpose of the matrix $(u_{k\xi_l})$, and for $\varepsilon=0$ it reduces to the identity.

To determine the function $S(x, \eta; \varepsilon)$ with the above properties, we expand it as a power series in ε and, using (26), compare coefficients in equation (25). In this way we will establish the existence of such a power series and at the same time obtain uniqueness under the added normalization

$$S(x, \eta; 0) = \sum_{k=1}^{n} x_k \eta_k \tag{28}$$

and

$$b_0^* = a_k^* = 0 \quad (k=1, \ldots, n), \tag{29}$$

where the asterisk again denotes the mean value taken with respect to the variables $x_1, \ldots, x_n$.

For $\lambda \geqq 0$ an integer, let the coefficients of ε^λ in $H, \Phi, S, \gamma, \ldots$ be denoted by $H^\lambda, \Phi^\lambda, S^\lambda, \gamma^\lambda, \ldots$, and assume that $S^0, \ldots, S^{\lambda-1}$ as well as $\gamma^0, \ldots, \gamma^{\lambda-1}$ have already been determined so that the coefficients of $\varepsilon^0, \ldots, \varepsilon^{\lambda-1}$ on both sides of (25), (26) agree and the normalization (28), (29) is satisfied. For $\lambda=1$ this will certainly be so if we start with (28) and set $\gamma^0 = H^0(0)$. Proceeding by induction, we compare the coefficients of ε^λ for $\lambda \geqq 1$ in (25) and obtain

$$\sum_{k=1}^{n} H^0_{y_k}(\eta) S^\lambda_{x_k}(x, \eta) - \Phi^\lambda(x, \eta) = g(x, \eta), \tag{30}$$

where $g(x, \eta)$ is a known real analytic function of period 2π in $x_1, \ldots, x_n$. The requirement that Φ satisfy (26), as far as the coefficient of ε^λ is concerned, leads further to the relation

$$\sum_{k=1}^{n} H^0_{y_k}(\eta) S^\lambda_{x_k}(x, \eta) - \gamma^\lambda = g(x, \eta) + O(|\eta|^2),$$

so that, expressing $S^\lambda(x, \eta)$ in the form

$$S^\lambda(x, \eta) = \sum_{k=1}^{n} c_k x_k + b(x) + \sum_{k=1}^{n} a_k(x) \eta_k$$

and introducing the notation

$$\omega_k = H^0_{y_k}(0), \qquad \Omega_{kl} = H_{y_k y_l}(0)$$

$$\partial = \sum_{k=1}^{n} \omega_k \frac{\partial}{\partial x_k}, \quad (c, \omega) = \sum_{k=1}^{n} c_k \omega_k$$

$$g_0(x) = g(x, 0), \qquad g_k(x) = g_{\eta_k}(x, 0),$$

we are led to the equations

$$(31)\qquad \begin{cases} \partial b + (c,\omega) - \gamma^\lambda = g_0 \\ \partial a_k + \sum_{l=1}^{n} \Omega_{lk}(b_{x_l} + c_l) = g_k \quad (k=1,\dots,n)\,. \end{cases}$$

A necessary condition for the existence of a multiply periodic solution b, a_k to (31) is that the mean value of both sides agree, leading to the relations

$$(c,\omega) - \gamma^\lambda = g_0^*$$

$$\sum_{l=1}^{n} \Omega_{lk} c_l = g_k^* \quad (k=1,\dots,n)\,,$$

which can be solved uniquely for $c_1, \dots, c_n$ and γ^λ since, by (20), the matrix (Ω_{kl}) is nonsingular. The equations (31) then reduce to

$$\partial b = g_0 - g_0^*$$

$$\partial a_k = h_k(x) - h_k^*$$

with

$$h_k = g_k - \sum_{l=1}^{n} \Omega_{lk} b_{x_l}\,,$$

and these partial differential equations are precisely of the form (4) discussed in the beginning of this section. Since the right-hand sides have mean values 0, these equations possess real analytic solutions $b, a_1, \dots, a_n$ of period 2π in $x_1, \dots, x_n$ and, moreover, these solutions are unique if we require that $b^* = a_k^* = 0$ $(k=1,\dots,n)$. This completes the induction step and proves the existence of a formal power series expansion for the desired function $S(x,\eta;\varepsilon)$.

As a simple application of the results we have derived, we consider the restricted three-body problem. This problem, we recall, is described by the Hamiltonian

$$E(q,p;\mu) = \tfrac{1}{2}(p_1^2 + p_2^2) + q_2 p_1 - q_1 p_2 - F\,,$$

where $F(q;\mu)$ is defined by (21; 29) with q_1, q_2 in place of x_1, x_2. For $\mu = 0$ this reduces to the Kepler problem in a rotating coordinate system, and the elliptical orbits appear as precessing ellipses. Analytically these solutions can be described in the form

$$q_1 + i q_2 = e^{-it} P(\alpha t)\,,$$

where $P(\theta)$ is a complex valued function of period 2π in θ, and the frequency α is related to the major axis $2a$ through Kepler's third law

$(\alpha+1)^2 a^3 = 1$. Clearly, for rational values of α these solutions are periodic, while for irrational α they are quasi-periodic of class $\mathfrak{Q}(1, \alpha)$. Our aim is to find such quasi-periodic solutions for small values of μ that approach the above solutions as $\mu \to 0$. For this it is convenient to introduce the so-called Delaunay variables x_1, x_2, y_1, y_2 in terms of which the Hamiltonian has period 2π in x_1, x_2 and takes the form $H^0(y_1, y_2) + O(\mu)$, so that the previous results apply.

To introduce the Delaunay variables [6], we begin with the Kepler problem described by the Hamiltonian $K = \frac{1}{2}(p_1^2 + p_2^2) - (q_1^2 + q_2^2)^{-\frac{1}{2}}$. For negative values of K the solutions are elliptical and up to rotation can be written in the form

$$q_1 = a(\cos u - e), \quad q_2 = a\sqrt{1-e^2}\,\sin u$$

with

$$t - t_0 = a^{\frac{3}{2}}(u - e\sin u).$$

Here $2a > 0$ is the major axis of the ellipse and $0 \leqq e \leqq 1$ is the eccentricity, while $K = -(2a)^{-1}$. For $0 \leqq e < 1$, the variable x_1 is now introduced by

$$x_1 = u - e\sin u\,.$$

Eliminating u, we can express the $q_k = \phi_k(x_1; a, e)$ $(k = 1, 2)$ as periodic functions of x_1 with period 2π, from which we can again obtain the solutions of the Kepler problem by setting $x_1 = a^{-\frac{3}{2}}(t - t_0)$. The remaining variables x_2, y_1, y_2 are introduced by setting

$$a = y_1^2, \quad y_1\sqrt{1-e^2} = y_2 \tag{32}$$

and expressing the most general elliptical solution of the Kepler problem in the form

$$\begin{cases} q_1 = \psi_1(x, y) = \phi_1 \cos x_2 - \phi_2 \sin x_2 \\ q_2 = \psi_2(x, y) = \phi_1 \sin x_2 + \phi_2 \cos x_2\,, \end{cases} \tag{33}$$

where x_2, y_1, y_2 are constant and $\dot{x}_1 = y_1^{-3}$. These equations also serve to define the functions ψ_1, ψ_2. We extend these formulas by adding the relations

$$p_k = y_1^{-3}\frac{\partial \psi_k(x, y)}{\partial x_1} \quad (k = 1, 2)\,, \tag{34}$$

and consider (33), (34) as a coordinate transformation of x_1, x_2, y_1, y_2 into q_1, q_2, p_1, p_2. To avoid collision orbits as well as circular orbits, on which x_2 becomes meaningless, we restrict ourselves to $a>0, 0<e<1$, or

$$0<y_2^2<y_1^2 .$$

It can be verified that this transformation is canonical and has period 2π in x_1, x_2, while the transformed differential equations are governed by the Hamiltonian $K=-(2a)^{-1}=-\frac{1}{2}y_1^{-2}$. We will restrict y_1 to positive values and admit both signs for y_2, so that

$$0<|y_2|<y_1 .$$

One sees that $y_2>0$ corresponds to counterclockwise motion with increasing t, and $y_2<0$ to clockwise motion. The variables x_1, x_2 are known in astronomy as the mean anomaly and the longitude of the perihelion, respectively.

We now apply the transformation (33), (34) to the restricted three-body problem. In order to avoid collisions with the particle of mass μ, we require that $a(1+e)<1$ or, in terms of y_1, y_2, that $y_1<(2-y_2^2)^{-\frac{1}{4}}<1$. Thus we restrict ourselves to the domain $\mathfrak{D}$ given by

$$0<|y_2|<y_1<(2-y_2^2)^{-\frac{1}{4}}<1 ,$$

which has two components, one corresponding to clockwise, the other to counterclockwise motion. It is useful to interpret the quantities $2a, e$ related to y_1, y_2 by (32) as the major axis and the eccentricity, respectively, of an osculating ellipse. The Hamiltonian $E(q, p; \mu)$ is seen to transform into

$$H(x, y; \mu)=H^0(y)+O(\mu), \quad H^0(y)=-\frac{1}{2y_1^2}-y_2 ,$$

where $H(x, y; \mu)$ is analytic in $\mathfrak{D}$ for μ sufficiently small. The additional term $-y_2$ in $H^0(y)$ is due to the uniform rotation of the coordinate system.

Thus $H(x, y; \mu)$ has the form required for the application of our previous results and, since $H^0(y)$ clearly satisfies (15), we conclude that on each energy surface $H(x, y; \mu)=h$ for which $H^0(y)=h$ intersects $\mathfrak{D}$ there exist quasi-periodic solutions with two frequencies. These solutions can be viewed as continuations of the precessing elliptical orbits that correspond to $\mu=0$. They clearly have the property that $|y_k(t)-y_k(0)|<c\mu$ $(k=1,2)$ for all real t, where c is a constant independent of t, μ. In other words, the major axis and the eccentricity of the osculating ellipses change little with time if μ is sufficiently small and, in particular, the solutions remain bounded and avoid collisions for all t.

An analogous statement holds for the other, possibly not quasi-periodic, solutions. More precisely, we will show that for each $\varepsilon > 0$ and any compact subdomain $\mathfrak{D}_1$ of $\mathfrak{D}$ there exists a positive $\mu_0 = \mu_0(\varepsilon, \mathfrak{D}_1)$ such that for $y(0) \in \mathfrak{D}_1$ and $0 \leqq \mu < \mu_0$ we have

$$|y_k(t) - y_k(0)| \leqq \varepsilon \quad (k = 1, 2)$$

for all real t. This again means that for a solution starting in $\mathfrak{D}_1$ the shape of the osculating ellipse changes very little with time if μ is small.

To prove the above statement, consider the collection of rectangles R defined by

$$(35) \qquad |y_1 - \eta_1| \leqq \varepsilon, \quad |y_2 - \eta_2| \leqq \varepsilon/2$$

for η in $\mathfrak{D}_1$, with ε chosen so small that all these rectangles lie in $\mathfrak{D}$ while all complex y_1, y_2 satisfying (35) lie in the domain of analyticity of $H(x, y; \mu)$ for $|\mu|$ small. We claim that if $\eta = y(0)$, where $x(t), y(t)$ is a solution of the given system, then $y(t)$ remains in the interior of the corresponding R for all real t, provided that μ is small enough. To verify this claim, we observe that along any solution the function $H(x, y; \mu)$ is equal to a constant, which we denote by

$$h = H(x(0), y(0); \mu) = H^0(y(0)) + O(\mu).$$

Along the curve $H_0(y) = h$ in $\mathfrak{D}$ we have

$$0 < \frac{dy_1}{dy_2} = y_1^3 < 1,$$

so that if y, y^* are two points on this curve we have

$$|y_1 - y_1^*| \leqq |y_2 - y_2^*|.$$

Similarly, for any two four-vectors x, y and x^*, y^* on the energy surface $H(x, y; \mu) = H(x^*, y^*; \mu) = h$ with y and y^* in $\mathfrak{D}$ we have

$$(36) \qquad |y_1 - y_1^*| \leqq |y_2 - y_2^*| + O(\mu).$$

Choosing μ so small that the last term in the above inequality can be estimated by $\frac{\varepsilon}{2}$, assume now that $\eta = y(0)$ and that our solution $y(t)$ escapes the corresponding rectangle R. Let t^* be the smallest positive value of t for which $y(t)$ lies on the boundary of R, so that $y(t)$ lies in the interior of R for $0 \leqq t < t^*$. Setting $y = y(t)$, $y^* = y(0)$ in (36), we have

$$|y_1(t) - \eta_1| < |y_2(t) - \eta_2| + \frac{\varepsilon}{2} \leqq \varepsilon$$

for $0 \leqq t \leqq t^*$, and it follows that

$$|y_2(t^*) - \eta_2| = \frac{\varepsilon}{2}.$$

In other words, $y(t)$ can escape R only through the sides of the rectangle that are parallel to the y_1-axis. But this escape, on the other hand, is blocked by the invariant tori we constructed earlier. Indeed, if we apply our earlier results to $H(x, y; \mu)$ with y restricted to R_+ or R_- defined by

$$|y_1 - \eta_1| < \varepsilon, \quad 0 < \pm(y_2 - \eta_2) < \frac{\varepsilon}{2},$$

we find that (15) is satisfied, and therefore each of the corresponding domains in the 4-dimensional (x, y)-space contains a 2-dimensional invariant torus which lies on the 3-dimensional energy surface $H(x, y; \mu) = h$. This torus serves as a barrier for our solution, and $y_2(t)$ can not pass from $y_2 = \eta_2$ to $y_2 = \eta_2 \pm \frac{\varepsilon}{2}$. Thus a first time of exit t^* actually does not exist, and $y(t)$ must remain in R, and therefore in $\mathfrak{D}$, for all $t \geqq 0$. The same argument applied to negative t now completes the proof.

In this discussion we had to exclude orbits of small eccentricity, but this was due only to our choice of coordinates and, in fact, the result in § 34 on stability of periodic solutions covers precisely that situation. Our previous argument did, however, depend crucially on the fact that the two-dimensional invariant tori formed the boundary of open sets on the three-dimensional energy surface. For problems with n degrees of freedom, for $n > 2$, the energy surface has $2n - 1$ dimensions and therefore requires $2n - 2$ dimensions for the boundary of an open set, while the invariant tori provided by the theory have only n dimensions. For this reason, there are no analogous stability theorems for more than 2 degrees of freedom.

Finally, for a problem of more than two degrees of freedom, we consider the three-body problem. If we restrict ourselves to the planar case, there are, at first, six degrees of freedom, but taking in account the center of mass integrals, we can reduce the problem to one of four degrees of freedom. Assuming that two of the masses, say m_1, m_2, are small compared to the third mass m_0, which can be normalized to be 1, Arnold [7] established the existence of quasi-periodic solutions with four rationally independent frequencies. More precisely, we assume that $m_k = \alpha_k \mu$ $(k = 1, 2)$ depend on a single small parameter μ, with α_1, α_2 fixed positive constants, and describe the relative state of the "planets" of masses m_1, m_2 with respect to the "sun" of mass $m_0 = 1$ by two position vectors and two velocity vectors, thus obtaining an eight-dimensional phase space. For small values of μ we would expect solutions for which the planets move in approximately elliptical orbits, at least for a limited time, and therefore it is appropriate to describe the points in the phase space by the geometrical quantities describing the osculating ellipses. If $2a_k$ denote the major axes and e_k the eccentricies of the corresponding ellipses, we consider the open set Ω of the form

$$c_k < a_k < C_k, \quad 0 < e_k < \varrho \quad (k=1,2)$$

in the eight-dimensional phase space. Arnold's statement then asserts that given constants α_k, c_k, C_k satisfying $0 < c_1 < C_1 < c_2 < C_2$, and $\varepsilon > 0$, there exists $\delta = \delta(\varepsilon) > 0$ such that for $0 \leqq \mu < \delta$, $0 < \varrho < \delta$ the set Ω contains a closed subset S with the following properties: every solution starting in S remains in S for all real t and is quasi-periodic with four independent frequencies; S is the union of four-dimensional invariant tori, each covered densely by a solution on it; and, the complement $\Omega - S$ has Lebesgue measure $m(\Omega - S) < \varepsilon m(\Omega)$. In particular, none of the solutions in the invariant set S ever encounters a collision.

In contrast to the restricted three-body problem, however, no assertion can be made about the solutions starting in the exceptional set $\Omega - S$. While for solutions in S the variables a_1, a_2, for example, vary only between constants differing by amounts of order at most μ, it is conceivable that there are exceptional solutions starting in $\Omega - S$ for which a_1, a_2 vary by amounts of finite order and eventually encounter a collision. Indeed, Arnold [8] constructed a remarkable example of a Hamiltonian system of three degrees of freedom in which precisely a phenomenon of this type occurs. The reason for the difference in behavior of the three-body problem and the restricted three-body problem is that in the former the four-dimensional invariant tori do not bound open domains in the seven-dimensional energy surface. Actually one could reduce this problem to one of three degrees of freedom by using the angular momentum integral, but the disparity in dimension would still remain. Very little is known about the long time behavior of solutions in the exceptional set, both for the three-body problem and other systems of more than two degrees of freedom.

Arnold extended these results on quasi-periodic solutions for the planar three-body problem to the general three-dimensional case, and even to the N-body problem for $N \geqq 3$. In principle the proofs are based on the ideas developed in the previous sections, although one encounters serious difficulties due to the degenerate character of the system for $\mu = 0$, leading to violation of the condition (20) as well as (23). Moreover, some of the frequencies of the quasi-periodic solutions approach 0 as $\mu \to 0$. These difficulties are overcome by carefully improved approximations, for which we refer to [5, 7]. For recent books discussing questions of stability we refer to [9, 10].

§ 37. The Recurrence Theorem

We begin with a system of differential equations

(1) $$\dot{x}_k = f_k(x) \qquad (k = 1, \ldots, v),$$

where the functions f_k need not be regular but are at least continuously differentiable in a real domain $\mathfrak{R}$ under consideration. In addition we assume that

(2) $$\sum_{k=1}^{v} f_{kx_k} = 0$$

throughout this domain, which in particular is true for Hamiltonian systems. If $x(t, \xi)$ again denotes the solution to (1) with $x(0, \xi) = \xi$, then according to § 21 the mapping S_t taking ξ into $x(t, \xi)$ is measure-preserving. From now on we will consider only those initial values ξ for which the orbit $x = x(t, \xi)$ lies in the domain $\mathfrak{R}$ for all real time t. If $\mathfrak{M}$ is any set of such orbits $x(t, \xi)$ then clearly $S_t\mathfrak{M} = \mathfrak{M}$, so that $\mathfrak{M}$ is invariant.

In what follows we will use some results from the theory of Lebesgue measure. For $\mathfrak{Q}$ an arbitrary set in v-dimensional Euclidean space, let $m^*(\mathfrak{Q})$ denote the outer Lebesgue measure of $\mathfrak{Q}$, and if $\mathfrak{Q}$ is measurable in the sense of Lebesgue, let $m(\mathfrak{Q})$ denote its measure. It will also be assumed that the set $\mathfrak{M}$ of orbits under consideration has a finite outer measure $m^*(\mathfrak{M})$. Let $\mathfrak{A}$ be a measurable subset of $\mathfrak{M}$ and let $\mathfrak{A}_n = S_{n\tau}\mathfrak{A}$ $(n = 0, \pm 1, \ldots)$, where τ is an arbitrary fixed positive number. We abbreviate $S_\tau = S$ and consider the sets

$$\mathfrak{B}_n = \bigcup_{k \leq n} \mathfrak{A}_n \qquad (n = 0, \pm 1, \ldots),$$

where k ranges over all integers $\leq n$. Clearly

$$S\mathfrak{B}_n = \mathfrak{B}_{n+1} = \mathfrak{B}_n \cup \mathfrak{A}_{n+1} \supset \mathfrak{B}_n,$$

and being a countable union of measurable sets the set $\mathfrak{B}_n$ itself is measurable, while being a subset of $\mathfrak{M}$ it has a finite measure. Since S is a measure-preserving mapping, we have $m(\mathfrak{B}_{n+1}) = m(S\mathfrak{B}_n) = m(\mathfrak{B}_n)$ and therefore the difference $\mathfrak{B}_{n+1} - \mathfrak{B}_n$ is a set of measure zero. Let $\mathfrak{B}_{-\infty}$ denote the intersection of all the $\mathfrak{B}_n$ $(n = 0, \pm 1, \ldots)$. From $\mathfrak{B}_n \subset \mathfrak{B}_{n+1}$ it follows that

$$\mathfrak{B}_0 - \mathfrak{B}_{-\infty} = \bigcup_{k<0} (\mathfrak{B}_{k+1} - \mathfrak{B}_k),$$

so that also $\mathfrak{B}_0 - \mathfrak{B}_{-\infty}$ is a set of measure zero. Finally, we set $\mathfrak{D} = \mathfrak{A} \cap \mathfrak{B}_{-\infty}$ and, in view of $\mathfrak{A} = \mathfrak{A}_0 \subset \mathfrak{B}_0$, obtain the relation

(3) $$\mathfrak{A} = \mathfrak{A} \cap \mathfrak{B}_0 = \mathfrak{D} \cup \{\mathfrak{A} \cap (\mathfrak{B}_0 - \mathfrak{B}_{-\infty})\}.$$

Since the intersection $\mathfrak{A} \cap (\mathfrak{B}_0 - \mathfrak{B}_{-\infty})$ has measure zero, by (3) also the difference $\mathfrak{A} - \mathfrak{D}$ has zero measure, so that we end up with

$$m(\mathfrak{A} - \mathfrak{D}) = 0 \quad (\mathfrak{D} = \mathfrak{A} \cap \mathfrak{B}_{-\infty}). \tag{4}$$

To interpret this result, let us consider all the images $\mathfrak{p}_n = S^n \mathfrak{p}$ of a point $\mathfrak{p}$ in $\mathfrak{M}$. For $\mathfrak{p}$ to lie in $\mathfrak{B}_n$ it is necessary and sufficient that there exist an integer $k \leqq n$ such that $\mathfrak{p} \in \mathfrak{A}_k$, or equivalently $\mathfrak{p}_{-k} \in \mathfrak{A}$. In particular, $\mathfrak{p}$ will lie in $\mathfrak{B}_{-\infty}$ if and only if there exists a sequence $k \to -\infty$ with this property, meaning that there exists a sequence of integers $l = l_1, l_2, \ldots$ with $l \to \infty$ such that $\mathfrak{p}_l \in \mathfrak{A}$. Consequently (4) says that:

Each measurable subset $\mathfrak{A}$ of $\mathfrak{M}$ contains a set $\mathfrak{D}$ of the same measure each of whose points $\mathfrak{p}$ has infinitely many image points $\mathfrak{p}_l$ ($l = l_1, l_2, \ldots$; $l \to \infty$) in $\mathfrak{A}$.

This is the recurrence theorem of Poincaré [1, 2]. For the case when $\mathfrak{M}$ itself is measurable, the theorem can be formulated in another way. Let $\mathfrak{C}_1, \mathfrak{C}_2, \ldots$ be a countable basis of open sets for the v-dimensional x-space, for example all spheres with rational coordinates for their centers and rational radii. The intersections $\mathfrak{M} \cap \mathfrak{C}_r = \mathfrak{A}_r$ ($r = 1, 2, \ldots$) are then again measurable, and we apply the recurrence theorem to $\mathfrak{A} = \mathfrak{A}_r$. Since a countable union of sets of measure zero again has measure zero, it follows that each point of $\mathfrak{M}$ outside a certain set of measure zero is an accumulation point of its images $\mathfrak{p}_n$ ($n = 1, 2, \ldots$).

For applications of the recurrence theorem it should be stressed that $\mathfrak{M}$ has to be an invariant set of finite outer measure in the domain of definition $\mathfrak{R}$. For example, if $f_1 = 1$, $f_2 = 0$, $v = 2$ we can take $\mathfrak{R}$ to be the whole plane, and the orbits are then parallel to the abscissa axis; however, for $m^*(\mathfrak{M})$ to be finite, $\mathfrak{M}$ itself must then have measure zero, and the assertion of the recurrence theorem becomes vacuous. To obtain an actual result for a given system (1) using the recurrence theorem, one must already know enough about the behavior of the orbits in the large to be able to prove existence of measurable invariant sets with a positive finite measure. Namely, one needs an invariant set $\mathfrak{M}$ having finite $m^*(\mathfrak{M})$ and containing a measurable set $\mathfrak{A}$ with $m(\mathfrak{A}) > 0$; but then it readily follows that the collection of orbits passing through the points of $\mathfrak{A}$ forms a measurable invariant set of finite positive measure. Examples are obtained from stationary incompressible flows in a closed container, or from having a stable equilibrium solution of (1), in which case for $\mathfrak{M}$ one can take an invariant neighborhood of this solution.

A deeper application arises in connection with the restricted three-body problem. As before, let the three points P_1, P_2, P_3 have the respective masses $m_1 = \mu$, $m_2 = 1 - \mu$, $m_3 = 0$, with $0 < \mu < 1$. The points P_1, P_2 rotate about their center of mass with angular velocity 1, and the coordinates of

P_1, P_2, P_3 in the rotating coordinate system are $(1-\mu, 0)$, $(-\mu, 0)$, (x_1, x_2) respectively, so that the respective distances P_3P_1, P_3P_2 and the distance of P_3 from the origin P_0 are given by

$$r_1 = \{(x_1+\mu-1)^2 + x_2^2\}^{\frac{1}{2}}, \quad r_2 = \{(x_1+\mu)^2 + x_2^2\}^{\frac{1}{2}}, \quad r = (x_1^2 + x_2^2)^{\frac{1}{2}}.$$

As noted in § 24, setting

$$E = \frac{1}{2}(y_1^2 + y_2^2) + x_2 y_1 - x_1 y_2 - \frac{\mu}{r_1} - \frac{1-\mu}{r_2}, \tag{5}$$

we can express the equations of motion for P_3 in the canonical form

$$\dot{x}_k = E_{y_k}, \quad \dot{y}_k = -E_{x_k} \quad (k=1,2). \tag{6}$$

From this it follows that $\dot{x}_1 = y_1 + x_2$, $\dot{x}_2 = y_2 - x_1$, so that upon elimination of y_1, y_2 the Hamiltonian function can be expressed in the form

$$E = \tfrac{1}{2}(\dot{x}_1^2 + \dot{x}_2^2) - G \tag{7}$$

with

$$G = \frac{1}{2}r^2 + \frac{\mu}{r_1} + \frac{1-\mu}{r_2}. \tag{8}$$

The function E, which is known as the Jacobian integral, is constant along each orbit.

As before, we associate with the system (6) the measure-preserving mapping S_t in the space of the four variables x_1, x_2, y_1, y_2, while for the domain of definition $\mathfrak{R}$ we take the set of all points that do not lie on the two-dimensional planes $x_1 = 1-\mu$, $x_2 = 0$ and $x_1 = -\mu$, $x_2 = 0$. Let $\mathfrak{L}$ denote the set of all points (x_1, x_2, y_1, y_2) for which the function E defined by (5) satisfies the inequality

$$c_1 < -E < c_2, \tag{9}$$

where c_1, c_2 are two positive constants with $c < c_1 < c_2$ for a sufficiently large positive c. Then $\mathfrak{L}$ is an open subset of $\mathfrak{R}$ and indeed invariant, since E is an integral. Along the orbits in $\mathfrak{L}$ we have according to (7), (9) the inequality

$$G = \tfrac{1}{2}(\dot{x}_1^2 + \dot{x}_2^2) - E > c_1 > c.$$

With G defined by (8), we consider for a fixed c the curve $G = c$, in the (x_1, x_2)-plane which forms the boundary of what is known as Hill's region. For a large c this curve consists of three simple closed curves $\mathfrak{K}_0, \mathfrak{K}_1, \mathfrak{K}_2$ which have equations of the respective forms

$$r = (2c)^{\frac{1}{2}} + O(c^{-\frac{1}{2}}), \; r_1 = \mu c^{-1} + O(c^{-2}), \; r_2 = (1-\mu)c^{-1} + O(c^{-2}) \; (c \to \infty)$$

and consequently can be approximated by the circles with radii $(2c)^{\frac{1}{2}}$, $\mu c^{-1}, (1-\mu) c^{-1}$ and centers P_0, P_1, P_2 respectively. The two-dimensional region $G > c$ thus decomposes into three disjoint regions, namely the exterior of $\mathfrak{K}_0$ and the interiors of $\mathfrak{K}_1$ and $\mathfrak{K}_2$, which we denote by $\mathfrak{F}_0, \mathfrak{F}_1, \mathfrak{F}_2$. Analogously, the four-dimensional region $\mathfrak{L}$ decomposes into three disjoint regions $\mathfrak{L}_0, \mathfrak{L}_1, \mathfrak{L}_2$ each of which remains invariant, since S_t is continuous in t. For the application of the recurrence theorem we will, in particular, choose $\mathfrak{M} = \mathfrak{L}_1$.

The above discussion requires an additional consideration in that there may also appear collision orbits. However, if a collision between P_1 and P_3 is regularized as in § 8 for the three-body problem, it follows that the collision orbits form only a set of measure zero in $\mathfrak{L}_1$, which for our purposes can be neglected.

The coordinates x_1, x_2 of the points from $\mathfrak{L}_1$ lie in the bounded set $\mathfrak{F}_1$, while over each fixed $(x_1, x_2) \neq (1-\mu, 0)$ in $\mathfrak{F}_1$ the function G is finite, with the admissible coordinates y_1, y_2 being restricted by the condition

$$2(G-c_2) < (y_1+x_2)^2 + (y_2-x_1)^2 < 2(G-c_1).$$

In the (y_1, y_2)-plane this condition defines an annulus whose area is $\leqq 2\pi(c_2-c_1)$ independently of x_1, x_2. Since also the two-dimensional area of $\mathfrak{F}_1$ is finite, it follows that $m(\mathfrak{L}_1)$ is finite, and the recurrence theorem now says that for almost all initial values in $\mathfrak{L}_1$ the point P_3 keeps returning after arbitrarily large time intervals arbitrarily closely to its initial values, both as to its position and velocity. The same can, of course, be said for $\mathfrak{L}_2$, and it is also readily seen that a corresponding assertion can be made for Hill's problem.

The ideas needed to prove the recurrence theorem were developed by Birkhoff [3] and others for ergodic theory. However, the possibility of applying this theory to a given system of differential equations is limited by difficulties that may prove to be more imposing than those in the problem of stability. In this connection we refer to a beautiful result of Denjoy [4–6].

In conclusion, we make still another observation about the n-body problem that goes back to Schwarzschild [7–9] and is derived from the central ideas of the recurrence theorem. Again we begin with the system (1), for which (2) is assumed to hold. Let $\mathfrak{A}$ be an open set in the domain of definition $\mathfrak{R}$, with the measure $m(\mathfrak{A})$ finite, and for each $\tau > 0$ let $\mathfrak{A}^\tau$ be the set of all points $\mathfrak{p}$ in $\mathfrak{A}$ for which the corresponding orbit remains in $\mathfrak{A}$ for the whole time interval $0 \leqq t \leqq \tau$, so that $\mathfrak{p}^t = S_t \mathfrak{p} \in \mathfrak{A}$ $(0 \leqq t \leqq \tau)$. Then certainly $\mathfrak{A}^{\tau_2} \subset \mathfrak{A}^{\tau_1}$ if $0 < \tau_1 < \tau_2$, and we denote the intersection of all the $\mathfrak{A}^\tau$ $(\tau > 0)$ by $\mathfrak{B}$. Since $\mathfrak{A}^\tau$ is an open subset of $\mathfrak{A}$, the set

$$\bigcap_{\tau > 0} \mathfrak{A}^\tau = \lim_{\tau \to \infty} \mathfrak{A}^\tau = \mathfrak{B}$$

is measurable and $m(\mathfrak{B})$ is finite. The points $\mathfrak{p}$ in $\mathfrak{B}$ are characterized by the property that the orbit $\mathfrak{p}^t$ remains in $\mathfrak{A}$ for all positive t, and we will express this by saying that $\mathfrak{B}$ is faithful to $\mathfrak{A}$ with respect to the future.

For each $\tau > 0$ we also define the set $\mathfrak{B}^\tau = S_\tau \mathfrak{B}$ for whose points $\mathfrak{p}$ we have $\mathfrak{p}^t \in \mathfrak{A}$ $(t \geqq -\tau)$. Consequently $\mathfrak{B}^\tau$ is a measurable subset of $\mathfrak{B}$, and again $\mathfrak{B}^{\tau_2} \subset \mathfrak{B}^{\tau_1}$ if $0 < \tau_1 < \tau_2$. It follows that also

$$\bigcap_{\tau > 0} \mathfrak{B}^\tau = \lim_{\tau \to \infty} \mathfrak{B}^\tau = \mathfrak{D}$$

is a measurable subset of $\mathfrak{B}$, while from the measure-preserving character of our mapping we can draw the same conclusion as in (4), namely

$$m(\mathfrak{B} - \mathfrak{D}) = 0\,. \tag{10}$$

The points $\mathfrak{p}$ in $\mathfrak{D}$ are characterized by the property that the whole orbit $\mathfrak{p}^t$ remains in $\mathfrak{A}$ for all real t, or in other words, $\mathfrak{D}$ is faithful to $\mathfrak{A}$ for all time. The formula (10) thus says that the set of points faithful to $\mathfrak{A}$ with respect to the future exceeds the set of points faithful to $\mathfrak{A}$ for all time only by a set of measure zero. For this assertion to have content, one must of course show in a particular case that $m(\mathfrak{B}) > 0$, and that may indeed be quite difficult.

We apply this result to the n-body problem, retaining the notation of § 5. Let q_k $(k = 1, \ldots, 3n)$ denote the rectangular coordinates of the n particles $P_1, \ldots, P_n$, ordered as before, and let p_k be the corresponding impulse coordinates. Again the center of mass may be assumed to be at rest, and in analogy to the relative coordinates for the three-body problem introduced in § 7, we set $x_k = q_k - q_{3n-3+\varkappa}$, $y_k = p_k$ $(k = 1, \ldots, 3n-3)$ with $\varkappa = 1, 2, 3$ congruent to k modulo 3. With the potential function U given by (5;2) and the kinetic energy T by (5;10), the Hamiltonian function becomes $H = T - U$ and the equations of motion are described by the corresponding canonical system in the $6n-6$ new coordinates x_k, y_k, so that also (2) holds. If r_{kl} $(k, l = 1, \ldots, n)$ again denotes the distance between the particles P_k, P_l $(k \neq l)$, the function H is regular provided that all $r_{kl} > 0$. For $s > 1$ sufficiently large, let $\mathfrak{A}(s)$ be the set of all points x, y in our $(6n-6)$-dimensional coordinate space for which

$$s^{-1} < r_{kl} < s \quad (1 \leqq k < l \leqq n)\,, \qquad -s < H < s\,. \tag{11}$$

This is an open set on which U and $T = H + U$ are bounded. From this it follows that $\mathfrak{A}(s)$ has finite measure, so that the above theorem applies, and we conclude that the set $\mathfrak{B}(s)$ of all points faithful to $\mathfrak{A}(s)$ in the future exceeds the set $\mathfrak{D}(s)$ of points faithful to $\mathfrak{A}(s)$ for all time only by a set of measure 0. Furthermore, for $s_1 < s_2$ we have $\mathfrak{A}(s_1) \subset \mathfrak{A}(s_2)$, $\mathfrak{B}(s_1) \subset \mathfrak{B}(s_2)$, $\mathfrak{D}(s_1) \subset \mathfrak{D}(s_2)$, so that we can construct $\lim\limits_{s \to \infty} \mathfrak{A}(s) = \mathfrak{A}$,

$\lim_{s\to\infty} \mathfrak{B}(s) = \mathfrak{B}$, $\lim_{s\to\infty} \mathfrak{D}(s) = \mathfrak{D}$ and obtain the corresponding result for $\mathfrak{A}, \mathfrak{B}, \mathfrak{D}$. Then $\mathfrak{B}$ is the set of points $\mathfrak{p}$ for which there exists a number $s > 1$, independent of t, such that the orbit $\mathfrak{p}^t$ remains in the region (11) for all $t \geqq 0$, while for $\mathfrak{D}$ the above must be true for all real t. The number s, however, may depend on $\mathfrak{p}$. Since H is an integral, this says that for $\mathfrak{B}$ all the distances r_{kl} remain between positive bounds for all future time, while for $\mathfrak{D}$ this is true for all future and past time; with the bound, however, permitted to depend on the initial point $\mathfrak{p}$. The orbits through the points of $\mathfrak{B}$ are said to be weakly stable with respect to the future, while those through $\mathfrak{D}$ are weakly stable with respect to all time. It then follows that almost all solutions of the n-body problem weakly stable with respect to the future are actually weakly stable with respect to all time.

Under the unproved assumption that the planetary system is weakly stable with respect to all time, we can draw the following conclusion. If the planetary system captures a particle coming in from infinity, say some external matter, then the new system with this additional particle is no longer weakly stable with respect to all time. Consequently, if we neglect a certain null set of exceptional initial values, the new system is also not weakly stable with respect to just future time, and it follows that the particle – or a planet, or the sun – must be again expelled, or else a collision must take place. For an interpretation of the significance of this result one must, however, consider that we do not even know whether for $n > 2$ the solutions of the n-body problem that are weakly stable with respect to all time actually form a set of positive measure.

If we assume that one mass is much larger than the others, we know from the remarks at the end of the last section that the quasi-periodic motions form a set of positive measure and hence so do the orbits which are weakly stable for all time. However, it seems pointless to use these deep investigations of Arnold in this connection since they yield more information about the nature of the orbits than the crude methods of this section.

In the recent work of Alekseev [10], it is shown that for the three-body problem such capture orbits actually do exist even though they form a set of measure zero. More precisely, he establishes the existence of solutions which are bounded for positive time but unbounded for $t \to -\infty$, provided the energy constant is negative and the masses are suitably restricted. These orbits have the cardinality of the continuum. This is a particular application of Alekseev's interesting more general investigations which also give information about other solutions of the three-body problem with prescribed asymptotic behavior for $t \to +\infty$ and $t \to -\infty$.

Bibliography

For § 5:

1. Lejeune Dirichlet, G.: Werke, vol. 2, p. 344. Berlin 1897.
2. Mittag-Leffler, G.: Zur Bibliographie von Weierstrass. Acta Math. **35**, 29—65 (1912).
3. Poincaré, H.: Sur le problème des trois corps et les équations de la dynamique. Acta Math. **13**, 1—271 (1890).
4. Sundman, K. F.: Mémoire sur le problème des trois corps. Acta Math. **36**, 105—179 (1913).
5. Bruns, H.: Über die Integrale des Vielkörper-Problems. Acta Math. **11**, 25—96 (1887—1888).

For § 6:

1. Levi-Civita, T.: Sur la régularisation du problème des trois corps. Acta Math. **42**, 99—144 (1920).

For § 12:

1. Sundman, K. F.: Recherches sur le problème des trois corps. Acta Soc. Sci. Fennicae **34**, Nr. 6 (1907).

For § 13:

1. Block, H.: Sur une classe de singularités dans le problème de *n* corps. Thèse pour le doctorat, Lund 1909.
2. Chazy, J.: Sur certaines trajectoires du problème des *n* corps. Bull. Astron. **35**, 321—389 (1918).
3. Siegel, C. L.: Der Dreierstoß. Ann. Math. **42**, 127—168 (1941).

For § 14:

1. Lagrange, J. L.: Oeuvres, vol. 6, p. 272—292. Paris 1873.
2. Euler, L.: De motu rectilineo trium corporum se mutuo attrahentium. Novi Comm. Acad. Sci. Imp. Petrop. **11**, 144—151 (1767).

For § 19:

1. Hill, G. W.: Researches in the lunar theory. Am. J. Math. **1**, 5—26, 129—147, 245—260 (1878).
2. Wintner, A.: Zur Hillschen Theorie der Variation des Mondes. Math. Z. **24**, 259—265 (1926).
3. Hopf, E.: Über die geschlossenen Bahnen in der Mondtheorie. Sitzber. Preuß. Akad. Wiss., phys.-math. Kl. **1929**, 401—413.

For § 20:

1. Siegel, C. L.: Über eine periodische Lösung im ebenen Dreikörperproblem. Math. Nachr. **4**, 28—35 (1950—1951).
2. Brown, E. W.: On the part of the parallactic inequalities in the moon's motion which is a function of the mean motions of the sun and moon. Am. J. Math. **14**, 141—160 (1892).
3. Moulton, F. R.: A class of periodic solutions of the problem of three bodies with application to the lunar theory. Trans. Am. Math. Soc. **7**, 537—577 (1906).
4. Perron, O.: Über eine Schar periodischer Lösungen des ebenen Dreikörperproblems (Mondbahnen). Sitzber. Bayer. Akad. Wiss., math.-nat. Abt. **1936**, 157—176.
5. Crandall, M. G.: Two families of periodic solutions in the plane four-body problem. Am. J. Math. **89**, 275—318 (1967).
6. Perron, O.: Über eine Schar periodischer Lösungen des ebenen Vierkörperproblems. Math. Ann. **113**, 95—109 (1937).
7. — Neue periodische Lösungen des ebenen Drei- und Mehrkörperproblems. Math. Z. **42**, 593—624 (1937).

For § 21:

1. Poincaré, H.: Les méthodes nouvelles de la mécanique céleste, vol. 1, chap. 3. Paris 1892.
2. Wintner, A.: Grundlagen einer Genealogie der periodischen Bahnen im restringierten Dreikörperproblem. I. Math. Z. **34**, 321—349 (1932).

For § 22:

1. Poincaré, H.: Sur un théorème de géométrie. Rend. Circ. mat. Palermo **33**, 375—407 (1912).
2. Birkhoff, G. D.: Proof of Poincaré's geometric theorem. Trans. Am. Math. Soc. **14**, 14—22 (1913).
3. Poincaré, H.: Les méthodes nouvelles de la mécanique céleste, vol. 3, chap. 22. Paris 1899.

For § 23:

1. Birkhoff, G. D.: Surface transformations and their dynamical applications. Acta Math. **43**, 1—119 (1922).
2. Rüssmann, H.: Über die Existenz einer Normalform inhaltstreuer elliptischer Transformationen. Math. Ann. **137**, 64—77 (1959).
3. — Über die Normalform analytischer Hamiltonscher Differentialgleichungen in der Nähe einer Gleichgewichtslösung. Math. Ann. **169**, 55—72 (1967).
4. Moser, J.: The analytic invariants of an area-preserving mapping near a hyperbolic fixed point. Comm. Pure Appl. Math. **9**, 673—692 (1956).
5. Siegel, C. L.: Vereinfachter Beweis eines Satzes von J. Moser. Comm. Pure Appl. Math. **10**, 305—309 (1957).

For § 24:

1. Birkhoff, G. D.: Nouvelles recherches sur les systèmes dynamiques. Mem. Pont. Acad. Sci. Novi Lyncaei (3) **1**, 85—216 (1935).
2. Moser, J.: Periodische Lösungen des restringierten Dreikörperproblems, die sich erst nach vielen Umläufen schließen. Math. Ann. **126**, 325—335 (1953).
3. Conley, C.: On some new long periodic solutions of the plane restricted three-body problem. Comm. Pure Appl. Math. **16**, 449—467 (1963).

For § 25:

1. Schröder, E.: Über iterierte Functionen. Math. Ann. **3**, 296—322 (1871).
2. Cremer, H.: Über die Häufigkeit der Nichtzentren. Math. Ann. **115**, 573—580 (1938).

For § 26:

1. Kolmogorov, A. N.: On conservation of conditionally periodic motions under small perturbations of the Hamiltonian. Dokl. Akad. Nauk SSSR **98**, 527—530 (1954) [Russian].
2. Arnold, V. I.: Proof of A. N. Kolmogorov's theorem on the preservation of quasi-periodic motions under small perturbations of the Hamiltonian. Usp. Mat. Nauk SSSR **18**, no. 5 (113), 13—40 (1963) [Russian].
3. — Singularities of differentiable mappings. Usp. Mat. Nauk SSSR **23**, 3—44 (1968), in particular, p. 28—29 [Russian].
4. Siegel, C. L.: Iteration of analytic functions. Ann. Math. **43**, 607—612 (1942).

For § 27:

1. Poincaré, H.: Oeuvres, vol. 1, p. 95—114. Paris 1951.
2. Dulac, H.: Détermination et intégration d'une certaine classe d'equations différentielles ayant pour point singulier un centre. Bull. Soc. Math. France (2) **32**, 230—252 (1908).
3. Frommer, M.: Über das Auftreten von Wirbeln und Strudeln (geschlossener und spiraliger Integralkurven) in der Umgebung rationaler Unbestimmtheitsstellen. Math. Ann. **109**, 395—424 (1934).
4. Saharnikov, N. A.: On Frommer's conditions for the existence of a center. Prikl. Mat. Meh. Akad. Nauk SSSR **12**, 669—670 (1948) [Russian].

For § 28:

1. Liapounoff, A.: Problème général de la stabilité du mouvement. Ann. Fac. Sci. Toulouse (2) **9**, 203—474 (1907).
2. Siegel, C. L.: Über die Normalform analytischer Differentialgleichungen in der Nähe einer Gleichgewichtslösung. Nachr. Akad. Wiss. Göttingen, math.-phys. Kl. **1952**, 21—30.

For § 29:

1. Lejeune Dirichlet, G.: Werke, vol. 2, p. 5—8. Berlin 1897.

For § 30:

1. Birkhoff, G. D.: Dynamical systems, chap. 3. New York 1927, revised edition 1966.
2. Lindstedt, A.: Beitrag zur Integration der Differentialgleichungen der Störungstheorie. Abhandl. K. Akad. Wiss. St. Petersburg **31**, Nr. 4 (1882).
3. Poincaré, H.: Les méthodes nouvelles de la mécanique céleste, vol. 2, chap. 9. Paris 1893.
4. Whittaker, E. T.: On the solution of dynamical problems in terms of trigonometric series. Proc. London Math. Soc. **34**, 206—221 (1902).
5. Cherry, T. M.: On the solution of Hamiltonian systems of differential equations in the neighbourhood of a singular point. Proc. London Math. Soc. (2) **27**, 151—170 (1928).

6. Siegel, C. L.: Über die Existenz einer Normalform analytischer Hamiltonscher Differentialgleichungen in der Nähe einer Gleichgewichtslösung. Math. Ann. **128**, 144—170 (1954).
7. Birkhoff, G. D.: Stability and the equations of dynamics. Am. J. Math. **49**, 1—38 (1927).
8. Siegel, C. L.: On the integrals of canonical systems. Ann. Math. **42**, 806—822 (1941).
9. Poincaré, H.: Les méthodes nouvelles de la mécanique céleste, vol. 1, chap. 5. Paris 1892.

For § 31:

1. Levi-Civita, T.: Sopra alcuni criteri di instabilità. Ann. Mat. pura appl. (3) **5**, 221—307 (1901).
2. Siegel, C. L.: Some remarks concerning the stability of analytic mappings. Univ. nac. Tucumán Rev. A **2**, 151—157 (1941).

For § 33:

1. Moser, J.: On invariant curves of area-preserving mappings of an annulus. Nachr. Akad. Wiss. Göttingen, math.-phys. Kl. **1962**, 1—10.

For § 34:

1. Moser, J.: Stabilitätsverhalten kanonischer Differentialgleichungssysteme. Nachr. Akad. Wiss. Göttingen, math.-phys. Kl. **1955**, 87—120.

For § 35:

1. Arnold, V. I.: The stability of the equilibrium position of a Hamiltonian system of ordinary differential equations in the general elliptic case. Dokl. Akad. Nauk SSSR **137**, 255—257 (1961) [Russian] or Soviet Math. **2**, 247—249 (1961).
2. Leontovitch, A. M.: On the stability of Lagrange's periodic solutions of the restricted three-body problem. Dokl. Akad. Nauk USSSR **143**, 525—528 (1962) [Russian].
3. Deprit, A., and Deprit-Bartolomé, A.: Stability of the triangular Lagrangian points. Astron. J. **72**, 173—179 (1967).

For § 36:

1. Bohr, H.: Fastperiodische Funktionen. Berlin 1932. English translation: Almost periodic functions. New York 1947.
2. Kolmogorov, A. N.: On conservation of conditionally periodic motions under small perturbations of the Hamiltonian. Dokl. Akad. Nauk SSSR **98**. 527—530 (1954) [Russian].
3. — Théorie générale des systèmes dynamiques et mécanique classique. Proc. of Int. Congress of Math. Amsterdam 1954, vol. 1, 315—333. Amsterdam 1957.
4. Arnold, V. I.: Proof of A. N. Kolmogorov's theorem on the preservation of quasi-periodic motions under small perturbation of the Hamiltonian. Usp. Mat. Nauk **18**, no. 5 (113), 13—40 (1963) [Russian].
5. — Small divisor problems in classical and celestial mechanics. Usp. Mat. Nauk **18**, no. 6 (114), 91—192 (1963) [Russian].
6. Poincaré, H.: Les méthodes nouvelles de la mécanique céleste, vol. 1, chap. 1. Paris 1892.
7. Arnold, V. I.: On the classical perturbation theory and the stability problem of planetary systems. Dokl. Akad. Nauk SSSR **145**, 481—490 (1962) [Russian].

8. Arnold, V.I.: Instability of dynamical systems with several degrees of freedom. Dokl. Akad. Nauk SSSR **156**, 9—12 (1964) [Russian].
9. — and Avez, A.: Problèmes ergodiques de la mécanique classique. Paris 1967. English translation: Ergodic problems of classical mechanics. New York 1968.
10. Sternberg, S.: Celestial mechanics, vol. I, II. New York 1969.

For § 37:
1. Poincaré, H.: Les méthodes nouvelles de la mécanique céleste, vol. 3, chap. 26. Paris 1899.
2. Carathéodory, C.: Über den Wiederkehrsatz von Poincaré. Sitzber. Preuß. Akad. Wiss. **1919**, 580—584.
3. Birkhoff, G.D.: Proof of the ergodic theorem. Proc. Nat. Acad. Sci. USA **17**, 656—660 (1931).
4. Denjoy, A.: Sur les courbes définies par les équations différentielles à la surface du tore. J. math. pures appl. (9) **11**, 333—375 (1933).
5. Kampen, E.R. van: The topological transformations of a simple closed curve into itself. Am. J. Math. **57**, 142—152 (1935).
6. Siegel, C.L.: Note on differential equations on the torus. Ann. Math. **46**, 423—428 (1945).
7. Schwarzschild, K.: Über die Stabilität der Bewegung eines durch Jupiter gefangenen Kometen. Astr. Nachr. **141**, 1—8 (1896).
8. Hopf, E.: Ergodentheorie. Ergeb. Math. **5**, 48 (1937).
9. Littlewood, J.E.: On the problem of *n* bodies. Meddel. Lunds Univ. mat. Sem., Suppl. M. Riesz, 143—151 (1952).
10. Alekseev, V.M.: Quasirandom dynamical systems I, II, III. Mat. Sbornik **76** (118), 72—134 (1968); **77** (119), 545—600 (1968); **78** (120), 3—50 (1969) [Russian]. English translation in Math. USSR Sbornik **5**, 73 – 128 (1968); **6**, 505 – 560 (1968); **7**, 1 – 43 (1969).

Subject Index

∞ CIM

M. Aigner Combinatorial Theory ISBN 978-3-540-61787-7
A. L. Besse Einstein Manifolds ISBN 978-3-540-74120-6
N. P. Bhatia, G. P. Szegő Stability Theory of Dynamical Systems ISBN 978-3-540-42748-3
J. W. S. Cassels An Introduction to the Geometry of Numbers ISBN 978-3-540-61788-4
R. Courant, F. John Introduction to Calculus and Analysis I ISBN 978-3-540-65058-4
R. Courant, F. John Introduction to Calculus and Analysis II/1 ISBN 978-3-540-66569-4
R. Courant, F. John Introduction to Calculus and Analysis II/2 ISBN 978-3-540-66570-0
P. Dembowski Finite Geometries ISBN 978-3-540-61786-0
A. Dold Lectures on Algebraic Topology ISBN 978-3-540-58660-9
J. L. Doob Classical Potential Theory and Its Probabilistic Counterpart ISBN 978-3-540-41206-9
R. S. Ellis Entropy, Large Deviations, and Statistical Mechanics ISBN 978-3-540-29059-9
H. Federer Geometric Measure Theory ISBN 978-3-540-60656-7
S. Flügge Practical Quantum Mechanics ISBN 978-3-540-65035-5
L. D. Faddeev, L. A. Takhtajan Hamiltonian Methods in the Theory of Solitons ISBN 978-3-540-69843-2
I. I. Gikhman, A. V. Skorokhod The Theory of Stochastic Processes I ISBN 978-3-540-20284-4
I. I. Gikhman, A. V. Skorokhod The Theory of Stochastic Processes II ISBN 978-3-540-20285-1
I. I. Gikhman, A. V. Skorokhod The Theory of Stochastic Processes III ISBN 978-3-540-49940-4
D. Gilbarg, N. S. Trudinger Elliptic Partial Differential Equations of Second Order ISBN 978-3-540-41160-4
H. Grauert, R. Remmert Theory of Stein Spaces ISBN 978-3-540-00373-1
H. Hasse Number Theory ISBN 978-3-540-42749-0
F. Hirzebruch Topological Methods in Algebraic Geometry ISBN 978-3-540-58663-0
L. Hörmander The Analysis of Linear Partial Differential Operators I – Distribution Theory and Fourier Analysis ISBN 978-3-540-00662-6
L. Hörmander The Analysis of Linear Partial Differential Operators II – Differential Operators with Constant Coefficients ISBN 978-3-540-22516-4
L. Hörmander The Analysis of Linear Partial Differential Operators III – Pseudo-Differential Operators ISBN 978-3-540-49937-4
L. Hörmander The Analysis of Linear Partial Differential Operators IV – Fourier Integral Operators ISBN 978-3-642-00117-8
K. Itô, H. P. McKean, Jr. Diffusion Processes and Their Sample Paths ISBN 978-3-540-60629-1
T. Kato Perturbation Theory for Linear Operators ISBN 978-3-540-58661-6
S. Kobayashi Transformation Groups in Differential Geometry ISBN 978-3-540-58659-3
K. Kodaira Complex Manifolds and Deformation of Complex Structures ISBN 978-3-540-22614-7
Th. M. Liggett Interacting Particle Systems ISBN 978-3-540-22617-8
J. Lindenstrauss, L. Tzafriri Classical Banach Spaces I and II ISBN 978-3-540-60628-4
R. C. Lyndon, P. E Schupp Combinatorial Group Theory ISBN 978-3-540-41158-1
S. Mac Lane Homology ISBN 978-3-540-58662-3
C. B. Morrey Jr. Multiple Integrals in the Calculus of Variations ISBN 978-3-540-69915-6
D. Mumford Algebraic Geometry I – Complex Projective Varieties ISBN 978-3-540-58657-9
O. T. O'Meara Introduction to Quadratic Forms ISBN 978-3-540-66564-9
G. Pólya, G. Szegő Problems and Theorems in Analysis I – Series. Integral Calculus. Theory of Functions ISBN 978-3-540-63640-3
G. Pólya, G. Szegő Problems and Theorems in Analysis II – Theory of Functions. Zeros. Polynomials. Determinants. Number Theory. Geometry ISBN 978-3-540-63686-1
W. Rudin Function Theory in the Unit Ball of $\mathbb{C}^n$ ISBN 978-3-540-68272-1
S. Sakai C*-Algebras and W*-Algebras ISBN 978-3-540-63633-5
C. L. Siegel, J. K. Moser Lectures on Celestial Mechanics ISBN 978-3-540-58656-2
T. A. Springer Jordan Algebras and Algebraic Groups ISBN 978-3-540-63632-8
D. W. Stroock, S. R. S. Varadhan Multidimensional Diffusion Processes ISBN 978-3-540-28998-2
R. R. Switzer Algebraic Topology: Homology and Homotopy ISBN 978-3-540-42750-6
A. Weil Basic Number Theory ISBN 978-3-540-58655-5
A. Weil Elliptic Functions According to Eisenstein and Kronecker ISBN 978-3-540-65036-2
K. Yosida Functional Analysis ISBN 978-3-540-58654-8
O. Zariski Algebraic Surfaces ISBN 978-3-540-58658-6